Neuroscience and Legal Responsibility

OXFORD SERIES IN NEUROSCIENCE, LAW, AND PHILOSOPHY

SERIES EDITORS

Lynn Nadel, Frederick Schauer, and Walter P. Sinnott-Armstrong

Conscious Will and Responsibility
Edited by Walter P. Sinnott-Armstrong and Lynn Nadel

Memory and Law
Edited by Lynn Nadel and Walter P. Sinnott-Armstrong

Neuroscience and Legal Responsibility
Edited by Nicole A Vincent

Neuroscience and Legal Responsibility

EDITED BY NICOLE A VINCENT

OXFORD
UNIVERSITY PRESS

Oxford University Press is a department of the University of Oxford.
It furthers the University's objective of excellence in research, scholarship,
and education by publishing worldwide.

Oxford New York
Auckland Cape Town Dar es Salaam Hong Kong Karachi
Kuala Lumpur Madrid Melbourne Mexico City Nairobi
New Delhi Shanghai Taipei Toronto

With offices in
Argentina Austria Brazil Chile Czech Republic France Greece
Guatemala Hungary Italy Japan Poland Portugal Singapore
South Korea Switzerland Thailand Turkey Ukraine Vietnam

Published in the United States of America by
Oxford University Press
198 Madison Avenue, New York, NY 10016

Library of Congress Cataloging-in-Publication Data
Neuroscience and legal responsibility/edited by Nicole A Vincent.
p. cm.—(Oxford series in neuroscience, law, and philosophy)
ISBN 978-0-19-992560-5
1 Law—Psychological aspects. 2. Neurosciences. 3. Forensic neurology. I. Vincent, Nicole A
K346.N48 2013
345'.04—dc23 2012019790

9 8 7 6 5 4 3 2 1
Printed in the United States of America on acid-free paper

CONTENTS

CONTRIBUTORS

Christoph Bublitz
Scientific Researcher
Faculty of Law
Institute of Criminal Sciences
Universität Hamburg
Hamburg, Germany

Adrian Carter
NHMRC Postdoctoral Fellow
UQ Centre for Clinical Research
The University of Queensland
Brisbane, QLD, Australia

Alicia Coram
Faculty of Arts
University of Melbourne
Melbourne, VIC, Australia

Jillian Craigie
Wellcome Trust Research Fellow in Biomedical Ethics
Department of Philosophy
University College London
London, UK

Leora Dahan Katz
Yale Law School, Doctoral Candidate; and
Philosophy Department, Graduate Student
Hebrew University of Jerusalem
Jerusalem, Israel

Paul Sheldon Davies
Professor
Department of Philosophy
College of William and Mary
Williamsburg, VA, USA

Colin Gavaghan
Associate Professor
Director of The New Zealand Law Foundation Centre for Law and Policy in Emerging Technologies
Faculty of Law
University of Otago
Dunedin, New Zealand

Wayne Hall
Professor and NHMRC Australia Fellow
Deputy Director (Policy)
UQ Centre for Clinical Research
The University of Queensland
Brisbane, QLD, Australia

Jeanette Kennett
Professor of Moral Psychology
Centre for Agency Values and Ethics
Department of Philosophy
Macquarie University
Sydney, NSW, Australia

Neil Levy
ARC Future Fellow
Florey Neuroscience Institutes
Melbourne, VIC, Australia

Anne Ruth Mackor
Professor of Professional Ethics
Faculties of Law, Philosophy and Theology
Rijksuniversiteit Groningen
Groningen, The Netherlands

Reinhard Merkel
Professor of Criminal Law and Philosophy of Law
Faculty of Law
Institute of Criminal Sciences
Universität Hamburg
Hamburg, Germany

Stephen J. Morse
Ferdinand Wakeman Hubbell Professor of Law & Professor of Psychology and Law in Psychiatry
University of Pennsylvania Law School
Philadelphia, PA, USA

Thomas Nadelhoffer
Assistant Professor of Philosophy
Department of Philosophy
Dickinson College
Carlisle, PA, USA

Katrina L. Sifferd
Associate Professor
Department of Philosophy
Elmhurst College
Elmhurst, IL, USA

Walter P. Sinnott-Armstrong
Chauncey Stillman Professor of Practical Ethics
Department of Philosophy and Kenan Institute for Ethics
Duke University
Durham, NC, USA

Nicole A Vincent
Research Fellow
Centre for Agency Values and Ethics
Department of Philosophy
Macquarie University
Sydney, NSW, Australia; and
Chief Investigator
Enhancing Responsibility Project
3TU Centre for Ethics and Technology
Philosophy Section
Delft University of Technology
Delft, The Netherlands

1

Law and Neuroscience

Historical Context[1]

NICOLE A VINCENT

The chapters in this volume investigate whether, and if so how, advances in the mind sciences—chiefly in neuroscience, psychology, and behavioral genetics—might affect the moral foundations of legal responsibility practices. These chapters have been arranged under five headings—responsibility and mental capacity; reappraising agency; assessment; disease and disorder; and modification—and in what follows, I survey how the individual chapters engage with issues that fall under these headings. But first is a brief historical interlude to help me explain why there is even reason to suppose that advances in the mind sciences might affect the moral foundations of legal responsibility practices.

LAW, SCIENCE, HARD DETERMINISM, LIBERTARIANISM, AND COMPATIBILISM

In a paper that charts the increasingly prominent role of psychology and forensic psychiatry in Victorian legal responsibility adjudication, Joel Eigen writes:

> Before the 13th century, the person who sat at the heart of English common law consisted of an agent capable of physical harm... it was the *doing*—not the contemplation—of evil that conferred responsibility.

> Beginning with Henri de Bracton, Henry II's legal scribe...one finds the first mention of a mental element required for ascribing legal accountability to a physical act. De Bracton's conception of the legally culpable actor incorporated existing notions of physical harm with a mental component to produce the common law's position on moral guilt...It was the sustained focus paid to the "mental element" that de Bracton transported from canon to common law, grounding mens rea in the Roman principle, *voluntas nocendi*: the will to harm. Explicit reference to the will dates at least to *Justinian's Digest* (535 ace) but de Bracton was the first English lawyer...to give contemplation equal footing with physical execution in arriving at the legal basis for criminal responsibility: "For a crime is not committed unless the will to harm be present...In misdeeds we look to the will and not the outcome." It was the *will to choose* that defined human action, and consequently, criminal responsibility. Choosing to do evil revealed a wicked will, and, after de Bracton's formulation: evil intent. For a person to be criminally culpable in the medieval common law, the act must have been willfully, intentionally committed. [But a]lthough sound as a moral principle and a criterion for legal guilt fastening, inferring an actor's capacity to choose has proven a daunting task for jurors ever since de Bracton...Medical practitioners since antiquity have employed a host of diagnostic terms to label the mental life of the deranged, most notably delirium, melancholia, and mania... But however familiar and accessible delirium, melancholia, and mania remained as medical...terms, they did not constitute a legal defense. The journey from the doctor's surgery to the judge's courtroom was a considerable one. (2004:397–398)

Had the law's focus in responsibility adjudication remained confined to investigating *physical harm*—that is, if responsibility consisted purely of the *physical doing* but not *willing* or *intending*—then the reach of forensic psychiatry today might have been less pronounced. But because of the increasing importance of this *mental element* in legal responsibility adjudication, with the passage of time, forensic psychiatrists have come to play an increasingly prominent role by helping courts to assess defendants' mental states, disorders, and mental capacities and by commenting on how these things might relate to legal criteria for responsibility. During this period, *medical* advances in the diagnosis and understanding of mental disorders were translated into increasingly more sophisticated *legal* tools for the assessment and conceptualization of legal responsibility by gradually reshaping legal views about human agency. Thus, it is in this context that advances in neuroscience, psychology, and behavioral genetics might again affect legal responsibility practices—that is, by providing further insights into the nature of human agency and by offering revamped diagnostic criteria and more powerful diagnostic and intervention tools with which to assess and to alter minds.

Thanks to the development of powerful new diagnostic brain imaging techniques that made it possible to study living human brains in situ without needing to reach for cranial saws, drills, and scalpels—for instance, computed axial tomography (CAT), single-photon emission computed tomography (SPECT), positron emission tomography (PET), and more recently, magnetic resonance imaging (MRI) and its functional equivalent (fMRI) as well as diffusion tensor imaging (DTI)—the past two decades have witnessed important advances in our understanding of brain structure and function. And although much empirical and conceptual work remains to be done before we can proclaim to know how the brain enables the mind, two points seem uncontroversial: (1) that the operation of the mind is closely linked to the operation of the brain[2] and (2) that like everything else subjected to scientific inquiry, as far as we can tell, the brain, too, operates according to the same laws of nature that govern the rest of the physical world. Put together, these two observations suggest that the mind is indeed a very complicated mechanism, but a mechanism nevertheless, and this has led some authors to start writing a eulogy for responsibility.

For instance, in a much-cited paper, Joshua Greene and Jonathan Cohen maintain that advances in neuroscience demonstrate that "[f]ree will, as we ordinarily understand it, is an illusion," and although they stop short of denying any place for responsibility in our mechanical universe, they nevertheless contend that the criminal law's backward-looking retributive aims should be replaced with forward-looking aims such as deterrence, prevention, and treatment (2004:1783). Similarly, Robert Sapolsky's discussion of the role that the frontal cortex plays in grounding our capacity for self-control is also a plea for "a world of criminal justice in which there is no blame [but] only prior causes" (2004:1794). Richard Dawkins takes a harder stance against what he sees as "the flawed concept of retributi[ve]" punishment—he argues that "as a moral principle [retribution] is incompatible with a scientific view of human behaviour" because, by his account, "a truly scientific, mechanistic view of the nervous system make[s] nonsense of the very idea of responsibility"—and so he, too, recommends that the law's proper aims should be deterrence, prevention, treatment, and removal of dangerous individuals from society, but certainly not retribution (2006). And more recently, Anthony Cashmore brings these various points together in an article in the *Proceedings of the National Academy of Sciences*:

> Many discussions about human behavior center around the relative importance of genes and environment, a topic often discussed in terms

> of nature versus nurture. In concentrating on this question of the relative importance of genes and environment, a crucial component of the debate is often missed: an individual cannot be held responsible for either his genes or his environment. From this simple analysis, surely it follows that individuals cannot logically be held responsible for their behavior. Yet a basic tenet of the judicial system and the way that we govern society is that we hold individuals accountable (we consider them at fault) on the assumption that people can make choices that do not simply reflect a summation of their genetic and environmental history. As de Duve has written, "If... neuronal events in the brain determine behavior, irrespective of whether they are conscious or unconscious, it is hard to find room for free will. But if free will does not exist, there can be no responsibility, and the structure of human societies must be revised." (2010:4499)

Subsequent reflection on the role that genetics, the environment, and stochasticism play in shaping human behavior leads Cashmore to assert that "we are mechanical forces of nature [which] have evolved the phenomenon of consciousness, which [in turn] has conferred upon us the illusion of responsibility," and thus he concludes that "it is time for the legal system to confront this reality... The reality is, not only do we have no more free will than a fly or a bacterium, in actuality we have no more free will than a bowl of sugar. The laws of nature are uniform throughout, and these laws do not accommodate the concept of free will" (2010:4503).

This is a radical approach to how advances in the mind sciences might affect the moral foundations of legal responsibility practices. Viewed against the backdrop of the story that Eigen tells about the increasingly prominent role of psychology and forensic psychiatry in legal responsibility adjudication, the above authors' arguments can be understood as follows. What began all the way back in the 13th century as a relatively benign addition of a *mental element* to legal responsibility criteria started a slow but gradual conceptual shift that bit-by-bit eroded away the traditional *legal* notion of agency and responsibility, allegedly based on the presupposition that humans have a metaphysically free will, and replaced it with a *psychologically* inspired notion of agency that has room for neither free will nor responsibility, and all that now remains to be done is for the law to step aside for science and medicine. Put bluntly, these authors' claim is not just that advances in the mind sciences should *affect* the moral foundations of legal responsibility practices but that legal responsibility practices should be *eliminated* altogether because they have no place in a scientifically informed approach to the regulation of society.

However, this is *not* how this volume approaches the questions of whether and how advances in the mind sciences should affect the moral foundations of legal responsibility practices.

One problem with the above arguments is their assumption that the cited empirical findings rule out the possibility that humans have a metaphysically free will. Although I do not endorse the claim that humans have a metaphysically free will, we must admit that much of what the empirical sciences study still remains unexplained, and one distinct possibility is that the as-yet-unexplained phenomena are indeed brought about by the operation of a metaphysically free will. More important, though, even if the sciences explained all of the observable phenomena, that state of affairs would still be compatible with the view that a metaphysically free will brings about those phenomena. In response, we could of course stipulate that a metaphysically free will must necessarily be temperamental, whimsical, irrational, irregular, and ultimately inexplicable to ensure that anything the sciences ever manage to explain will by definition not qualify as a candidate for "free will"—but that would make the claim that science provides evidence against free will trivial. Thus, although this volume does *not* presuppose a libertarian approach to how the mind sciences affect the moral foundations of legal responsibility practices, the possibility that the will is metaphysically free is *not* ruled out by findings from the empirical sciences.

Another problem with those arguments is the looseness with which they employ terms like "determinism," "indeterminism," and "free will." Precisely how the thesis called "determinism" should be understood—candidates include causation, necessitation, sufficiency, entailment, explanation, prediction, inevitability, fate, and predestination (this list is intended to be demonstrative not exhaustive)—and thus precisely how determinism is meant to pose a threat to responsibility are both matters of debate. Similarly, precisely how the term "indeterminism" should be understood—for example, in terms of probability, chance, chaos, or something else—and thus precisely how indeterminism is meant to pose a threat to responsibility are also matters of debate. Finally, the literature on free will is vast, and although some authors tie the will's freedom to its alleged existence outside of the physical world's causal order, others ground its freedom in indeterministic principles, and some even claim that free will in any meaningful sense requires determinism (I shall come back to this last point shortly). That the arguments cited in the previous section do not engage with this immense theoretical diversity is understandable—to do so would require engagement with libraries' worth of scholarship, and no article or even book can do that. This is not an ideal to which anyone should be expected to aspire. It is, however, regrettable that the above arguments barely acknowledge this diversity and fail to respond accordingly by setting their targets more narrowly.

The above arguments are also not new. For instance, in bygone years, much ink was spilled debating whether *God's omniscience* posed a challenge to the propriety of tormenting sinners in hell for their sinful actions. After all, if God knows everything, then God surely knows what each of us will ever do, which entails that what we will do is already set, and that in turn makes it seem terribly unfair to torment sinners in the depths of hell when what they do is the only thing they *can* do. Comparable quantities of ink are still being spilled debating whether determinism (whatever that amounts to) undermines responsibility, and the structure of the above authors' arguments closely tracks the structure of these older arguments. Now, my aim here is not to score points by deriding the above authors' arguments as "old hat," but rather to highlight that despite centuries of debate, little progress has been made in settling these questions. This point is salient because it surely provides at least some reason to resist the temptation to scaffold new arguments on top of the same basic argumentative structure that has so far failed to yield firm conclusions, or to start composing a eulogy for responsibility. Unless the older forms of these arguments present a compelling case against responsibility, then without addition of new argumentative components, I do not see how these newer versions of those older arguments change anything. And if the claim is that the new neuroscience will make old arguments more vivid, then the reply must surely be that unless this vividness also carries with it the authority of sound argument, then these new arguments should hold no more sway than those older arguments.

For such reasons, I find the above approach to the questions of whether and how advances in the mind sciences might affect the moral foundations of legal responsibility practices unattractive. Additionally, if we wish to say anything more interesting in this volume about how the mind sciences might affect the moral foundations of legal responsibility practices than just that responsibility should be eliminated, then I see no other option than to set aside the above approach. But perhaps most important, I believe that a better conceptual approach is available—one that provides the mind sciences with a genuine opportunity to enrich the legal understanding of agency and to inform legal responsibility practices, rather than just attempting to eliminate them and in effect to change the topic.

As I point out elsewhere (Vincent and van de Poel 2011:1), it has been well over a decade since John Fischer and Mark Ravizza (1998), and before them Jay Wallace (1994) and Daniel Dennett (1984), defended responsibility from the threat of determinism.[3] On their compatibilist accounts, determinism does not pose a threat to moral responsibility, and moral responsibility does not require free will either. Rather, responsible people are those who have the right mental capacities in the right degree—that is, the mental capacities required for moral agency. What justifies holding some people responsible for what they do is that their actions issue from mechanisms[4] that bestow upon them mental capacities like the ability to perceive the world without delusion, to think clearly, to guide their actions by the light of their judgments, and to resist acting on mere impulse.

This brief sketch of what matters for moral responsibility on the compatibilist approach closely resembles what H. L. A. Hart said about what matters for legal responsibility—namely, "[t]he capacities . . . of understanding, reasoning, and control of conduct: the ability to understand what conduct legal rules or morality require, to deliberate and reach decisions concerning these requirements, and to conform to decisions when made" (1968:227). Given these parallels between compatibilist moral responsibility theory and Hart's claims about what matters for legal responsibility, I find it puzzling that authors like Greene and Cohen, Sapolsky, Dawkins, and Cashmore insist that the mind sciences pose a critical threat to responsibility and that retribution can only be defended under a libertarian theory that they find wanting, when another approach is available that does not reject or overlook the possibility that the world is deterministic and that does not require radical legal reform.

RESPONSIBILITY AND MENTAL CAPACITY

A central compatibilist premise that informs the approach taken by the chapters of this volume is the capacitarian idea that responsibility tracks mental capacity.

In chapter 2, Stephen J. Morse not only argues that legal responsibility practices sit comfortably in a deterministic setting but also explains why compatibilism provides the most attractive foundation for those practices—more attractive than either libertarianism or hard determinism. On the first point, Morse argues that determinism poses no challenge to establishing the elements of criminal responsibility and liability—that is, the existence of an *action* accompanied by a *mental state* appropriate to the *circumstances* of the case, as well as the existence of *causation* between the action and the *result*—and of positive *defenses* and *excuses* (e.g., duress and insanity). Whether the universe turned out to be deterministic or not, some people's behavior would still be classified as *actions,* whereas others' (e.g., epileptic tremors and convulsions, or the unconscious ambulation of a sleepwalker) would not. Some people would have performed their actions with the required *mental states* (e.g., purpose, intent, knowledge), whereas others would not. Determinism poses no problem for the existence of *results* or *causation*, and whether someone could *justify* their behavior or establish *duress* would still, as always, depend on the circumstances of the case—that is, on whether given what was at stake, their actions reflected a reasonable choice in a hard-choice situation. And finally, whether the person concerned could cite the defense of *insanity* would depend unsurprisingly on whether that person had sufficiently grave deficits in the requisite mental capacities. On Morse's account, scientific findings that reveal (or hint at) the physical implementation of our agency-relevant mental capacities do not pose a threat to legal responsibility—as he has pointed out previously, to think otherwise is to commit the "fundamental psycho-legal error," the all-too-common tendency to confuse causation with excuse (Morse 2003:289, 2006:399). Rather, the main challenge to legal responsibility by his account comes from the claim that recent findings from the mind sciences challenge the law's "folk psychological view of the person and behaviour"—the idea that people can be motivated to act by being presented with reasons, such as those provided by legal rules and related threats of sanction. However, Morse maintains that given "how little we know about the brain-mind-action connections," it would be the height of "neuroarrogance" to insist that the folk psychological view is outdated and should thus be abandoned. On the second point, Morse notes that although the above picture of the criteria of criminal responsibility, liability, justifications, and defenses is also compatible with libertarianism, he finds libertarianism unattractive. He argues:

> I consider this position extremely implausible in the modern, scientific age. Human beings, as complex as they are, are still part of the physical universe and subject to the same laws that govern all phenomena. In short, I believe that libertarianism does not furnish a justifiable foundation for an institution that is essentially about blaming and punishing culpable agents. If determinism or something quite like it is true, as I assume it is, then only compatibilism provides a secure basis for criminal responsibility.

The reason he rejects hard determinist justifications for responsibility practices—namely, ones similar to the sorts of forward-looking justifications that are cited by Greene and Cohen and the other authors whom I mentioned above—is that not only does hard determinism fail to accommodate a genuine notion of agency, but it also discards notions like desert, autonomy, and dignity, which are important parts of the moral landscape and play a crucial role in the law's protection of rights and justice. Strikingly, though not surprisingly,[5] it's business as usual under Morse's common law compatibilist approach to how the mind sciences can inform the moral foundations of legal responsibility practices. Namely, they can do this (to the extent that they *can* indeed do so, given the current state of science) by helping us to establish the elements of criminal responsibility and liability in largely the same way as psychology and forensic psychiatry have been doing for quite some time.

In chapter 3, Anne Ruth Mackor explains how the mind sciences can and cannot contribute to our understanding of the mental capacities required for fully responsible moral agency, and what they can contribute to legal responsibility adjudication practices. She does this in the context of responding to Dawkins' earlier-cited claim about neuroscience's radical relevance to legal responsibility practices, and to Michael Gazzaniga's opposite claim that neuroscience is irrelevant to legal responsibility practices. In response to Dawkins' claim that "a truly scientific, mechanistic view of the nervous system make[s] nonsense of the very idea of responsibility" and that retributive punishment is as pointless as flogging a broken car in an attempt to make its engine start (2006), Mackor argues that neuroscience shows neither that our responsibility practices are conceptually flawed nor that they are pointless because these practices presuppose only that humans possess the capacity for practical reason, and the neuroscientific evidence that Dawkins cites says nothing to challenge this supposition. On the other hand, Gazzaniga argues that "neuroscience has nothing to say about concepts such as free will and personal responsibility" (2006:141), and hence he maintains that neuroscientists should refrain from commenting on such topics. In response, Mackor cites empirical work on the role of affect in moral reasoning to demonstrate that neuroscience can alter our conception of moral reasoning—that is, that neuroscientific evidence can change our understanding of precisely what sorts of mental capacities are required for responsible moral agency—and she also suggests that in some cases, neuroscientific techniques and technologies might be epistemically better positioned to help us assess whether a particular individual possesses a given mental capacity or not. On Mackor's account, neuroscientific findings do not pose "a threat to the continued existence of our accountability practices[, and] neuroscience... can suggest new ways of looking at both the content and the application of our conceptions" of responsibility.

Finally, Jillian Craigie and Alicia Coram's contribution in chapter 4 helps to round out the message of the first section of this volume by drawing attention to a range of normative and conceptual reasons that explain why neuroscience cannot answer all of the important questions that criminal responsibility adjudications hinge upon. First, Craigie and Coram point out that although on the surface mental capacity evaluations inform rationality assessments in two similar-looking legal contexts—namely, in *medical competence* assessments to determine a person's ability to consent or to refuse medical treatment, and in *criminal responsibility* adjudications to determine whether a conviction can be secured—because different things are at stake in these two contexts, it would be a mistake to suppose that a person with a given set of mental or brain features would be deemed rational concurrently in both of these contexts. In neither context can neuroscience settle the critical question of how much of which kind of reasoning capacity might be required for a person to be considered fully or sufficiently rational—that is, how much of which mental capacities are required for moral agency—because this is a meta-ethical and a normative issue rather than an empirical neuroscientific issue. Second, it is also far from clear that there is indeed such a thing as generic moral agency—that is, that a single group of mental capacities will always suffice for rationality in every context across the board—because agency may be inherently situational or context bound. Third, when this is coupled with Morse's observation that assessments of rationality are ultimately decided on the basis of *behavioral* observations rather than *neurological* data (e.g., Morse 2011:32–35), this seriously reigns in the degree to which advances in the mind sciences (and specifically in neuroscience) can affect legal responsibility practices. Craigie and Coram's contribution rounds out the message of this first section by challenging Mackor's claim that neuroscience can alter our conception of moral reasoning and that it can help us to determine whether a given person is rational. In addition, Craigie and Coram's discussion of neuroscience in the context of medical law, and their comparison of this to the criminal responsibility context, also offers a valuable foray into an area that has so far received insufficient attention in the field of neurolaw, which tends to focus on issues of criminal law.

REAPPRAISING AGENCY

Having sketched some important details of a broadly compatibilist approach to neurolaw (i.e., of how the mind sciences might and might not bear on the moral foundations of legal responsibility practices), before making use of these conceptual foundations, the next group of chapters examines issues that bear on whether Morse is right to contend that recent findings from the mind sciences do not challenge the law's folk psychological view of the person and behavior—the idea that people can be motivated to act by being presented with reasons, such as those provided by legal rules and related threats of sanction. Put another way, this group of chapters investigates whether evidence from the mind sciences about even nonpathological ways of being—that is, about the mental capacities of the normal, ordinary, and common person, or the proverbial man on the Clapham omnibus—might provide reasons to revise our understanding of human agency and corresponding legal responsibility practices.

In chapter 5, Paul Sheldon Davies argues that there is sufficient scientific reason to be skeptical about the role that consciousness plays in regulating our behavior, and this, he contends, exposes serious problems with how criminal responsibility is conceptualized and assessed. By his account, criminal responsibility presupposes voluntary control: the Model Penal Code postulates that we are only responsible for what we do voluntarily or for what can be traced back to a prior voluntary act, and we also generally assume that our consciousness clearly flags when our actions are voluntary and when they are not. However, Davies argues that recent work in the fields of cognitive and social psychology, as well as neuroscience—in particular, the work of Daniel Wegner whom Morse also cites—shows that we are simply not the sorts of agents who we think we are. Despite our convictions that we consciously control our actions and that we are consciously aware of when our actions were voluntary in this sense, our actions are in fact caused by lower level brain processes of which we have little or no awareness, and what appears in our consciousness is little more than post hoc confabulation. By his account, "we cannot justifiably claim to know from the first-person perspective the causes of our actions, nor can we justifiably claim to know of some other person that he 'determined' his own action," and he believes that this entails that "[t]he characterization of criminal responsibility given in the Model Penal Code cannot serve its intended function." If Davies is right, then the way in which criminal responsibility is conceptualized and how it is assessed both presuppose an empirically inaccurate account of agency, and this might in turn entail that our responsibility practices are in need of reform.

Although Davies' paper challenges the almost universally held assumption that we are responsible for our *consciously controlled* actions, the following two chapters target responsibility for *negligent* actions.

In chapter 6, Leora Dahan-Katz argues that the objective reasonable person standard upon which attributions of responsibility in negligence are based presupposes a particular kind and degree of human rationality that, contrary to widely held assumptions, humans simply do not possess. By her account, the heuristics and biases tradition of cognitive psychology shows that humans are not like what we think we are because we have various blind spots, weaknesses, and infirmities in our heuristic-based reasoning processes. She argues that this growing body of work "has demonstrated that human reasoning processes often rely upon inaccurate rules of thumb, or heuristics, and are subject to a wide variety of identifiable biases [that often] lead to [unavoidable] errors in judgment," and she holds "that when heuristic reasoning has [this] impact upon human judgment and decision-making, this fact can ultimately negate moral culpability." On Dahan-Katz's account, this entails that we should either reduce the sorts of expectations associated with the objective reasonable person standard or, preferably, replace the objective standard with a subjective standard of negligence. In her view, "[t]he current reasonable person standard imposed in criminal offences is...entirely unjustified," and it "must be modified or removed from the criminal law."

And in chapter 7, Neil Levy argues that empirical studies like Benjamin Libet's and Daniel Wegner's only challenge the moral justification of negligence-based responsibility ascriptions, but that they leave unscathed our ascriptions of responsibility for people's intentional actions. On the former point, he argues that actions that stem from unconscious processes express neither our will nor our evaluative agency, and so agents cannot be legitimately held responsible for them. On the latter point, he argues that "[i]t simply does not matter, for the purposes of moral responsibility, whether I consciously initiate my actions, or whether my consciousness that I have initiated an action is itself caused by mechanisms that initiate actions. What matters is the content of the mental states to which these mechanisms are responsive." According to Levy, such empirical studies further vindicate the intuition that people are at best responsible only for what they do intentionally, not for what they do unintentionally. One consequence of accepting his argument would be that (barring other considerations) negligence should cease to function as a legitimate basis for attributions of legal responsibility.

It is crucial to notice the following feature of these chapters' arguments. The challenges to legal responsibility that they discuss do not surface from the mere fact that our behavior is caused, determined, produced through mechanisms, fully predictable, completely explainable, or anything else of a similar ilk. Rather, these challenges are meant to stem from the fact that given what mounting empirical evidence is telling us about the nature of human agency—about the mental capacities of paradigmatically normal agents—we have reason to suppose that we are not the sorts of agents whom we have taken ourselves to be—that the man on the Clapham omnibus is not the man we took him to be because in some ways he is more frail, though in other ways, he might be more capable. In effect, contrary to what Morse says, these authors claim that, in at least some contexts, we already have sufficient evidence and reason to abandon the law's folk psychological view of the person and behavior. And if they are right that our agency is so radically different from what we have hitherto imagined—and, importantly, if the way that we are cannot be changed, for instance, by trying harder to live up to the standards of agency presupposed by our current legal responsibility doctrines, or by threatening people with even more severe sanctions should they fail to comply with those standards (Vincent 2011a:327)—then this *may* require us to adopt concomitant legal responsibility doctrines and criteria that take into account what can actually be expected of us given this more accurate picture of what we are actually like as agents. I say "may" rather than "will" because there may still be valid policy reasons to hold people to a higher standard of care than what they are actually capable of living up to—for instance, because being more realistic and thus expecting less of people may provide insufficient deterrence for some.

ASSESSMENT

Taking solace from Levy's suggestion that what matters for responsibility is that the mechanisms from which our actions issue are responsive to reasons, and from the fact that it is still far from settled whether the studies cited by the earlier chapters indeed show what some take them to show, the next two chapters look at how neuroscience might be used to shed light on assessments of responsibility. A very practical way in which the mind sciences could at least in principle affect legal responsibility practices would be by providing new techniques and tools with which to assess defendants' responsibility—for instance, to more accurately evaluate their true mental capacities[6] or to predict people's propensities toward irresponsible or even dangerous behavior—and this is the topic of the next two chapters.

In chapter 8, Katrina L. Sifferd argues that one way in which neuroscience could help the law would be by providing additional evidence pertaining to an accused person's brain's capacity for executive function. Assessments of defendants' mental capacity are usually based on expert testimony from forensic psychiatrists or from lay people who are acquainted with the defendant—their neighbors, teachers, employers, family, and so forth. However, this means that *behavioral observations* are ultimately what informs legal assessments of defendants' mental capacity, and this creates at least two potential problems. One, cunning defendants can put on an act in order to convince the court that their cognitive and volitional mental capacities are impaired when in fact they are not. Two, lay people's observations may either be mistaken or intentionally distort the picture that they present when giving testimony—for instance, if they feel compassion for the defendant (e.g., their parents, siblings, or friends), then they might claim that they have always been impaired, or if they dislike the defendant (e.g., relatives of the victim), then they might insist that they have always been competent. Given these drawbacks of *behavioral* mental capacity assessment, it would be useful if other sources of data could be used to supplement such assessments, and here Sifferd suggests a role for diagnostic *neuroimaging* techniques. By Sifferd's account, something like H.L.A. Hart's "capacity responsibility"—that is, the earlier-cited "ability to understand what conduct legal and moral rules require, to deliberate and reach decisions concerning these requirements, and to conform to decisions when made" (Hart 1968:227)—is a mental condition of criminal responsibility. But even though the law normally relies upon behavioral criteria to determine whether someone has capacity responsibility or not, Sifferd claims that there is no principled reason why neuroscientific evidence could not be used to more individually assess people's capacity responsibility. Furthermore, in Sifferd's view, the sorts of capacities that Hart describes bear striking resemblance to the brain's executive functions—in particular, to the control, attention, and planning functions of the prefrontal areas of the frontal lobe of the human brain—and hence she suggests that the degree and type of a particular individual's capacity responsibility could be at least partially ascertained by considering neuroimaging evidence about that individual's brain's capacities for executive function. To demonstrate the utility of her position, she finishes off her paper by explaining how this approach can be helpful in the context of assessing juvenile and mentally retarded defendants' responsibility.

In chapter 9, Colin Gavaghan argues that although recent neuroimaging techniques might provide more accurate and less degrading techniques with which to assess the degree of danger a person might pose to society, we must be mindful to clearly distinguish between techniques that measure mere appetite and those that measure propensity to act. Gavaghan's discussion begins with the description of a device called the *penile plethysmograph* (PPG), which measures:

> ...changes in penile circumference or volume occasioned by images of persons who vary in age and sex, or audiotaped stories concerning sexual interactions with persons who vary in age and sex. An increase in either penile circumference or volume is assumed to indicate sexual arousal, thereby indicating sexual desire.

This device has been used around the world in a range of contexts, some of which seem dubious at best. For instance, in some countries where homosexuals are prohibited from entering military service, the PPG has been used in attempts to confirm whether given individuals who have been drafted into the army and who claim to be homosexual *really* are homosexual or whether they are just *pretending* in order to escape military service. In other contexts, though, it is at least imaginable that a plausible prima facie case could be constructed in favor of allowing the PPG to be used. For instance, if the PPG could accurately detect whether a given person is a pedophile, then perhaps it should be permissible for child care centers to make it a condition that potential child care employees submit themselves to a PPG test before being offered a job. Alternatively, it is also conceivable that its use might be justified when courts consider whether a sex offender should be incarcerated (to confirm whether they are likely to reoffend and should thus be incarcerated for public protection), or when parole boards consider whether to release a convicted sex offender from prison (the question of whether an inmate still poses a threat to society plays an important role in parole boards' decisions). Because of the highly invasive, deeply degrading, and undignified nature of these tests, support for the PPG has been relatively limited. However, recent techniques have been developed that use fMRI to measure sexual desire. Given that these techniques are even more accurate than the PPG and that they are less invasive and degrading because they do not involve direct genital contact, this raises the question of whether it might be reasonable to use these newer brain-based techniques in cases in which PPG use would have been impermissible.

Gavaghan argues that although at least some of the worries about the PPG are circumvented by these brain-based sexual appetite measuring techniques, this helps to foreground a different kind of worry—namely, that if such techniques actually measure only a person's sexual appetites, but not their ability to resist acting on them or their intent, then this could have two highly controversial consequences. First, some people might be denied their liberty merely because they harbor certain desires, even though they may actually pose no threat to anyone because despite having those desires, they would not have acted on them because they can now control their actions and genuinely intend to exercise that control. Second, and perhaps more worryingly, this would also eliminate the space in which people exhibit genuinely virtuous human agency—that is, when in spite of inclinations to the contrary, they still choose to act as they believe is right (or perhaps as they believe is the most prudent)—and in which convicted offenders can attempt to redeem themselves in the eyes of society by demonstrating that they have indeed reformed.

It is easy to be virtuous when no temptation presents itself and when no effort must be expended on doing what is right, and for this reason, truly responsible human action is exemplified in moments when we resist our base inclinations and temptations in favor of doing what is right. Those working on brain-based prediction techniques and technologies and those who might consider their use should take heed of Gavaghan's acute observation to ensure that people are not stigmatized as irresponsible menaces to society merely because they harbor inclinations that they would never in fact act upon and that, if questioned about, they would distance themselves from rather than identifying with.

DISEASE AND DISORDER

The next group of chapters investigates whether recent evidence about the brain-based correlates of psychopathy and drug addiction warrants viewing these conditions as mental disorders capable of at least in principle diminishing guilt or mitigation—that is, whether this evidence supports moving away from the moralized and stigmatized views under which psychopaths are bad and addicts are weak and self-indulgent.

In chapter 10, Thomas Nadelhoffer and Walter Sinnott-Armstrong cite mounting empirical evidence suggesting that not only do psychopaths experience a range of significant cognitive, volitional, affective, and attention deficits, but also that these deficits appear to correlate with distinct neural and neurochemical dysfunctions, which in turn appear to have a strong genetic basis. Normally, this kind of evidence might provide compelling support for a defense of insanity, especially if we think that the mental capacities in which psychopaths have deficits are required for moral agency. However, in many jurisdictions, defendants are only entitled to raise the insanity defense if their actions were brought about by a condition classified as a "disease of the mind" or as a "mental disease or defect." But because psychopathy is often moralized and stigmatized—that is, because psychopaths are often viewed as more bad than mad, as evil rather than insane—this appears to prevent them from citing what seems like otherwise-pertinent empirical evidence to raise and support an insanity defense. In effect, when it comes to psychopathy, this kind of evidence is at best ambiguous, and at worst it is treated as evidence of psychopaths' character flaws—as evidence that condemns them for who they are—rather than as evidence of mental incapacities that at least in principle could diminish their degree of guilt for what they do or mitigate at sentencing. After all, who in their right minds would suggest that the law should excuse bad people for the bad things they do simply because they are bad, as demonstrated by neuroscientific evidence of the brain-based implementation of their badness?

For this kind of reason, Nadelhoffer and Sinnott-Armstrong survey five different accounts of mental illness—namely, eliminativist, social constructivist, biomedical, harmful dysfunction, and objective harm accounts—and although they do not themselves endorse any of these accounts, they point out that on any of the plausible accounts, psychopathy *would* qualify as a disease or as mental illness. I have previously argued that once evidence of morally relevant mental incapacity is cited to support a claim of diminished responsibility, nothing further would be gained by pointing out that the incapacity in question was an instance of a disease or disorder (Vincent 2008). Elsewhere, I have also argued that without appropriate criteria to distinguish the correlates of character flaws from mental incapacities, it is unclear whether such evidence as Nadelhoffer and Sinnott-Armstrong cite should excuse or condemn psychopaths (Vincent 2011b). This chapter provides some reasons to reconsider my previous claims.

The next two chapters continue the disease/disorder theme, but their topic shifts to the neuroethics of drug addiction.

In chapter 11, Jeanette Kennett argues that although behaviors associated with drug addiction are not like muscular twitches, epileptic tremors, or ticks of people with Tourette syndrome, this does not entail that they are therefore voluntary behaviors in a full-responsibility-conferring sense either because mounting empirical evidence suggests that in at least a portion of drug users, addiction significantly impairs their capacity for synchronic and diachronic self-control. Gene Heyman has recently argued that addictive behavior can be explained in terms of the same universal principles of motivation and choice that also explain other non-addiction-related behavior—that is, that like anybody else, addicts also choose to do what they most want to do, albeit their choices reveal a bias against long-term and in favor of short-term rewards. This seems to obviate the need for the medical model according to which addiction is a chronic relapsing brain disease—a medical condition in Nadelhoffer and Sinnott-Armstrong's sense that, at least in principle, could mitigate or diminish addicts' responsibility—while lending support for the moral model that views drug use and criminal acts to support it as freely chosen behavior for which offenders may be held responsible.

If taken to heart, Heyman's argument dislodges the medical model of addiction and re-entrenches the moral model that has traditionally stigmatized addicts as weak, self-indulgent individuals and denied them recourse to diminished responsibility defenses. Kennett offers four responses to Heyman. First, she points out that Heyman's examples—for instance, that a substantial portion of addicts spontaneously cease their drug use—fail to lend support to the claim that addiction is not a disease. After all, most people also recover spontaneously from the common cold, but nobody takes that to show that the common cold is not a disease. Second, she points out that Heyman's discussion fails to acknowledge the fact that *some* drug users will fail to cease using drugs even when drug use brings no rewards of any sort but only visits suffering upon them, and even when the drug users themselves are at a loss to explain what could possibly count in favor of their continued drug use. The significance of this point on Kennett's account—and this now brings me to her third point—is that we must distinguish at least two importantly different categories of drug user: those who *will stop* using drugs when presented with sufficient incentives, and those who *will continue* despite acknowledging that continued drug use will visit upon them significant costs that they would rather avoid and deprive them of what they value. These two categories of drug user must be distinguished from one another, according to Kennett, because while the former's behavior evidently issues from mechanisms that are moderately reasons responsive in Fischer and Ravizza's (1998) sense and that thus bestow upon them sufficient mental capacity (in this case, self-control) to warrant treating them as responsible agents—after all, they cease using drugs when presented with sufficient reason to stop—the latter's behavior seems instead to issue from mechanisms that respond only to biological causes and not to reasons. Finally, Kennett cites mounting empirical evidence suggesting that, in at least some users, addiction compromises synchronic and diachronic self-control capacities—it monopolizes attention, the mere perception of stimuli (e.g., familiar environments associated with drug use) evokes drug-use action plans, and these in turn not only present further obstacles in themselves but also over time diminish addicts' will power, which appears to be a limited resource that can become depleted by chronic strain. This leads Kennett to recommend that future research should focus on investigating whether neuroscientific evidence might help courts to distinguish these two very different types of drug users from one another.

In chapter 12, Wayne Hall and Adrian Carter do three things. First, they contrast the medical and moral models of addiction that Kennett also discusses which dominate the debate about whether addicted individuals should be held legally responsible for their drug use and crimes committed to fund it (moral model), or whether their responsibility is at least partly diminished (medical model). Second, they describe current legal practice in dealing with addicted offenders, and summarize empirical data on the effectiveness of different addiction treatment options, including coerced treatment, as well as outline the criminological arguments offered in favor of coerced treatment. Third, they consider three related questions: (1) How plausible is the brain disease model of addiction? (2) How does current legal practice fit with the moral and medical models of addiction? and (3) How might current practice be improved in light of research evidence on the effectiveness of different types of treatment and findings from neuroscience research? Their answer to the first question is that the moral and brain disease models each miss the point: the former ignores the effects that chronic drug use has on addicted individuals' capacity to refrain from continuing to use drugs, but the latter overstates the degree of impaired control in most addicted individuals. This leaves a theoretical vacuum—an alternative account is needed that does justice to both of these views—however, rather than proposing such an account, Hall and Carter only point out the necessity of finding one. In light of the above, their answer to the second question is that current legal practice represents a defensible pragmatic compromise between the brain disease and moral models—namely, that addicted individuals' control over their drug use is impaired but not eliminated, and hence an appropriate alternative to imprisonment is coerced addiction treatment in the community. Finally, in response to the third question, they suggest that current practice could be improved in two ways. One, courts (especially in the United States) should stop using ineffective abstinence-oriented treatment and instead avail themselves of more effective drug-assisted treatments such as opioid substitution treatment for heroin addiction. Although abstinence-oriented approaches might appear more virtuous when viewed from the perspective of the moral model, as indicated above, empirical evidence suggests that appropriate responses to drug addiction should find a midway position somewhere between the moral and medical models. Two, they also recommend that courts should consider trialing Kleiman's proposal to use behavioral triage to decide what form of addicted treatment each offender requires, perhaps with coerced abstinence as the first option and court-ordered treatment reserved for those in whom that option fails.

MODIFICATION

Discussions in the field of neurolaw often stem either from *discoveries* that are made by the mind sciences or from potential opportunities created by new (ways of employing) *diagnostic* techniques and technologies. However, neuroscientific *intervention* techniques are also improving—for instance, transcranial magnetic stimulation, gamma ray scalpels, and improved psychopharmaceuticals. These techniques and technologies, and the opportunities that they create—to treat, to enhance, and generally to modify people's minds—arguably raise an even greater number of vexing and important issues. The final two chapters in this volume investigate how neuroscientific intervention techniques might change the shape of our legal responsibility practices, and whether mental manipulations through direct brain interventions should give rise to a special legal defense.

In chapter 13, I argue that for the same reason responsibility diminishes when mental capacities are lost and it is restored as they are subsequently regained, so, too, responsibility would become enhanced if mental capacities were enhanced, for instance, through the use of cognitive enhancement medications. The ability to stay focused and to think quickly, efficiently, and accurately is critical in professions in which people's lives are at risk (e.g., surgeons, pilots, and soldiers). Minor errors and oversights can have catastrophic effects, and professionals who jeopardize their mental acuity by working while intoxicated can lose their license and even face civil and criminal sanctions. However, recent research suggests that drugs originally designed to treat mental disorders (e.g., methylphenidate, bromocriptine, donepezil, and modafinil) can significantly improve mental performance when taken by healthy individuals. This raises the questions of whether in some circumstances (e.g., surgeons working long shifts in the hospital), professionals in socially important roles might have a moral and maybe even a legal duty to enhance themselves, and of whether their failure or refusal to do this would be a form of negligence or even recklessness. Where technology exists to minimize a substantial risk, if its costs are sufficiently low and its benefits are sufficiently high, then we should normally use that technology. But should this thinking also extend to cognitive enhancement, which involves expecting professionals to use these drugs? May we expect people in socially important roles to enhance themselves to minimize the risk for harm to others? May we expect this even if they would rather not enhance themselves? Although current literature considers issues such as safety, effectiveness, coercion, and justice, these drugs' effects on people's responsibility have barely begun to be investigated, and this leaves society in a morally and legally uncertain position.

The structure of my argument in support of affirmative answers to the above questions is straightforward. After we develop sufficiently effective cognitive enhancement techniques that have few or no significant negative side effects, we could impose on some people a responsibility to cognitively enhance themselves—after all, little would be lost, and potentially great gains could be made. However, once these people have become enhanced, it might then become legitimate to expect them to fulfill new responsibilities, and to be blamed, made liable, and maybe even punished when they fail to live up to these new responsibilities—that is, when they fail to live up to this new cognitively enhanced standard of care.

I take the above to be interesting for at least two reasons. First, because it suggests that the capacitarian idea that responsibility tracks mental capacity holds true even for cases in which mental capacities are enhanced—that is, capacitarianism seems to be vindicated in situations that involve mental capacity enhancement. Second, this scenario highlights the fact that progress made in one area—in this case, in the field of psychopharmacology—might have repercussions that reshape the legal relations between people in other walks of life. In essence, the idea here is that merely by creating new techniques and technologies—in this case, ones that affect mental function—we may bootstrap new legal obligations into existence first by expecting people to avail themselves of those technologies, and then by inflating the standard of care such that it comes into line with people's technologically enhanced capacities. This phenomenon is not new—the introduction in the 1950s of pagers for use by physicians is one prior example, and the development, refinement, and proliferation of cost-effective diagnostic techniques like the x-ray is another. But what *is* new is that, on this occasion, the technology involved would be used not as an external tool but rather as something with which to sharpen our brains.

Finally, in chapter 14, Christoph Bublitz and Reinhard Merkel investigate whether the responsibility of agents whose minds have been tampered with by others—for instance, those whose preferences have been nonvoluntarily or involuntarily changed through direct brain-intervention-based techniques—would be diminished, and if so, then why, and if not, then why not, and whether in light of this, the law should recognize a brainwashing defense.

In the philosophical compatibilist literature, it is often claimed that agents whose minds have been manipulated by others would not be responsible for what they do as a consequence of those manipulations until they have made those manipulations *their own*. For instance, according to Fischer and Ravizza's influential account of guidance control, people are only responsible for those actions that issue from *their own* moderately reasons-responsive mechanisms, or when they are responsible for the fact that those mechanisms are not moderately reasons responsive (1998:227). That is, in addition to needing the right *mental capacities* and satisfying the *tracing condition*, it is crucial that this *ownership condition* is met. Although until recently this may have sounded like a largely academic debate, several examples can be cited to explain why advances in the mind sciences have made this a very practical and urgent topic. Consider an example provided by Carter and colleagues:

> There are cases...in which DRT [dopamine replacement therapy] induces behaviour that individuals claim [is] authentic. For example, one male who became fascinated with anal sex following DRT claimed that he had these desires prior to DRT treatment but was too embarrassed to act on them. The medication allowed him to "realise these desires." His interest in these sexual behaviours stopped following a change in his medication, and he later expressed regret at his behaviour. (Carter, Ambermoon et al. 2011:95–96)

Given that in some cases patients taking DTR commit crimes (e.g., see Tasmania v. Martin 2011), this raises the question of whether patients who have been affected like the patient whom Carter and colleagues mentioned should be held responsible for what they do, and if not, then whether this is because the medications have undermined the ownership condition. Similar issues are raised by the reported effects that deep brain stimulation (DBS) sometimes has on patients (Klaming and Haselager 2010), and by the effects that the drug methylphenidate (Ritalin), used to treat attention deficit/hyperactivity disorder as well as allegedly for cognitive enhancement purposes, has reportedly had (Bolt and Schermer 2009). In a recent opinion piece, Hank Greely provides another pertinent example:

> As neuroscience learns more about the causes of human behaviors, it will give us new ways to change those behaviors. When behaviors are caused by "brain diseases," effective actions that intervene directly in the brain will be readily accepted, but what about direct brain interventions that treat brain-based causes of socially disfavored behaviors that are not generally viewed as diseases? (2012:163)

Putting aside the interesting and troubling issues that Greely wishes to highlight, one might legitimately worry about what would happen to the responsibility of involuntarily treated criminals. Would they be responsible for their actions upon their release from custody subsequent to successful treatment, or would the fact that their minds were involuntarily altered undermine the ownership condition and thus their responsibility for what they subsequently do? Finally, an example from Bublitz and Merkel:

> Rumors have it that some Casinos secretly spray the odorless substance Oxytocin in gambling halls to increase players' trust in their luck. Suppose Oxytocin disposes gamblers to be overconfident and to take higher risks—would this be permissible, or could players claim their stakes back? What if they commit illicit acts under Oxytocin's influence—are they responsible?

Bublitz and Merkel argue that although in some cases we might indeed have reason to be more lenient toward people whose minds have been tampered with in such ways, this lenience should not be viewed as an instance of diminished responsibility but rather as mitigation of liability, and that what explains why mitigation might be appropriate is not that such manipulations might affect these people's mental capacity ownership but rather that they might significantly alter legal relations between people in the sociolegal domain. Bublitz and Merkel's legal background helps them to notice an important point that may otherwise be easily overlooked, namely "the contemporary tendency to locate the source of mental illness, life problems, and even the current financial crises in individual deficits and psychological or neuronal processes. But with this perspective we run the risk of losing sight of social and structural levels" of explanation. Consequently, they recommend against the introduction or recognition of a brainwashing defense because, after all, the *tracing condition* already suffices to explain the intuition that lenience toward manipulated parties might be appropriate.

CONCLUSION

The addition of a mental element to the criteria for legal responsibility all the way back in the 13th century opened up a dialogue between science and law about the law's understanding of human agency and about its related notion of responsibility. The rich story told by the chapters in this volume about the myriad ways in which advances in the mind sciences have the potential to affect the moral foundations of legal responsibility practices testifies to the fact that far from *ending* this dialogue, advances in the fields of neuroscience, psychology, and behavioral genetics have the potential to *enrich* and *extend* this dialogue even further.

NOTES

1. My work on this volume was supported by funding from plural sources: the Netherlands Organization for Scientific Research (NWO), the 3TU Centre for Ethics and Technology, Delft University of Technology, and Macquarie University. The first draft of this volume, which emerged from papers presented at

the *Moral Responsibility: Neuroscience, Organisation and Engineering* conference held in Delft in August 2009, was put together while I worked at Delft University of Technology on Gert-Jan Lokhorst's project, *The Brain and the Law*, a sub-project of his parent project, *Neuroethics: Ethical, Legal and Conceptual Aspects of Neuroscience and Neurotechnology* (NWO Dossiernr: 360-20-180). The second draft of this volume, which saw substantial revisions as well as the addition of new chapters, was put together while I worked on my Macquarie University–funded project, *Reappraising the Capacitarian Foundation of Neurolaw: On the Assessment, Restoration and Enhancement of Responsibility.* I am deeply grateful to everyone who has helped me during the process of editing this volume both for refereeing and providing useful comments on all of the chapters and also for their emotional support. Special thanks go to Walter Sinnott-Armstrong for suggesting that this volume should be put together in the first place; to Gert-Jan Lokhorst and Jeanette Kennett whose faith in me, encouragement, and moral support gave me the strength to bring this project through to completion; and most important, to the wonderful authors whose chapters populate this volume because without their work, this volume would not exist. NAV.

2. I say "closely linked to" rather than something stronger like "equivalent to" or "nothing more than" in order to leave conceptual room for the thesis that cognition is *embodied* and *extended*.
3. I set aside earlier compatibilist innovations for the sake of brevity.
4. These may be *brain* mechanisms, but given that the mind might be *embodied* or *extended*, this should not be stipulated as a necessary condition.
5. Not surprisingly given that compatibilism aims to preserve our intuitions about responsibility.
6. Given what has been written on the topic of the alleged *directness* and *objectivity* of neuroimages (e.g. see Logothetis 2008; or Roskies and Sinnott-Armstrong 2011), I use the words "objectively" and "directly" with some trepidation. These words should thus be understood to express an *ideal* rather than as descriptions of what the current neuroimaging techniques and technologies are capable of delivering.

REFERENCES

Bolt, I., and M. Schermer (2009). "Psychopharmaceutical enhancers: Enhancing identity?" *Neuroethics* 2(2): 103–111.

Carter, A., P. Ambermoon, et al. (2011). "Drug-induced impulse control disorders: A prospectus for neuroethical analysis." *Neuroethics* 4(2): 91–102.

Cashmore, A. R. (2010). "The Lucretial swerve: The biological basis of human behavior and the criminal justice system." *Proceedings of the National Academy of Sciences of the United States of America* 107(10): 4499–4504.

Dawkins, R. (2006). "Let's all stop beating Basil's car." Retrieved December 2, 2007, from http://www.edge.org/q2006/q06_9.html—dawkins.

Dennett, D. C. (1984). *Elbow Room: The Varieties of Free Will Worth Wanting.* Cambridge, MA, MIT Press.

Eigen, J. P. (2004). "Delusion's odyssey: Charting the course of Victorian forensic psychiatry." *International Journal of Law and Psychiatry* 27(5): 395–412.

Fischer, J. M., and M. Ravizza (1998). *Responsibility and Control: A Theory of Moral Responsibility.* Cambridge, UK, Cambridge University Press.

Gazzaniga, M. S. (2006). Facts, fictions and the future of neuroethics. In: *Neuroethics: Defining the Issues in Theory, Practice, and Policy.* J. Illes. Oxford, UK, Oxford University Press, pp. 141–148.

Greely, H. T. (2012). "Direct brain interventions to treat disfavored human behaviors: Ethical and social issues." *Clinical Pharmacology and Therapeutics* 91(2): 163–165.

Greene, J., and J. D. Cohen (2004). "For the law, neuroscience changes nothing and everything." *Philosophical Transactions of the Royal Society of London* 359(1451): 1775–1785.

Hart, H. L. A. (1968). IX. Postscript: Responsibility and retribution. In: *Punishment and Responsibility.* Oxford, UK, Clarendon Press, pp. 210–237.

Klaming, L., and P. Haselager (2010). "Did my brain implant make me do it? Questions raised by DBS regarding psychological continuity, responsibility for action and mental competence." *Neuroethics* OnlineFirst: 13 pages.

Logothetis, N. K. (2008). "What we can do and what we cannot do with fMRI." *Nature* 453(7197): 869–878.

Morse, S. J. (2003). "Diminished rationality, diminished responsibility." *Ohio State Journal of Criminal Law* 1: 289–308.

Morse, S. J. (2006). "Brain overclaim syndrome and criminal responsibility: A diagnostic note." *Ohio State Journal of Criminal Law* 3: 397–412.

Morse, S. J. (2011). NeuroLawExuberance: A plea for neuromodesty. In: *Technologies on the Stand: Legal and Ethical Questions in Neuroscience and Robotics.* B. van den Berg and L. Klaming. Nijmegen, The Netherlands, Wolf Legal Publishers, pp. 23–40.

Roskies, A., and W. Sinnott-Armstrong (2011). Brain images as evidence in the criminal law. In: *Law and Neuroscience: Current Legal Issues, Volume 13.* M. Freeman. Oxford, UK, Oxford University Press, pp. 97–114.

Sapolsky, R. M. (2004). "The frontal cortex and the criminal justice system." *Philosophical Transactions of the Royal Society of London* 359: 1787–1796.

Tasmania v. Martin, No. 2, 2011 TASSC 36 (Australia).

Vincent, N. (2008). "Responsibility, dysfunction, and capacity." *Neuroethics* 1(3): 199–204.

Vincent, N. (2011a). "Legal responsibility adjudication and the normative authority of the mind sciences." *Philosophical Explorations* 14(3): 315–331.

Vincent, N. (2011b). Madness, badness and neuroimaging-based responsibility assessments. In: *Law and Neuroscience, Current Legal Issues, Volume 13.* M. Freeman. Oxford, UK, Oxford University Press, pp. 79–95.

Vincent, N., and I. van de Poel (2011). Introduction. In: *Moral Responsibility: Beyond Free Will and Determinism.* N. Vincent, I. van de Poel, and J. van de Hoven. New York, NY, Springer, pp. 1–14.

Wallace, R. J. (1994). *Responsibility and the moral sentiments.* Cambridge, MA, Harvard University Press.

PART 1

Responsibility and Mental Capacity

2

Common Criminal Law Compatibilism

STEPHEN J. MORSE

INTRODUCTION

The common criminal law is a beautiful, albeit sometimes ramshackle, institution devoted to blaming and punishing culpable agents that has been developing for well over half a millennium. In this chapter, I shall use the term "common law" more broadly to include its usual definition as judge-made law, and also penal legislation, and the interaction between legislatures (influenced by politics, of course) and judges that further develops the penal law.[1] Although all US jurisdictions have a statutory penal code, the criminal law can be fully comprehended only if the accretion of interpretations by the courts is considered. Indeed, most penal codes are based on preexisting common law. Many legislative changes are responses to common law innovations, changes that will then be interpreted by the courts. So it goes. The common law is the product of an immense number of penal statutes and judicial decisions over the years, and it has stood the test of time as the product of human trial and error. We common lawyers like to think that it is impossible to produce an ex ante watertight criminal code and that the bottom-up, "organic" methodology of the common law process will ultimately produce reasonably coherent and just, but not perfect, criminal law.

The thesis of this chapter is simple and straightforward. The criminal law is a thoroughly folk psychological enterprise that is completely consistent with the truth of determinism or universal causation. It does not claim that judges and legislators throughout the centuries of development of modern criminal law explicitly adopted the compatibilist position in the metaphysical debate about determinism, free will, and responsibility. Far from it. Most criminal justice actors are probably implicitly libertarian and believe that we somehow have contra-causal free will. To be sure, criminal law doctrine and practice are also fully consistent with metaphysical libertarianism, but I consider this metaphysical view extremely implausible in the modern, scientific age.[2] Human beings, as complex as they are, are still part of the physical universe and subject to the same laws that govern all phenomena. In short, I believe that libertarianism does not furnish a justifiable foundation for an institution that is essentially about blaming and punishing culpable agents. If determinism or something quite like it is true, as I assume it is, then only compatibilism provides a secure basis for criminal responsibility.

I do not aim to argue for the truth of compatibilism. It is sufficient that it is one of the two plausible positions in the metaphysical debate—the other being hard determinist incompatibilism—and, in one form or another, it is probably the position held by the vast majority of professional philosophers.[3] Thus, I will simply assume that compatibilism is a thoroughly acceptable, albeit contestable, metaphysics. I am not suggesting that compatibilism is monolithic; there are many compatibilisms that differ widely in details. But all have in common the possibility of genuine responsibility in a deterministic or universally caused universe, and this is a plausible, defensible position. Based on it, I will try to demonstrate that it provides an accurate positive account of criminal law and that compatibilist criminal law is normatively desirable.

I begin with a brief account of desert–disease jurisprudence that explains under what conditions the state may constrain a person's liberty. The purpose is to situate in legal context the place of criminal justice as a means of social control. Next, the chapter offers an introduction to the criminal law's implicit psychology and view of the person that underpins desert–disease jurisprudence. Following that, the chapter offers a nutshell summary of foundational assumptions of criminal law and of criminal law doctrine. The next section demonstrates that the doctrines and practices we have are fully consistent with the truth of determinism. Then, the chapter turns to the external challenges to compatibilist criminal law and demonstrates why they do not succeed and would lead to normatively undesirable doctrines and practices. The final section addresses the normative desirability of compatibilist criminal law.

DESERT–DISEASE JURISPRUDENCE

At present, the state's ability to deprive people of their liberty is constrained by desert–disease jurisprudence. The state may imprison people in the criminal justice system if they deserve punishment for crimes they have committed,[4] and it may civilly commit dangerous people if they are not responsible agents—usually because they have a mental abnormality, such as a major mental disorder.[5] Otherwise, with rare, limited exceptions, such as rejecting bail for dangerous agents,[6] the state must leave people at liberty even if they arepotentially dangerous. It is crucial to recognize that the criminal law is distinguishable from noncriminal regulation of undesirable conduct because it is essentially communicative and "speaks with a distinctly moral voice."[7]

The concern with justifying and protecting liberty that produces the desert–disease constraints is deeply rooted in the conception of rational personhood. Only human beings self-consciously and intentionally decide how they should live; only human beings have projects that are essential to living a good life. Only human beings have expectations of each other and require justification for interference in each other's lives that will prevent seeking the good. If liberty is unjustifiably deprived, a good life is impossible. In sum, both the criminal and the medical-psychological systems of behavior control require a justification in addition to public safety—desert for wrongdoing or nonresponsibility (based on disease)—to justify the extraordinary liberty infringements that these systems impose.

Virtually all criminals are rational, responsible agents, and according to the dominant story, the deprivation imposed on them—punishment—is premised on considerations of desert. No agent should be punished without desert for wrongdoing, which exists only if the agent culpably caused or attempted prohibited harm. The threat of punishment for a culpable violation of the criminal law is itself arguably a form of preventive infringement on liberty, but it is an ordinary, "base-rate" infringement that requires no special justification. After all, no one has a right to harm other people unjustifiably. In our society, the punishment for virtually all serious crimes, and thus for dangerous criminals, is incapacitation, which is preventive during the term of imprisonment. But criminals must actually have culpably caused or attempted harm to warrant the intervention of punishment. We cannot detain them unless they deserve it, and desert requires culpable wrongdoing. In the interest of liberty, we leave potentially dangerous people free to pursue their projects until they actually offend, even if their future wrongdoing is quite certain. We are willing to take great risks in the name of liberty.

For people who are dangerous because they are disordered or because they are too young to "know better," the usual presumption in favor of maximal liberty yields. Because the agent is not rational or not fully rational, the person's choice about how to live demands less respect, and the person is not morally responsible for his or her dangerousness. The person can therefore be treated more "objectively," like the rest of the world's dangerous but nonresponsible instrumentalities, ranging from hurricanes to microbes to wild beasts.[8] In brief, agents with sufficiently compromised rationality do not actually have to cause or attempt harm to justify nonpunitive intervention. We can take preemptive precautions, including broad preventive detention, with nonresponsible agents based on an estimate of the risk they present. Justified on consequential grounds, such deprivation will be acceptable if the conditions of deprivation are both humane and no more stringent than is necessary to reduce the risk of harm. Such deprivations are forms of greater or lesser quarantine and may include "treatment," but in theory they are not punishment, and they should never have a punitive justification or effect.[9]

In sum, the normative basis of desert–disease jurisprudence is that it enhances liberty and autonomy by leaving people free to pursue their projects unless they responsibly commit a crime or unless they are nonresponsibly dangerous through no fault of their own. This scheme maximizes respect for liberty, dignity, and autonomy. Responsible agents are left free on the theory that a rational agent may always recognize the wrongness and danger to oneself of criminally infringing the legitimate interests of others. Therefore, the state may not intervene unless the agent has attempted or committed a crime. If agents are not responsible for their danger, then the usual presumptions in favor of liberty and autonomy yield because they are based on rational agency.

This is a satisfyingly neat account, but desert–disease jurisprudence leaves a "gap." It provides no mechanism to restrain dangerous people who may not be punished because they have committed no crime or have completed their sentence and who may not be civilly committed because they are responsible agents. The state tries to fill the gap by expanding desert and disease jurisprudence. For example, desert jurisprudence is widened by lengthening sentences or by recidivist sentencing laws. Disease jurisprudence is expanded by widening the definition of nonresponsibility, most egregiously in sex offender civil commitment statutes. Both types of expansion are open to substantial normative and practical problems, however, so the gap inevitably remains. There is no seamless set of alternative means to preventively detain dangerous agents. Nonetheless, legislatures continue to try, and the Courts usually approve.[10]

THE IMPLICIT PSYCHOLOGY AND CONCEPT OF THE PERSON IN CRIMINAL LAW

What is the law's implicit psychology and concept of the person that underpins desert–disease jurisprudence and criminal justice more generally? Lawyers take the law's implicit psychology for granted because there is seldom any need to identify or to question it. It is crucial to recognize it consciously, however, in order to understand the doctrines and practices of criminal law.

The law presupposes the folk psychological view of the person and behavior. "Folk psychology" is the term used to identify any psychological theory that causally explains behavior *in part* by mental states such as desires, beliefs, intentions, and plans. Biological, other psychological, and sociological variables also play a role, but folk psychology considers mental states fundamental to a full explanation of human action. Human behavior cannot be adequately understood if mental state causation is completely eliminated. Lawyers, philosophers, and scientists do, of course, argue about the definitions of mental states and about theories of the nature of human action, but that does not undermine the general claim that mental states are fundamental. Indeed, the arguments and evidence disputants use to convince others presuppose the folk psychological view of the person. Brains do not convince each other; people do.

Folk psychology does not presuppose the truth of free will, it is perfectly consistent with the truth of determinism, it does not hold that we have minds that are independent of our bodies (although it, and ordinary speech, sound that way), and it presupposes no particular moral or political view. It does not claim that all mental states are necessarily conscious or that people go through a conscious decision-making process each time that they act. It allows for "thoughtless," automatic, and habitual actions and for nonconscious intentions. The definition of folk psychology being used here does not depend on any particular bit of folk wisdom about how people are motivated, feel, or act. Any of these bits—for example, that people intend the natural and probable consequences of their actions—may be wrong. The definition of folk psychology presupposes and insists only that human action can at least be partially explained by mental state explanations or that it will be responsive to reasons, including incentives, under the right conditions.

Brief reflection should indicate that the law's psychology must be a folk psychological theory, a view of the person as a conscious (and potentially self-conscious) creature who forms and acts on intentions that are the product of the person's other mental states, such as desires, beliefs, and plans. We are practical reasoners, the sort of creatures that can act for and respond to reasons. The law does not treat persons generally as nonintentional creatures or mechanical forces of nature. As a system of rules and standards that guides and governs human interaction, law tells citizens what they may and may not do, what they must or must not do, what abilities are required competently to perform certain tasks, and what consequences will follow from their conduct. Law is primarily action guiding[11] and could not guide people ex ante and ex post unless people could use rules and standards as premises in their reasoning about how they should behave. Otherwise, law as an action-guiding normative system would be useless, and perhaps incoherent.[12] Law can directly and indirectly affect the world we inhabit only by its influence on human beings who can potentially use legal rules to guide conduct. Unless people were capable of understanding and then using legal rules to guide their conduct, law would be powerless to affect human behavior.[13]

The legal view of the person does not hold that people must always reason or consistently behave rationally according to some preordained notion of what rationality entails. Rather, the law's view is that people are capable of acting for reasons and are capable of minimal rationality according to predominantly conventional, socially constructed standards. What the law requires is the ordinary person's common-sense view of rationality, not the technical notion that might be acceptable within the disciplines of economics, philosophy, psychology, computer science, and the like. Virtually everything for which people deserve to be praised, blamed, rewarded, or punished is the product of mental causation[14] and, in principle, is responsive to reason. People are intentional agents with the potential to act—to violate expectations of what they owe each other and to do wrong.

Many scientists and some philosophers of mind and action consider folk psychology to be a primitive or prescientific view of human behavior.[15] For the foreseeable future, however, the law will be based on the folk psychological model of the person and behavior. Until and unless scientific discoveries convince us that our view of ourselves is radically wrong, a claim discussed in the penultimate section of this chapter, the basic explanatory apparatus of folk psychology will remain central. It is vital that we not lose sight of this model lest we fall into confusion when various claims about criminal responsibility are made. Any evidence allegedly relevant to criminal responsibility must be translated into the folk psychological framework.

CRIMINAL LAW ASSUMPTIONS

The predominant view is that criminal law is generated by the harm and fault principles. The harm principle guides which actions and omissions the criminal law should prohibit. There is great controversy, of course, about what types of harms should be criminalized, but virtually all criminal law theorists believe that the decision to criminalize behavior should be based on the harm that behavior causes or risks.[16] The fault principle guides judgments about which agents may fairly be blamed and punished and how much they should be punished. This is the focus of this chapter. As we shall see, indicia of fault are part of the definition of virtually all crimes.

Criminal blame and punishment entail the intentional infliction of pain by the state and therefore require substantial moral justification. Although the Supreme Court of the United States has never held that the Constitution requires a jurisdiction to adopt any particular justification of punishment, there is widespread agreement that retributivism and consequentialism (especially deterrence and incapacitation) are the leading justifications. Retributivism is a form of deontological justice that holds that offenders should receive their just deserts. No one should be punished unless an agent deserves such treatment, and no one should be punished more than they proportionately deserve for their criminal conduct. Retribution is not revenge, of course, and contrary to what many people think, retribution is silent about how much punishment is deserved in an ultimate sense. Unless one is a moral realist who thinks that there is in principle a right answer to how much punishment of what type each offender deserves, one can be either a tough or tender retributivist. Retributivism is not committed to harsh punishments. Finally, a retributivist can be either obligatory or permissive. The former argues that the state is obliged to give offenders their just deserts. The latter thinks that the state has permission to impose just deserts, but need not do so if it has good reason not to. This is an important distinction, but for the purposes of this chapter, it is not relevant.

Consequentialists think that the major goal of punishment is to prevent crime by general and specific deterrence and by incapacitating offenders. A purely consequential theory would not punish even a blameworthy offender if no social good would be achieved by punishment, and it would apportion punishment to maximize crime prevention. In its pure form, the offender's desert would be neither a necessary nor a limiting principle because desert bears no necessary relation to the goal of crime prevention. Notions of fault would simply be epistemic guides for deciding when prevention is maximized by punishing an offender.

Most criminal law theorists are not pure consequentialists, but a substantial number are pure retributivists. The majority are probably mixed theorists who believe that desert is a necessary condition for just punishment and that desert should set a proportionality limit on how much punishment an offender receives. Within the range of justifiable punishment that desert sets, the punishment imposed may be adjusted to achieve consequential goals. Thus, among most criminal law theorists, desert plays a crucial role because it is both a necessary and limiting condition on punishment.

If one examines the penal codes and appellate court decisions that explicitly address the justifications for punishment, the picture is less clear. Many penal codes contain a provision setting forth the jurisdiction's justifications, but these are usually just vague lists of retributive and consequential concerns and give no guidance about how they should be weighed, balanced, and applied. Nonetheless, there has been an apparent and major shift in penal policy in the United States beginning in the 1960s. For much of the 20th century,

American sentencing schemes strongly favored indeterminate sentencing with very wide ranges for most serious crimes. Although in theory and to some degree in practice, desert set a cap to the permissible ranges, the justification for this practice was largely a rehabilitation model. Deciding when to release the prisoner would depend on the professional judgment of the correctional officials concerning the prisoner's rehabilitation progress. At the maximal term, the prisoner would of course have to be released even if he had not been rehabilitated and was still dangerous, but the maximum was typically quite long. Thus, lengthy incapacitation was possible, and many otherwise unrehabilitated prisoners would simply "age out" of future violent conduct.

Although the indeterminate sentencing model was conceptually coherent, in practice it was subject to severe problems that led to its demise in many jurisdictions. First, it led to unprincipled and arbitrary discretion. Correctional officials had virtually unreviewable authority to make release decisions, but they had neither a coherent conceptual approach to such problems nor accurate predictive validity about who would be dangerous if released. The alleged rehabilitative programs available in the prisons were either paltry or unvalidated. In a word, the officials were "flying blind," and this produced differential treatment that could not possibly be justified and was widely believed to be racially biased. In addition, many believed that the sentences actually served and the maximal terms bore insufficient relation to the punishment that offenders deserved.

In the late 1960s and early 1970s, the critiques of indeterminate sentencing reached an unusually bipartisan crescendo. Critics and politicians from across the political spectrum called for a new regime that would tie offenders' punishments to their just deserts, and the legislatures responded.[17] Simultaneously, and in response to concerns about arbitrary discretion and unjustifiably unequal treatment, legislatures also began to adopt determinate sentences with relatively limited ranges for most crimes and sentencing guidelines that limited judicial sentencing discretion to various degrees. Although the new regime also faces substantial criticisms, it has been adopted in a considerable number of jurisdictions, including the federal criminal jurisdiction, and it exerts influence on those jurisdictions that still retain more indeterminate sentencing. Most recently, the influential American Law Institute's revision of the general Model Penal Code provisions on the purposes of sentencing replaced the previously consequential aims with a balance of just deserts proportionality and consequential goals.[18]

In sum, among criminal law theorists and in the positive law, desert plays a central role in the justification for criminal blame and punishment. The final assumption, addressed previously when considering desert–disease jurisprudence, is that only offenders who are genuinely culpable, who are genuinely at fault, deserve punishment. Let us therefore turn to the criminal law's criteria for blame and punishment.

CRIMINAL LAW DOCTRINE

Crimes are defined by what lawyers call their "elements," the criteria the prosecution must establish beyond a reasonable doubt to prove prima facie guilt. If the prosecution cannot prove any requisite element, the defendant must be acquitted (but may be convicted of another similar crime that does not require the missing element). Even if a defendant is prima facie guilty, the defendant may still avoid criminal liability by establishing what is known as an affirmative defense.[19] Let us consider the meaning of all these terms and their relation to culpability.

There are five types of elements that define crimes: acts (sometimes referred to as "conduct" elements), an accompanying mental state (termed *mens rea*), attendant circumstances, results, and causation (in cases in which there is a result element). Here is an example of a crime that involves all five: "the intentional killing of a police officer knowing that the victim is a police officer." The act element is any type of intentional killing conduct, such as shooting, stabbing, bludgeoning, and the like. The accompanying *mens rea* is the intent to kill. If the agent shoots with the intent only to wound or to scare the victim, the defendant lacks the intent to kill (but may be guilty of a different type of homicide crime if the victim dies). The attendant circumstance is that the victim is a police officer and this element has its own *mens rea*: the defendant must know the victim is an officer. If the victim is not a police officer or the defendant does not know that the victim is an officer, the defendant cannot be convicted of this particular homicide crime. The result element in this crime is the victim's death. If the victim does not die, at most the defendant can be convicted of an attempt to commit this crime. Finally, the defendant's action must be the legal cause of the death. This is a very complicated requirement, but, in brief, it means that how death occurred should not be a remote or unforeseeable consequence of the defendant's action. For example, suppose the defendant inflicts what would be a mortal wound, but the emergency department doctor commits gross malpractice, and the victim dies immediately as a result of the doctor's error. In such a case, the so-called causal chain between the defendant's conduct and the victim's death might be "cut," and once again, the defendant could be convicted only of attempting to kill the police officer.

All criminal statutes require an act element.[20] Although the meaning of this element is often not clear in statutes and the cases that interpret them, here is a rough approximation of the meaning. An act is an intentional bodily movement (or omission) performed by an agent whose consciousness is reasonably intact. Thus, a reflex or convulsive movement of the body would not satisfy the act requirement even if it causes harm to another because it is not an action at all. In most US jurisdictions, an act performed in a substantially dissociated state of consciousness, such as sleepwalking, is also not considered an action, although it may appear responsive to the environment and goal directed. The rationale for an act requirement for criminal guilt should be apparent. If the defendant's movements were not his actions, they cannot be imputed to him as an agent and be the foundation for culpability. The harm caused is simply not the defendant's fault, and it would be unjust to blame and punish the agent.

Mens rea terms, such as intention, purpose, knowledge, recklessness (conscious awareness of a substantial risk that another element is satisfied and disregard of the risk), and negligence (the failure to be aware of a substantial risk that a reasonable person should have been aware of) have their ordinary language, common-sense meanings. No degree of commitment or rationality is included in the definitions of *mens rea*. An act committed ambivalently for irrational reasons is considered intentional if it was done on purpose.

An agent who does not act is acquitted outright. Similarly, if the agent lacks a requisite mental state, the agent is also not prima facie criminally responsible and must be acquitted outright of the crime requiring that mental state. Let us return to the example concerning the intentional killing of a police officer. In Clark v. Arizona,[21] defendant Eric Clark shot and killed a police officer who was in full uniform and made a traffic stop of Clark's vehicle using the officer's marked police vehicle. Clark indisputably suffered from paranoid schizophrenia and was delusional. He claimed that he delusionally believed that the officer was a space alien impersonating a police officer. If Clark's testimony was believed, the shooting was intentional action, but he did not intend to kill a human being knowing the victim was a police officer. He intended to kill a space alien. Thus, he could not be guilty of intentional homicide of any sort, but he might have been guilty of a lesser offense such as negligent homicide because his belief that the officer was a space alien was unreasonable. Although a person with severe mental disorder perhaps cannot be expected to behave reasonably, the standard for negligence is objective and based on how a hypothetical reasonable person should have behaved.[22]

Even if the agent is fully prima facie responsible, however, the agent ultimately may still not be criminally responsible if an affirmative defense can be established. A defendant raising an affirmative defense is claiming that although his behavior satisfied the definitional elements of the crime, there is some further reason why the defendant should not be held responsible. Affirmative defenses can be either justifications or excuses. In the case of justifications, the defendant was a responsible agent, but his behavior in the specific circumstances was the permissible or perhaps right thing to do.

Self-defense, in which an agent intentionally uses otherwise unlawful force proportionately to prevent a wrongful attack on the agent, is the classic example. In cases of excuse, the defendant has done the wrong thing, but for some reason should not be held responsible. The two primary examples in US and English criminal law are legal insanity (essentially a rationality defect) and duress (a compelling hard-choice situation, such as a "do-it-or-else" threat at gunpoint).

Like the definitions of crimes, affirmative defenses also have specific criteria. Consider first the insanity defense, using the Model Penal Code test as an example.[23] To be found legally insane, at the time of the crime the defendant must have been suffering from a mental disorder and, as a result, lacked substantial capacity to appreciate the criminality of his action or to conform his action to the requirements of the law. Thus, if the defendant is not sufficiently disordered to meet the law's definition of mental disorder, or did not lack the requisite substantial capacity, a legal insanity defense will fail. The justification for the excuse is that a defendant who does wrong but is sufficiently irrational at the time is not a rational agent who deserves blame and punishment. Note, by the way, that the criteria for mental disorder in a statute are legal criteria and they need not and often do not reflect established psychiatric and psychological diagnostic criteria.

Now consider the Model Penal Code standard for the affirmative defense of duress.[24] Duress is established if the defendant is threatened with wrongful force unless he harms another and a person of reasonable firmness would have yielded in this situation. The common law standard was even stricter, granting the defense only if the threat was death or grievously bodily harm, and it did not apply to homicide crimes. The justification for the excuse is that demanding agents to accept harm to themselves rather than to harm another asks too much of agents in some circumstances. The choice is simply too hard. The defense will fail if the threat was not of sufficient magnitude, say, a threat to destroy only the defendant's valued property, or if a person of reasonable firmness would not have yielded, say, killing five people to save one's own life.

In cases of excuse, the defendant will be acquitted if the defense is established, but defendants who have been acquitted by reason of legal insanity may be committed to a secure hospital if they are still mentally disordered and dangerous. The Supreme Court has affirmed that such commitments are not punishment[25]—the defendant cannot be punished because he has been acquitted—and that the commitments must end if either criterion for the commitment, disorder or dangerousness, is no longer present.[26]

Criminal law has many wrinkles that have not been addressed,[27] and the details vary from jurisdiction to jurisdiction, but the foregoing is a modal account of the essential conditions for criminal culpability. The defendant will be culpable for the crime charged if all the elements are proved and no affirmative defense is established. In contrast, the defendant will be found not guilty of the crime charged if either the elements cannot be proved beyond a reasonable doubt or an affirmative defense is established.

For the purpose of this chapter, it is crucial to recognize that libertarian free will is not an element of any crime or of any affirmative defense. To establish prima facie guilt, the prosecution *never* needs to prove that the defendant had free will. To establish an affirmative defense, the party with the burden of persuasion *never* needs to prove the presence or absence of free will. To defeat the prosecution's prima facie case, the defendant must negate the elements of conscious, intentional action and *mens rea*; to establish an affirmative defense, the defense must introduce sufficient evidence of the criteria for the defense. People will often say that a defendant who acted under duress or who was legally insane lacked free will. In such cases, however, free will is simply a confusing and conclusory way of saying that the legal criteria for excuse were met. Lack of free will independent of the behavioral legal criteria for excuse does no work whatsoever in explaining why such a defendant is excused.

For a final confirmation of the thesis that free will plays no role in the positive criteria for criminal responsibility, consider the US Supreme Court's recent decision, Clark v. Arizona, in which the Court had one of its rare opportunities to clarify the relation between *mens rea* and legal insanity.[28] The questions presented were whether Arizona's unusually narrow insanity defense test, which asked only if the defendant could distinguish between right and wrong, violated substantive due process and whether an Arizona rule that excluded virtually all expert evidence concerning mental disorder offered for the purpose of negating *mens rea* violated procedural due process. Although legal insanity and the presence of *mens rea* are probably the criminal law issues to which free will is allegedly most relevant, and although there was extensive discussion of the history of legal insanity and of the role of *mens rea*, the Court did not so much as mention free will to decide the issues. There were many problems with the Court's analysis, especially of the *mens rea* issue,[29] but failure to discuss free will was not among them. In short, free will or lack of it is not a criterion for criminal responsibility or nonresponsibility. Once again, it is irrelevant to the actual practice of criminal law.

It is true that many lawyers, including Supreme Court justices, lower court judges, practicing lawyers, and some law professors, talk as if free will were important in criminal law, but this is clearly wrong as a matter of positive law.[30] As noted, when they use the locution, it is simply a confused proxy for the conclusion that some genuine culpability doctrine was or was not present. They sometimes mean, however, that free will is a necessary foundational justification for responsibility, even if it is not a criterion in any legal doctrine.

The next section, which discusses the metaphysical free will problem, demonstrates that this assumption is not necessary to justify responsibility doctrines and practices under a compatibilist account of the relation of determinism, free will, and responsibility.

THE COMPATIBILITY OF CRIMINAL LAW AND COMPATIBILISM

Readers of this chapter will be familiar with the metaphysical determinism–free will–responsibility debate, so I will not rehearse the various views. It is true that many people think that libertarian free will is necessary for criminal responsibility. As I noted earlier in the chapter, I do not consider libertarianism a plausible metaphysics; with P. F. Strawson, I believe it is "panicky."[31] If those who think that libertarianism is true are incorrect, then many legal doctrines and practices, especially those relating to responsibility, may be entirely incoherent. But, as we shall see, on the most natural reading of what ordinary people mean by claiming that free will is foundational, it is not necessary even for them, and determinism is not inconsistent with current responsibility doctrines and practices.

Based on my reading of the cases and legal commentary that mention free will and from talking with numerous judges and lawyers about this question, I believe that most criminal justice system participants and commentators have not even thought about and certainly do not have genuine understanding of the metaphysical debate and do not situate their views in that debate. I believe that most confuse libertarian free will with freedom of action, the ability to do what one wants. I recognize that one interpretation of hard determinism is that agents cannot do what they want because there is no alternative possibility other than the action that was performed. But this position is contested, and it is entirely consistent with determinism that people can act freely in the sense of doing what they choose to do based on their reasons for action and doing so without compulsion. Determinism is not fatalism.[32] As I have already mentioned, when adherents of foundational free will say a defendant should be excused because the defendant lacked free will, this is usually just shorthand for a conclusion based on the usual criteria for folk psychological irrationality, such as legal insanity, or for folk psychological compulsion, such as duress. If pressed, some will say things like the defendant could not help himself, but, again, they are not referring to the principle of alternative possibilities or the like. They are speaking loosely of standard folk psychological excusing conditions. If pressed further, they will recognize that all people have or do not have free will, and they really cannot coherently distinguish those they wish to excuse on the ground that those people lack libertarian freedom. In short, most nonphilosopher adherents of foundational free will as a necessary component of just blame and punishment are not committed to this view in any principled fashion, and it has no practical consequences in the practice of criminal law.

Now let us turn to some facts about the criteria for criminal responsibility—the prima facie elements of crime, and the criteria for the justifications and excuses—that cannot reasonably be denied, even if determinism is true. Most bodily movements that appear to satisfy the act element of criminal prohibitions are intentional, but some, like reflexes or movements performed while sleepwalking, are not. Most criminal defendants had some *mens rea*, but some did not because they made a mistake or were not paying attention or as a result of other perfectly ordinary reasons. The otherwise criminal acts that defendants commit may in fact be justified, but most often they are not. Some defendants commit crimes while legally insane or acting under duress, but most do not. Again, these are true facts about the world that no one who accepts the importance of agency and the partial explanatory power of folk psychology could reasonably deny.[33] Indeed, they cannot be deinied even if one rejects folk psychology and denies the possibility of genuine agency.

There are two striking things to notice about these facts. First, they are entirely consistent with the truth of determinism or universal causation. Indeed, the partial explanatory power of folk psychology, the psychology that is foundational for criminal responsibility, is perfectly consistent with the truth of determinism. It may not be true that folk psychology is genuinely partially causally explanatory of behavior, but if it is false, it is not because determinism is true. If it is false, it is false because we will have discovered that mental states do not play a causal role. Determinism is consistent with both folk psychology and epiphenomenalism about mental states even if they are not consistent with each other. And to be more specific about doctrine, most bodily movements may be determined to be intentional, and some will be determined not to be. Most defendants will be determined to form *mens rea*, and some will be determined not to form it. Determinism will also account for the presence and absence of the criteria for justifications and excuses. The distinctions the criminal law draws are perfectly consistent with the truth of determinism per se.

The second striking aspect of these facts is that the distinctions drawn matter morally and legally and are consistent with principles of justice that virtually all adopt. All of them are consistent with retributive, consequential, and mixed theories of the justification of punishment. For example, defendants who caused harm in the absence of action do not deserve punishment, and they cannot be deterred. Defendants who lack *mens rea* do not demonstrate the types of attitude and lack of concern for others that are crucial to fault and desert. For similar reasons, they may not present a danger to the community and do not need to be restrained. Of course, in some cases, retributive and consequential aims may be at odds. Consider a defendant with mental abnormalities that may compromise desert but that may also render the defendant particularly dangerous. The pure retributivist or consequentialist is not bothered by such a case, but even a mixed theorist who is bothered by this example is working with a set of considerations—culpability and danger—that everyone agrees are normatively crucial for deciding what to do. Criminal law assuredly operates imperfectly, but it is driven by conceptions of justice that are at the heart of Western morality.

In short, libertarian freedom is not even foundational for criminal responsibility. The criminal law holds most defendants responsible and excuses some. This is fully consistent with the compatibilist view of the relation between determinism and responsibility if one stays internal to the positive criminal law we have.[34] Compatibilism is the only metaphysical position that is consistent with both the criminal law's robust conception of responsibility *and* the contemporary scientific worldview. If one is not a compatibilist criminal lawyer, one is forced to adopt the panicky metaphysics of libertarians or to abandon the concept of responsibility altogether. Neither prospect is inviting, and neither is necessary. As this section has shown, the criminal law is consistent with the truth of determinism or something like it, and as the next section indicates, no external challenge to the very coherence of criminal responsibility is successful.

EXTERNAL CHALLENGES TO CRIMINAL RESPONSIBILITY

The two external challenges to criminal responsibility are hard determinism (hard incompatibilism)[35] and the even more radical claim and hope (of some) that neuroscience or other sciences will cause a paradigm shift in criminal responsibility by demonstrating that we are "merely victims of neuronal circumstances"[36] or of our genes, or some similar claim that denies human agency altogether. Both simply deny that human beings can be genuinely responsible. Many proponents claim further that these challenges, if successful, would lead to a pure prediction–prevention system of social control unmoored from considerations of responsibility, desert, and punishment that would be far kinder, gentler, and more rational than current criminal justice. In this section, I first address each challenge separately and then consider criticisms common to both.

The hard determinist challenge is completely familiar. The intuition that fuels it—roughly, that if all events in the universe are causally determined by prior events subject to the universe's causal laws, then no one could ever do otherwise; there never are genuine alternative possibilities—is profound. After all, how could anyone be praised or blamed, punished or rewarded, if what they did was in some ultimate sense not "up to them"? Therefore, if determinism or something like it is true, how can anyone rationally be held responsible? Many compatibilists claim that they do not feel the force of this intuition, but I feel its tug. On the other hand, most professional philosophers are compatibilists,[37] suggesting at the least that there are many good arguments in favor of compatibilism, and that there is no resolution of the metaphysical dispute in sight. Moreover, the technical progress made in the professional free will literature often depends on intuitive reactions to complicated, unrealistic hypotheticals, intuitions that are not well-grounded in ordinary experience and that are not remotely related to the intuitions that ground criminal doctrine and practice.[38]

The criminal law cannot wait for the doctors of metaphysics to arrive and to cure the philosophical ailments that bedevil the responsibility literature. Even if compatibilism does provide a strong, satisfying metaphysical basis for responsibility, as I believe it does despite the hard determinist tug, criminal justice agents really do not care about the genuine metaphysical dispute. Indeed, there is empirical evidence, admittedly nondispositive, that ordinary people are intuitive compatibilists,[39] and my hunch is that even the most committed hard incompatibilists actually live as if they were compatibilists. Finally, despite claims to the contrary,[40] it seems clear that interpersonal life would be exceptionally impoverished if concepts of responsibility, including genuinely deserved praise and blame, were extirpated from our lives, even assuming that this is possible. If people came to be treated as objects to be manipulated rather than full agents, much that we most treasure would be lost.[41] What reactive attitudes would still be justified, if any would be, would be pallid compared with those that enrich our lives now. Such losses may be the price one pays for living in truth, but it is inconceivable that a practical institution such as the criminal law would transform itself in apparently negative ways as the result of an insoluble theoretical, metaphysical debate. I am not claiming that reactive attitudes that fuel retributive desert, such as indignation and anger, cannot have negative consequences. They can and do. The issue is whether a more pallid world is preferable. I think not, and as this section argues, we need not seek to attain it.

It appears that most hard determinists do not necessarily deny that human beings have mental states and that those mental states may play a partial causal role in explaining behavior. What distinguishes them from compatibilists is whether such causal determination is consistent with responsibility. The hard determinist need not deny, however, that human behavior presents at least a simulacrum of genuine agency. But what if we are not agents at all? What if our mental states are simply epiphenomenal or the like?

At present, the law's "official" position—that conscious, intentional, rational, and uncompelled agents may properly be held responsible—is justified unless and until neuroscience, genetics, or any other discipline demonstrates convincingly that humans are not the type of creatures that we think we are. The law's implicit theory of action presupposes that its subject is an agent who acts based on his or her mental states, such as desires, beliefs, and intentions.[42] Agents are praised and blamed, rewarded and punished. Because it is an agent who acts, it makes sense to ask that person to give an account of his or her behavior and to be held accountable. Asking another animal without practical reason or a mechanistic force that does not act to answer to charges does not make sense. If humans are not intentional creatures who act for reasons and whose mental states play a causal role in their behavior, then the foundational facts for responsibility ascriptions are mistaken. If it is true that we are all automata, then no one is an agent and no one can be responsible. If the concept of mental causation that underlies folk psychology and current conceptions of responsibility is false, our responsibility practices, and many others, would appear unjustifiable. Why not move, then, to a system of pure preventive detention for dangerous agents and perhaps to mandated early detection and intervention—screen and intervene—for those who will become dangerous later?

This claim is not a strawperson. Here is a lengthy quote from a widely noticed article by neuroscientists Joshua Greene and Jonathan Cohen that expresses the mechanistic conception:

> [A]s more and more scientific facts come in, providing increasingly vivid illustrations of what the human mind is really like, more and more people will develop moral intuitions that are at odds with our current social practices... Neuroscience has a special role to play in this process for the following reason. As long as the mind remains a black box, there will always be a donkey on which to pin dualist and libertarian intuitions... What neuroscience does, and will continue to do at an accelerated pace, is elucidate the "when," "where" and "how" of the mechanical processes that cause behaviour. It is one thing to deny that human decision-making is purely mechanical when your opponent offers only a general, philosophical argument. It is quite another to hold your ground when your opponent can make detailed predictions about how these mechanical processes work, complete with images of the brain structures involved and equations that describe their function... At some further point... [p]eople may grow up completely used to the idea that every decision is a thoroughly mechanical process, the outcome of which is completely determined by the results of prior mechanical processes. What will such people think as they sit in their jury boxes?... Will jurors of the future wonder whether the defendant... *could have done otherwise*? Whether he really *deserves* to be punished...? We submit that these questions, which

> seem so important today, will lose their grip in an age when the mechanical nature of human decision-making is fully appreciated. The law will continue to punish misdeeds, as it must for practical reasons, but the idea of distinguishing the truly, deeply guilty from those who are merely victims of neuronal circumstances will, we submit, seem pointless.[43]

Greene and Cohen are not alone among thoughtful people in making such claims. The seriousness of science's potential challenge to the traditional foundations of law and morality is best summed up in the title of an eminent psychologist's recent book, *The Illusion of Conscious Will*.[44]

If our mental states play no role in our behavior and are simply epiphenomenal, then traditional notions of responsibility based on mental states and on actions guided by mental states would be imperiled. But is the rich explanatory apparatus of intentionality simply a post hoc rationalization that the brains of hapless *Homo sapiens* construct to "explain" what their brains have already done (and to whom and by what process if this view is correct)? Will the criminal justice system as we know it wither away as an outmoded relic of a prescientific and cruel age? If so, not only criminal law is in peril. What will be the fate of contracts, for example, when a biological machine that was formerly called a person claims that it should not be bound because it did not make a contract? The contract is also simply the outcome of various "neuronal circumstances." Compatibilism will not save us from this challenge because compatibilism presupposes that we are agents, a view the radical challenge denies.[45]

Given how little we know about the brain–mind–action connections,[46] to claim based on neuroscience that we should radically change our picture of ourselves and our legal doctrines and practices is a form of neuroarrogance. Although I predict that we will see far more numerous attempts to use neuroscience in the future as evidence in criminal cases and to affect criminal justice policy, I have elsewhere argued that for conceptual and scientific reasons, there is no reason at present to believe that we are not agents and that our mental states do not matter.[47] It is possible that we are not agents and that mental states are epiphenomenal, but the current science does not remotely demonstrate that this is true. The burden of persuasion is firmly on the proponents of the radical view. There is simply no present justification for concluding that the foundational folk psychological view of agency upon which criminal law is based is incorrect.

Now let us turn to two considerations that apply to both external challenges. First, neither view, and especially not the radical position, entails any normative agenda, despite the claim sometimes made that each entails consequentialism. If it is unjustified according to hard determinism to say to a harm-doer, "you should not have done that," because no alternative is possible, what is the justification for saying to anyone, "you should/should not do that in the future," because there are no alternative possibilities in the future either.[48] Moreover, when the hard determinist allegedly engages in normative reasoning, is it really reasoning, or is it simply a simulacrum of deliberation? And, if we cannot genuinely be guided by reason, if what we come to believe is simply the product of causal variables, what reason do we have as agents for adopting any normative view? And if we can be guided by reason, as compatibilists claim, then why does the hard determinist deny the possibility of responsibility?

This problem is even direr for the radical challenge. Suppose we were convinced by the mechanistic view that we are not intentional, rational agents after all. (Of course, the notion of being "convinced" would be an illusion, too. Being convinced means that we are persuaded by evidence or argument, but a mechanism is not persuaded by anything. It is simply neurophysically transformed.) What should we do now? We "know" (another mental state, but never mind) that it is an illusion to think that our deliberations and intentions have any causal efficacy in the world. We also know, however, that we experience sensations such as pleasure and pain and that we care about what happens to us and to the world. We cannot just sit quietly and wait for our brains to activate further, for determinism to happen. We must and will deliberate and act.

If we still thought that the radical view was correct and that standard notions of genuine moral responsibility and desert were therefore impossible, we might nevertheless continue to believe that the law would not necessarily have to give up the concept of incentives. Indeed, Greene and Cohen concede that we would have to keep punishing people for practical purposes.[49] Such an account would be consistent with "black-box" accounts of economic incentives that simply depend on the relation between inputs and outputs without considering the mind as a mediator between the two. For those who believe that a thoroughly naturalized account of human behavior entails complete consequentialism, such a conclusion might not be unwelcome.[50]

On the other hand, coming to conclusions and making decisions about whether or not to maintain a practice of punishment (and why should it be punishment as opposed to some other nonpunitive form of social control?) are mental acts that presuppose folk psychology. Moreover, the radical's concession seems to entail the same internal contradiction just explored. What is the nature of the "agent" that is discovering the laws governing how incentives shape behavior? Could understanding and providing incentives via social norms and legal rules simply be epiphenomenal interpretations of what the brain has already done? How do we "decide" which behaviors to reward or punish? What role does "reason"—a property of thoughts and agents, not a property of brains—play in this "decision"?

If the truth of pure mechanism is a premise in deciding what to do, this premise yields no particular moral, legal, or political conclusions. It will provide no guide to how one should live or how one should respond to the truth of reductive mechanism.[51] Normativity depends on reason, and thus the radical view is normatively inert. Neurons and neural networks do not have reasons; they do not have a sense of past, present, and future; and they have no aspirations. Only people do. If reasons do not matter, then we have no genuine, nonillusory reason to adopt any morals or politics or any legal rule, or to do anything at all. Thus, this view does not entail consequentialism or a pure preventive scheme of social control.

The second major criticism of the external challenges concerns the assumption that a social control system that abandons notions of agency, desert, and punishment will be kinder and gentler than current practice. This is an open, empirical prediction, but the current situation suggests that the opposite will be true. I agree with critics of current American criminal justice that many of our doctrines and practices are too harsh. They lock up too many people for too long. But although retribution has played an increasingly large role in theory in justifying criminal punishment since the 1960s, most of the most draconian aspects of punishment have been motivated by consequential concerns. Striking examples are recidivist sentencing enhancements, the approval of strict liability crimes, the "war on drugs," the vast proliferation of regulatory crimes, and mandatory minimum sentences. None of these can be retributively justified, and all punish disproportionately to desert. They are mainly justified by considerations of general deterrence, specific deterrence, and incapacitation, all of which are consequential justifications.

Now, it is open to consequentialists to argue that the legislatures have misunderstood the costs and benefits of what they have done and that they are unnecessarily creating welfare losses, which are a consequential form of "disproportion." But if we cannot be sure of the net social effects of our consequential policies, why does it matter if we get it wrong if people are not agents and do not have rights. After all, it was Bentham who claimed that rights were nonsense on stilts. Moreover, why wait for criminal conduct to occur to intervene in the kinder, gentler world? Why not adopt a regime of "screen and intervene" if we are convinced that we could increase social safety? In fact, it is the compatibilist view of agency, coupled with notions of desert and retribution, that are the foundation of proportionality and the protection of human rights within our desert–disease system of social control.

It is also not clear that a social control system without agency and responsibility would be more rational than a system that took desert seriously. A prediction–prevention, screen–intervene system would be rational if social control were our main goal and we had the technology to achieve it, but it would not be kinder and gentler. More important, if we took desert seriously, including being willing to take the risks in the name of liberty to respect desert, that system, too, would be rational. The question would then be which type of rational world one wished to live in: the brave new world or the messy current world that takes people seriously as agents capable of being responsible and taking responsibility.

CONCLUSION: COMMON CRIMINAL LAW COMPATIBILISM REDUX

There is simply no question that the criminal law we have is completely consistent with the truth of determinism, avoids a panicky metaphysics, and employs moral and political concepts that we have reason to endorse. Like all human institutions, it is messy and imperfect, but in principle, it takes people seriously and attempts to maximize liberty, autonomy, and dignity.

There are external challenges, of course, and hard determinists will not be persuaded that genuine responsibility and desert are possible. The criminal law is the product of centuries of development, however, and has the attractive features this chapter has described. The burden of persuasion is surely on the external challengers. They must not only mount a theoretical argument but also provide a practical, concrete proposal of what their social control system would look like, including how it would be administered by imperfect creatures like ourselves who will virtually always have inadequate information to make fully informed decisions. The critics must also provide an account of why, in concrete terms, human welfare would be enhanced in the brave new world. The record of social engineering in history is unappealing, to say the least, which simply enhances the externalist's burden of persuasion.

Ultimately, I believe that the vision of the person, of interpersonal relations, and of society the externalist endorses bleaches the soul. I have no incontrovertible argument to prove that the externalist view is incorrect or would lead to dire consequences, but neither does the externalist have an incontrovertible argument against the internalist. In the concrete and practical world we live in, however, we must be guided by our values and a vision of the good life. I do not want to live in the externalists' world that is stripped of genuine agency, desert, autonomy, and dignity. For all its imperfections, common criminal law compatibilism is more respectful and humane.

NOTES

1. A classic example is the federal law of legal insanity. Before 1984, the federal criminal code did not contain an insanity defense. Instead, each of the circuit courts of appeal had made the law of legal insanity by adopting one test or

another for use in that circuit. By 1984, all but the First Circuit had adopted the Model Penal Code test, which excuses a defendant if, at the time of the crime, as a result of mental disorder, the defendant lacked substantial capacity either to appreciate the wrongfulness of his action or to conform his conduct to the requirements of the law. American Law Institute, Model Penal Code, §4.01(1) (1962). In response to the unpopular Hinckley verdict, which was based on the MPC test in the US Court of Appeals for the District of Columbia, in 1984 Congress passed the Insanity Defense Reform Act, which imposed a uniform and much more limited "cognitive" test for legal insanity that applies in all federal trials nationwide. 18 USC17(a). Since then, the federal courts have been interpreting the federal test.

2. For a particularly trenchant critique, *see*, Hilary Bok, *Freedom and Responsibility* (1998).
3. *See,* http://philpapers.org/surveys/results.pl (reporting the results of a 2009 survey of professional philosophers that showed that a majority are either compatibilists or tend toward compatibilism).
4. The notion of desert I am employing is simply the traditional retributive conclusion that if an offender's behavior satisfies the elements of a charged offense and no justification or excuse obtains, then the offender is culpable and deserves the ensuing blame and punishment. All these concepts are discussed in more detail below.
5. I have explored the civil–criminal distinction as a basis for confinement elsewhere and will therefore provide only the briefest sketch here. *See generally*, Stephen J. Morse, *Neither Desert Nor Disease*, 5 Legal Theory 266, 269 (1999); Stephen J. Morse, *Uncontrollable Urges and Irrational People*, 88 Virginia L. Rev. 1025 (2002) [hereinafter Morse, *Irrational People*]; Stephen J. Morse, *Blame and Danger: An Essay on Preventive Detention,* 76 B.U. L. Rev. 113, 116–122 (1996).
6. United States v. Salerno, 481 U.S. 739, 741, 755 (1987).
7. A. P. Simester and Andreas von Hirsch, *Crimes, Harms and Wrongs: On the Principles of Criminalisation* 10 (2011).
8. Such people of course continue to deserve enhanced concern and respect by virtue of their being human, and they can never be completely objectified.
9. The civil and nonpunitive characterization of such interventions often justifies lesser procedural protections for the potential subject. *See, e.g.*, Allen v. Illinois. 478 U.S. 364, 374–375 (1986) (holding that the Fifth Amendment guarantee against compelled self-incrimination does not apply in a proceeding to determine whether a person is a "sexually dangerous person" because the proceeding is not "criminal").
10. Ewing v. California, 538 U.S. 11, 14–15, 30 (2003) (upholding enhanced sentences under California's three strikes law).
11. *See,* George Sher, *In Praise of Blame* 123 (2006) (stating that although philosophers disagree about the requirements and justifications of what morality requires, there is widespread agreement that "the primary task of morality is to guide action"); John R. Searle, *End of the Revolution*, 43 N.Y. Rev. of Books, at 33, 35 (2002).

12. Scott J. Shapiro, *Law, Morality, and the Guidance of Conduct*, 6 Legal Theory 127, 131–132 (2000).
13. *Id.* at 131–132. This view assumes that law is sufficiently knowable to guide conduct, but a contrary assumption is largely incoherent. As Shapiro writes: "Legal skepticism is an absurd doctrine. It is absurd because the law cannot be the sort of thing that is unknowable. If a system of norms were unknowable, then that system would not be a legal system. One important reason why the law must be knowable is that its function is to guide conduct. *Id.*"I do not assume that legal rules are always clear and thus capable of precise action guidance. If most rules in a legal system were not sufficiently clear most of the time, however, the system could not function. Further, the principle of legality dictates that criminal law rules should be especially clear.
14. I do not mean to imply dualism here. I am simply accepting the folk-psychological view that mental states—which are fully produced by and realizable in the brain—play a genuinely causal role in explaining human behavior.
15. *E.g.*, see the chapters in Part I of this volume.
16. *See generally*, Joel Feinberg, *The Moral Limits of the Criminal Law: Harm to Others* (Vol. 1,1987); *Offense to Others* (Vol. 2, 1988); *Harm to Self* (Vol. 3, 1986); *Harmless Wrongdoing* (Vol. 4, 1990) (providing a complete account of the limits of criminalization based on the harm principle).
17. *See* American Friends Service Committee, *Struggle for Justice: A Report on Crime and Punishment in America* (1971); Marvin Frankel, *Criminal Sentences: Law Without Order* (1972); Twentieth Century Fund Task Force on Criminal Sentencing, *Fair and Certain Punishment* (1976); Ernest van den Haag, *Punishing Criminals: Concerning a Very Old and Painful Question* (1975). There were also proposals to use selective incapacitation more effectively. *See* James Q. Wilson, *Thinking About Crime*, ch. 8 (1975). This was not coupled, however, with calls for extensive lengthening of sentences. *See, e.g.*, Uniform Determinate Sentencing Act, ch. 1139; 1976 Cal. Stat. 5062.
18. American Law Institute, Model Penal Code: Sentencing §1.02(2) (*Tentative Draft No. 1*, 2007).
19. I use the locution, "establish an affirmative defense," because the US Supreme Court has made it clear that a jurisdiction may shift the burden of persuasion to the defendant for affirmative defenses. See, Leland v. Oregon, 343 U.S. 790 (1952) (permitting shifting the persuasion burden for legal insanity to the defendant); Rivera v. Delaware, 429 U.S. 877 (1976) (same); Dixon v. United States, 126 U.S. 2437 (2006) (permitting shifting the persuasion burden for duress to the defendant). Many jurisdictions have taken advantage. For example, in federal insanity defense cases, the defendant must prove that he or she was legally insane by clear and convincing evidence. Of course, a jurisdiction can also place the burden of persuasion on the prosecution. In practice, wherever the burden of persuasion is placed for affirmative defenses, the defendant will have the burden of producing sufficient evidence to warrant submitting the defense to the finder of fact.
20. This element is sometimes referred to as the "voluntary act" requirement, but this is simply a legal label affixed after the criteria for it have been established. If the criteria are met, the movement is referred to as a "voluntary act;" if they are not,

the movement is not an action. The term, "voluntary," does not have its ordinary language meaning in this case.

21. 548 U.S. 735 (2006).
22. Many commentators believe that such objective standards cannot properly be the basis for criminal liability because agents who do not advert to risks are not genuinely to blame. *See, e.g.*, Michael S. Moore and Heidi M. Hurd, *Punishing the Awkward, the Stupid, the Weak, and the Selfish: The Culpability of Negligence*, 5 Crim. L. & Phil. 147 (2011). Nonetheless, the criminal law continues to employ an objective negligence *mens rea*, and it plays a large role in some of the most serious crimes, such as homicide and rape.
23. Model Penal Code, §4.01(1).
24. *Id.*, §2.09(1).
25. Jones v. United States, 563 U.S. 354 (1983).
26. Foucha v. Louisiana, 504 U.S. 71 (1992).
27. For example, there are affirmative defenses such as the statute of limitations or jurisdictional limits that are irrelevant to culpability but that may lead to acquittal if they are established.
28. *Clark*, *supra*, note 21.
29. See Stephen J. Morse and Morris B. Hoffman, *The Uneasy Entente Between Insanity and Mens rea: Clark v. Arizona*, 97 J. Crim. L. & Criminol. 1071 (2008).
30. There are exceptions, of course. *See, e.g.*, United States v. Moore, 486 F.2d 1139, 1145–1146 (D.C. Cir. 1973) (opinion of Wilkey, J.) (recognizing that lack of "free will" means simply loss of control capacity).
31. P. F. Strawson, *Freedom and Resentment*, in G. Watson, Ed., Free Will 59, 80 (1982).
32. Daniel Dennet, *Elbow Room: The Varieties of Free Will Worth Wanting* 104 (1984).
33. Even those theorists who believe that genuine agency is impossible and that folk psychological explanations are primitive and wrong, such as epiphenomenalists about mental states, act as if they are agents and primarily use folk psychological concepts to explain the behavior of others and themselves and to understand and shape their interpersonal interactions. Skeptics about agency and folk psychology may believe that they have no real alternative at present—a situation future science and conceptual work might alter—but they live their daily lives as folk psychologists. At present, even they would have to concede as a practical matter that the facts adduced above are indeed facts about the world.
34. An external critique of the concept of criminal responsibility is of course possible, but then one is no longer trying to explain and to normatively justify criminal law. Instead, one is seeking to abolish all notions of responsibility. This chapter turns to such challenges in the next section.
35. *See generally*, Derk Pereboom, *Living Without Free Will* (2001) (explaining hard compatibilism, contrasting it with compatibilism and libertarianism, and suggesting that human welfare would be improved by the whole-hearted embrace of hard compatibilism).

36. *E.g.*, Joshua Greene and Jonathan Cohen, *For the Law, Neuroscience Changes Nothing and Everything*, *in* Law and the Brain 207, 217–18 (Semir Zeki and Oliver Goodenough, Eds., 2006), at 218.
37. *See*, survey results, *supra* note 3.
38. I thank Gideon Yaffe for a brief, helpful discussion of this point.
39. *See*, Adina Roskies and Shaun Nichols, *Bringing Moral Responsibility Down to Earth*, CV J. Phil. 371 (2008); Eddy Nahmias, D. Justin Coates, and Trevor Kvaran, *Free Will, Moral Responsibility, and Mechanism: Experiments on Folk Intuitions*, XXXI Midwest Studies in Philosophy 214 (2007); Thomas Nadelhoffer et al., *Folk Retributivism: In Theory and Practice* (in press).
40. *E.g.*, Pereboom, *supra* note 35, at 199–213.
41. *See*, C. S. Lewis, *The Humanitarian Theory of Punishment*, 6 Res Judicatae 224 (1953).
42. Stephen J. Morse, *Lost in Translation?: An Essay on Neuroscience and Law*, *in* Law and Neuroscience 529, 540 (Michael Freeman, Ed., 2011), at 529–533. *See generally*, Robert Audi, *Action, Intention and Reason*, 109–178 (1993) (providing an account of practical reason); Katrina L. Sifferd, *In Defense of the Use of Commonsense Psychology in the Criminal Law*, 25 Law & Phil. 571 (2006) (providing an extensive defense of the use of folk psychology that underpins criminal law and claiming that criminal law does not need a substitute).
43. Greene and Cohen, *supra* note 36, at 217–218 (internal citation omitted).
44. Daniel M. Wegner, The Illusion of Conscious Will (2002); *see also* Daniel M. Wegner, *Précis of the Illusion of Conscious Will*, 27 Behav. & Brain Sci. 649 (2004). The *précis* is followed by open peer commentaries and a response from Professor Wegner. *Id.* at 679. In more recent work, Professor Wegner appears to have softened the radical interpretation of his claim, which is that we, as agents, are not really "controllers" whose mental processes cause action. Daniel M. Wegner, *Who Is the Controller of Controlled Processes?*, *in* The New Unconscious 19, 32 (Ran R. Hassin et al., Eds., 2005) ("This theory is mute on whether thought does cause action."). On the other hand, Professor Wegner seems ambivalent and unwilling to give up the radical interpretation. *See id.* at 27 (arguing that the "experience of conscious will is normally a construction" and referring to mental causation as "apparent"). This apparent ambivalence is present in the work of others.
45. *See* Stephen J. Morse, *The Non-Problem of Free Will in Forensic Psychiatry and Psychology*, 25 Behav. Sci. & L. 203, 214–216 (2007) (discussing the meaning of compatibilism and its relation to criminal law).
46. For example, we have no idea how the brain enables the mind. Paul R. McHugh & Phillip R. Slavney, The Perspectives of Psychiatry 11–12 (2d ed. 1998).
47. *See* Stephen J. Morse, *Determinism and the Death of Folk Psychology: Two Challenges to Responsibility from Neuroscience*, 9 Minn. J. L. Sci. & Tech. 1, 19–34 (2008); Morse, *supra* note 34, at 543–554; *see also*, Mario Beauregard, *Mind does really matter: Evidence from neuroimaging studies of emotional self-regulation, psychotherapy, and placebo effect*, 81 Progress in Neurobiology 218 (2007).
48. Mitchell Berman, *Punishment and Justification*, 118 Ethics 258, 271 n. 34 (2008).
49. Greene and Cohen, *supra* note 36, at 218.

50. A clear implication of the disappearing person argument is that it would not be fair to punish dangerous people because no one deserves punishment (or anything else). We might need to deprive dangerous people of their liberty to ensure social safety, but then we would be morally bound to compensate for the unfairness of restraining liberty by making the conditions of confinement (or other restraints) sufficiently positive. In other words, a regime of "funishment" would be morally required. Saul Smilansky, *Hard Determinism and Punishment: A Practical Reductio*, 30 Law & Phil. 353, 355 (2011). I leave the perverse incentives this would create to the reader's imagination, but defenders of the disappearing person view would be required to institute funishment and somehow to avoid the perverse incentives. Smilansky's argument is directed at hard determinists, but it is a fortiori directed at those who make the disappearing person claim, as long as the latter can make any moral claims at all. *See infra,* note 42 and accompanying text.
51. I was first prompted to this line of thought by a suggestion Mitch Berman made in the context of a discussion of determinism and normativity. *See* Mitchell Berman, *supra,* note 48.

3

What Can Neurosciences Say About Responsibility?

Taking the Distinction Between Theoretical and Practical Reason Seriously[1]

ANNE RUTH MACKOR

INTRODUCTION

Richard Dawkins has argued that concepts such as "responsibility" and "punishment" have been superseded because human behavior is determined by physiology, heredity, and the environment. If we look at the nervous system from a scientific perspective, we have to acknowledge that lawsuits about the guilt or the diminished responsibility of human beings are just as absurd as lawsuits against cars, or so Dawkins claims (2006:19). Michael Gazzaniga, on the other hand, has argued with just as much ardor that "neuroscience has nothing to say about concepts such as free will and personal responsibility..." (2006:141).

Two famous scientists, two radically opposed views. How can we tell who is right? Is Dawkins right to claim that, as (neuro)scientists learn more about our brains, about our genes, and about the influence of the environment on our behavior, there will come a time that we eliminate our concepts of responsibility and punishment and only talk in scientific terms about human behavior? Or is Gazzaniga right in claiming that the idea that neurosciences can contribute anything to our views about responsibility and punishment is fundamentally mistaken? In this chapter, I argue that both Dawkins and Gazzaniga are wrong.

Four Distinct Questions

The question of whether neurosciences have anything to say about responsibility is the topic of this chapter. Before I can answer this question, however, I must give a more precise formulation of it. I argue that we have to distinguish among at least four questions: two external and two internal.

External Questions

The first and most radical interpretation of the question is whether neuroscientific developments can have implications for the continued existence of our moral, legal, and other practices within which "responsibility" is a, if not the most, central notion. Such a question is posed from an external perspective and looks at these practices as a whole. We can distinguish two versions of the external question.

Presuppositions

The first version of the external question asks whether neuroscientific research can show that those practices are fundamentally untenable because they rest on false presuppositions. Dawkins and others argue that they rest on the assumption that we have a free will, which (neuro)sciences can show to be an illusion. For example, Libet's famous experiments are sometimes claimed to be part of such "hard" neuroscientific evidence. Libet's experiments have shown that activities in the motor cortex related to the action of pushing a button precede the conscious decision to do so. They are taken to imply that unconscious and determined "decisions" of our brain precede and determine our conscious and "free" decisions.

Implications

The second version of the external question asks whether neuroscientific research can show that those practices, whether or not they rest on untenable assumptions, are pointless to the extent that they cannot fulfill the goal they are intended to fulfill. Dawkins seems to deal with this question, too, when he argues that lawsuits against human beings are just as absurd as those against cars. He seems to imply not only that it is absurd to assume that human beings have a free will but also that holding them responsible and punishing them will not gain us anything. Thus, our second external question is whether neurosciences can offer evidence to support the claim that holding people responsible is pointless.[2]

In short, the external and most radical version of the question asks whether neurosciences can play a role in showing that practices within which "responsibility" and "punishment" play a central role are either fundamentally untenable because their presuppositions are false (1a) or pointless in that they fail to contribute to the realization of their own "pointe" (1b).

Internal Questions

In a second and more modest version of the question, the defensibility of legal and moral practices is not under debate. The practices are accepted as tenable and useful, be it sincerely or only for the sake of the argument.[3] In contradistinction to the first question, we might say that this question is posed from an internal perspective. The aim of such a question would most likely be to improve moral and legal practices. We can also distinguish at least two internal questions.[4] In order to understand the way in which these two questions differ, I must first briefly discuss the distinction between concepts and conceptions.

A *concept* is the meaning of a particular term, whereas a *conception* states the criteria for the application of the concept. The term *justice*, for example, is ambiguous and can refer to different concepts, such as distributive, retributive, compensatory, and procedural justice. Obviously, each concept of justice has its own conception, that is, its own criteria of application. However, matters are more complicated because for each concept of justice, several competing conceptions exist. Rawls' two principles of justice, for example, are just one, be it famous, conception of the concept of distributive justice.[5] I shall argue that the same holds for the term *responsibility*—that it can refer to different concepts and that each of those concepts in turn can be characterized by means of different conceptions.

The distinction between concepts and conceptions allows us to distinguish between the claim that neurosciences can change the concept of responsibility and the claim that they can change the conception. Again, take the concept of distributive justice. If distributive justice would no longer mean something like the fair distribution of goods and services among people, then the practices within which that concept plays a central role would become another kind of practice. By my account, a fundamental change of a concept would count as an external critique. Changing the conception of distributive justice (e.g., by adopting Rawls' two principles) will also bring about some changes in the practice. However, in this case, the purpose and in that sense the nature of the practice will remain the same. The same holds for the concept of responsibility. Both the elimination and a radical change of the concept of responsibility will result in a change of the nature of the practice of holding each other responsible, and it thus counts as an external critique. A change of the conception of responsibility, on the other hand, would count as an internal critique.[6] Therefore, internal questions only deal with the conception of responsibility.

Conception

The first version of the internal question asks whether findings of the neurosciences can change conceptions of responsibility. Stated differently, the question is whether neurosciences are capable of changing the criteria for application of one or more concepts of responsibility.

Application

The second internal question is the most modest of all four. It accepts not only the concept but also the conception of responsibility as these are used within certain practices. It only asks whether neurosciences can help to show that specific (categories of) persons whom we used to think that fit the criteria of application, in fact do not fit the criteria, or, conversely, that specific (categories of) persons whom we thought not to fit the criteria, turn out to fit the criteria after all. Thus, this question is not about the content of the conception of responsibility, but only about its proper application.

Structure of the Chapter

We have just seen that Dawkins argues that lawsuits about responsibility and guilt are nonsensical. In doing so, he offers an affirmative answer to the external question, in particular to 1a, that practices within which we hold each other responsible make no sense and thus should be eliminated. Gazzaniga, on the other hand, does not only give a negative answer to the external question. He also, and more controversially, seems to offer a negative answer to the internal question, in particular to 2a.[7] In this chapter, I deal with all four questions. Given its complexity, I spend most time on answering the first external question.

The structure of this chapter is as follows. In section 2, I first go deeper into the notion of responsibility. I argue that the term "responsibility," like "justice," refers to several concepts that should be carefully distinguished before we can understand which concepts or conceptions of responsibility might or might not be eliminated or changed by neurosciences.

This analysis takes us in section 3 to the external questions—that is, whether neuroscientific findings can make our practices of holding oneself and others responsible untenable or pointless. I shall argue that Dawkins falls prey to several confusions. The core of my argument is that "free will," "responsibility," and "punishment" are notions within the realm of practical reason and that neuroscientific findings, which belong to the realm of theoretical reason, cannot eliminate these concepts from the realm of practical reason. Moreover, I shall argue that Dawkins cannot argue for the elimination of moral and legal practices within which we hold each other responsible without contradicting himself.

In the remainder of the chapter, I start from the assumption that neuroscientific findings cannot eliminate our responsibility practices. In section 4, I discuss both internal questions, that is, whether neuroscientific findings can change the conception of responsibility or its application. The argument made in section 3 might seem to result in the conclusion that neuroscientific findings, because they belong to the realm of theoretical reason, have nothing to say about responsibility. However, this argument fails. Responsibility belongs to the realm of practical reason, and therefore neurosciences can never have a final say in changing the concept or the conception. I argue, however, that neurosciences can nevertheless contribute arguments for and against changing the conception of responsibility. A fortiori, or so it is argued, neurosciences can contribute arguments for and against the application of the conception to particular persons and categories of persons.

In section 5, finally, I briefly discuss the well-known worry that even if my argument that neurosciences cannot result in the elimination of responsibility holds, neuroscientific findings might nevertheless result in a drastic reduction of the categories (and probably also the number of) people who are held responsible. I argue that the opposite worry, that neurosciences will result in a future of hyperresponsibility, seems just as plausible.

RESPONSIBILITY

Responsibility, like most other fundamental notions, has the peculiar property that as long as no one asks us to explicate it, we seem to be able to apply it correctly. However, as soon as we have to explicate its meaning and elucidate its presuppositions, we get stuck. At least part of the explanation of this problem is that in everyday use we apply the notion of responsibility in different meanings. Hart (1968) makes an often-cited distinction between four senses of responsibility. Other philosophers distinguish eight or even 10 different concepts. We need not concern ourselves here with all possible concepts, however, because scientists like Dawkins and Gazzaniga seem to have only a few of them in mind when they discuss the relevance of neurosciences for responsibility. In this section, I discuss four concepts.

Causal Responsibility

The term responsibility is sometimes used in a purely causal sense. Hurricane Katrina is said to be responsible for the damage done to New Orleans just as the deer is said to be responsible for the traffic accident. In these examples, no more is said than that the hurricane and the deer are the cause of (i.e., an important link in the chain of events that resulted in) the damage. As a consequence, in such examples, the term responsibility can be replaced without loss of meaning by the term "cause."[8]

Neither Dawkins nor Gazzaniga refers merely to causal responsibility. Although they both make use of the concept of causal responsibility, it is not the most important concept. What other concepts do they have in mind? Let us look at an example. If we say that the hooligan is responsible for the damage to the bus shelter, we not only make the claim that he caused the damage, we also make a moral or legal claim. More specifically, we are likely to make at least three further claims: that the hooligan has the capacity to be responsible, that he therefore is accountable for his behavior, and that he is liable for his behavior.[9]

When Dawkins and Gazzaniga refer to responsibility, they are likely to have all three concepts in mind, most specifically the concept of capacity-responsibility. In this chapter, I focus on the question of what neurosciences have to say about our capacity for responsibility and what they have to say about accountability. Before I do so, I will go a bit deeper into all three concepts.[10]

Accountability

The concept of accountability is the central concept of frameworks within which we hold ourselves and others responsible. Claiming that someone was accountable for an event is different from saying he was merely causally responsible. However, holding someone accountable does presuppose that he was—in a direct or remote sense—causally responsible for that event, that is, that his behaving or forbearing played a role in the chain of events that resulted in the event that he is held accountable for.

Some readers might want to object to my claim that accountability presupposes causal responsibility. To support their view, they might refer to cases in which we are accountable for the behavior of others. Parents, for example, can be accountable for the behavior of their child. In such a case, it seems that it is only the child, not the parent, who is causally responsible. This seems to be true for cases of strict liability. In such cases, the causal role of the parent, in particular the fact that he did his utmost to prevent the damage and thus seems to have played a proper role in the causal chain of events, does not count as a justification or exculpation and thus does not count as an argument against his liability. In such cases (absence of) causal responsibility is indeed irrelevant for one's accountability (and liability). However, this argument holds for all cases of strict liability, not merely for those cases in which we are accountable for the behavior of others. In other cases, parents who are held accountable for the behavior of their child are required to answer the question of whether they had done enough to prevent their child from behaving the way he did. Accordingly, as a first step in their answer, they will have to elucidate what role they played or failed to play in the *causal* chain of events that resulted in the event they are held accountable for.[11]

This being said, let us now look at what it means to say that someone is accountable. To be accountable is to be answerable. That is to say, if someone holds himself accountable for his behavior or is held accountable by others,[12] he must offer a justification for his behavior. To offer such a justification is to offer reasons for one's behavior. This does not imply that others must accept or approve of his reasons, it only implies that his reasons must be understandable to others. Stated yet differently, to say someone is accountable for his behavior is to say he is fitting to be subject to the reactive attitude of demanding an answer to the question of why he behaved the way he did.[13] We might also say that someone who is accountable is under the obligation to offer justificatory reasons for his behavior. If we do not accept his reasons, this may give us reason to hold him liable for his behavior. Below, I go deeper into the concepts of accountability and liability, but I first will explicate the concept of capacity-responsibility.

Capacity-Responsibility

Holding someone accountable for his behavior presupposes that he is capable of being held accountable. In saying the hooligan is accountable for his behavior, we presume that he is capable of giving an account. In other words, we presume that he is capable of justifying his behavior in terms of reasons. That is to say, we ascribe to him another type of responsibility: capacity-responsibility. Note that in most cases we do so implicitly. We normally assume adults to have this capacity unless there are indications that someone does not have it, such as when someone shows symptoms that he suffers from acute psychosis or that he is severely mentally retarded.

What is special about this capacity is that it is claimed to distinguish normal adult human beings from inanimate things, plants, and animals and also from infants and young children as well as from mentally retarded and mentally insane adults. A key question is why would we want to make this distinction in the first place. To answer this question, we have to understand in more detail what it means to hold someone accountable.[14]

It implies first and foremost that we can argue with the person about the reasons for his do's and don'ts. Moreover, we can be angry with him, can be ashamed about him, and can forgive him.[15] Although we can be afraid of inanimate objects and of nonhuman creatures, and although we can pity animals and human beings that lack the capacity, we cannot be ashamed of them or forgive them. Put differently, with beings that we cannot hold accountable, we can fight physically, not verbally. Most important, only beings that we can hold accountable are beings that we can blame and praise (Strawson 1962). This brings us to our last concept of responsibility: liability.

Liability

Let us return to the hooligan and assume he was held accountable and came up with a justification. If the hooligan was causally responsible for the damage to the bus shelter, if he had capacity-responsibility, and if he was held accountable but failed to offer a convincing justification for his behavior, we can make the further claim that he is liable for his behavior. To be liable is, in the first place, to be subject to blame or praise for the things one has done. Only if someone is blameworthy can the question arise of whether he should undergo punishment and, if so, which kind of punishment. The same holds for being praiseworthy. Only when the fact has been established that someone is praiseworthy can we ask whether this brings about an entitlement to a reward and, if so, what kind of reward.

It will depend on the criteria for liability (plus the particular features of your behavior) whether your justification is fully accepted or whether you will be held liable for your behavior. In the case of the hooligan, his stating, "I vandalized the bus shelter because the authorities refuse to give us a youth center," would count as his justification, but it would probably not be accepted as a *good* justification. That is, it would not necessarily prevent him from being liable.[16] If he is considered liable, the next step would be to determine what punishment should be enforced.

Some readers might want to object that punishment and reward do not depend on being blameworthy or praiseworthy. They might argue that we also punish and reward animals and infants even though they do not belong to Strawson's circle of beings who are capable of justifying their behavior. It should be noted, however, that these kinds of punishments and rewards can only consist in *measures* such as beating or confinement and sweets or a pet, respectively. Among others, these punishments and rewards lack the retributive connotation characteristic of punishment and rewards that typically follow blame and praise. In this section, I have used the terms *punishment* and *reward* in their retributive sense.

Summary

Accountability presupposes both that the agent was causally responsible for his behavior and that he had capacity-responsibility. If the agent lacks this capacity, he cannot be held accountable for his behavior and as a consequence cannot be held liable and cannot be punished. Accountability implies that someone should offer reasons for his behavior. When the justification that the subject offered is satisfactory, the discussion stops there; no further discussion or action is needed. Only when his reasons are not (fully) satisfactory must the question of liability be addressed, that is, whether he was blameworthy or praiseworthy and whether he deserved punishment or a reward.

In this section, I have also briefly discussed liability and punishment to show how they presuppose accountability. They are not the topic of this chapter, however. In section 3, I restrict myself to a discussion of the presuppositions and the tenability of accountability practices. An analysis of the implications of the findings of section 3 for the tenability of liability practices is beyond the scope of this chapter.[17]

CAN NEUROSCIENTIFIC FINDINGS GIVE US REASON TO ELIMINATE OUR ACCOUNTABILITY PRACTICES?

The analysis of the different concepts of responsibility in section 2 makes it clear that Dawkins' critique seems to revolve around the claim that not only cars and animals, nor just the insane and the mentally retarded, but all human beings lack capacity-responsibility. In this section, I argue that Dawkins falls prey to several fundamental confusions. Most important, I show that Dawkins fails to grasp the distinction between theoretical and practical reason. I further argue that Dawkins contradicts himself, that the concepts of capacity-responsibility and accountability do make sense, and that accountability practices are neither untenable nor useless.

Although Dawkins does not distinguish between different concepts of responsibility, and although he does not use the term *freedom of will* or *capacity-responsibility*, it is clear from his text that on his deterministic account, it makes no sense to hold people accountable for their behavior *because* nobody has the capacity for responsibility *because* nobody has a free will.[18] Stated yet differently, it seems that according to Dawkins, the concepts of freedom of will and capacity-responsibility are empty concepts in that they lack application to any being in a deterministic world. As a consequence, the concept of accountability lacks application, too, and the practices of holding each other accountable are untenable. Below, I substantiate these claims in detail.

One might think that, because Dawkins' claim revolves around the concepts of freedom of will and capacity-responsibility, we must deal with those concepts first and foremost. However, we first have to know what is the purpose of holding oneself and others accountable before we can understand what capacity is needed for these practices. Because a capacity is always a capacity for something, we need to know exactly what is that "something" before we can understand what capacity is needed.

Practical Versus Theoretical Reason

I have argued that demanding and offering an account of one's behavior takes place within specific kinds of practices the point of which is to ask each other for justifications for our behavior (Strawson 1962). A way of understanding these practices is to say that they imply that we look at behavior from the viewpoint of practical reason as opposed to the viewpoint of theoretical reason (Kant 1974:16[AA 14]).

First note that both the theoretical and the practical point of view are guided by reason. However, the purpose for which reason is used within the practical realm is fundamentally different from that of theoretical reason. The aim of theoretical reason is the description and explanation and sometimes also the prediction of events. The aim of practical reason is not to describe and to explain and thereby to understand states of affairs or events in the world, but to answer the question, "*What should I do?*"

A few preliminary remarks with respect to this question are in order. First, although the question of what one should do seems to have a moral flavor, it need not be a moral question. The question need not be taken in its Kantian sense, implying that the interest of the practical use of reason exists in the determination of the will *in relation to the final and complete end.*[19] The question can also be posed from a prudential, legal, or some other non-overarching point of view. Second, although the question is future oriented, it asks, "What *should* I do?," not "What *shall* I do?" Again, the difference is not that the question "What should I do?" necessarily has a moral flavor. It rather refers to the fact that the former does, whereas the latter does not, presuppose more or less determinate standards to be used in answering the question. As a final remark, it should be noted that both the question of what *others* should do as well as the question of what I *should have done* are derivative from the question of what I should do. In section 3.4, the claim that the question of what *I* should do is the most fundamental question will be argued for. With respect to the latter, it seems that we have reason to bother about what we should have done only because our deliberations about our past decisions and actions, as our deliberations about our character, can help us make better decisions in the future (Bok 1998, chapter 4).

This being said, let us note that in order to describe and explain, theoretical reason takes human beings as objects that are subject to causes and their behavior as events that can be causally explained. Practical reason, on the other hand, takes human beings as subjects that are capable of choosing between and acting on reasons. Although theoretical reason, like practical reason, can refer to reasons, theoretical reason takes reasons to be motives, that is, causes that can influence human behavior and that scientists can refer to in order to explain and predict behavior. Thus, whereas the theoretical point of view understands relations between events in terms of causal necessity, the practical point of view understands them in terms of rational necessity.[20] Whereas theoretical reason aims to describe what we did or will do and explain why we did it or will do it, practical reason aims to prescribe what we ought to do and to justify or criticize what we did or intend to do.

Accountability practices belong to the realm of practical reason. Participants address themselves and others, not as objects of investigations, but as subjects that (at some point) have to answer the first-person question, "What should I do?" With the distinction between theoretical and practical reason in mind, we can now look more closely at Dawkins' central assumption that the claim that the world is deterministic and the claim that human beings are free and responsible cannot both be true.

Determinism and Freedom

Although Dawkins does not explicate his argument, it seems clear that his claim rests on the assumption that accountability, liability, and punishment presuppose freedom of will—that because determinism is true, freedom of will is an illusion, and that therefore accountability, liability, and punishment are illusions, too.

This is, in a nutshell, Dawkins' position in the classic debate about the question of whether freedom of will and determinism are compatible. There are roughly three standpoints in the debate. Libertarians are incompatibilists who claim that freedom of will exists and that therefore the assumption of determinism must be false. Dawkins defends the opposite position of deterministic incompatibilism, which holds that determinism is true and that therefore freedom of will cannot exist. The third group in this debate claims that freedom of will and determinism can coexist. They are called *compatibilists.*

If freedom is or presupposes the capacity to act noncausally or contracausally, incompatibilists seem correct to argue that freedom and determinism conflict.[21] However, according to philosophers such as Kant, Korsgaard, Habermas, and Dennett, freedom does not consist of the capacity to act contra-causally,[22] but rather of the capacity to deliberate about and to act on reasons. That is, to be free is to be capable of deliberating about ourselves, about our aims and values, our norms, our reasons, our choices, and our actions for the purpose of deciding what to do and to be capable of acting on the decision that is made based upon those reasons. It is widely believed that the capacity to think about all these things, more in particular our capacity to think about them in hypothetical and counterfactual terms,[23] distinguishes normal adult human beings not only from inanimate objects but also from animals, young children, and adults with mental restrictions.[24]

I agree with compatibilist philosophers who claim that we should understand freedom roughly in these terms. However, it is even more important to stress the fact that we can look at this capacity both from the theoretical and the practical point of view. Anyone who, like Dawkins, does not distinguish between both points of view runs into serious conceptual confusions.[25] When we look at the capacity to reason and to act upon reasons from a theoretical perspective, we can acknowledge that this capacity is distinctive of normal adult human beings without, however, thereby denying that the having and exercising of the capacity is subject to causal influences and might be fully determined. Thus, from a theoretical perspective, it can make sense to distinguish human beings in terms of this capacity. However, the fact that they have this capacity does not allow us to infer that they are fundamentally different from other beings, more specifically that they have a free will in any theoretically interesting sense. Stated differently: whereas the term "capacity to think about and to act upon reasons" makes sense within the realm of theoretical reason, "freedom of will" does not.

Practical Properties as Evaluative Supervening Properties

It is only within the realm of practical reason that we are allowed to infer, from the theoretical fact that a particular person has the capacity to act on reasons, the practical fact that he thus has freedom of will and that he can therefore be held accountable. In other words, freedom of will and capacity-responsibility are practical, more specifically evaluative, properties that supervene on the theoretical property of being able to act on reasons.[26]

What exactly is implied by the fact that practical properties are supervening evaluative properties? Examples of evaluative properties are *free*, *responsible*, and *good*, but also properties like *tasty*, *ugly*, and *pleasant*. To say a practical property supervenes upon theoretical properties is to say an object, event, or state of affair that has this practical property has particular more or less determinate[27] theoretical properties, that is, properties that belong to and can thus be assessed within the realm of theoretical reason.

However, the fact that the supervening property belongs to the practical realm brings with it that it cannot be equated with or reduced to the underlying properties that belong to the theoretical realm.[28] For example, when we say that this is a good or tasty orange, we make a claim, which can be empirically tested within the theoretical realm, that it is juicy, is not too sour, or has several other empirically verifiable characteristics. On top of it, however, we also make the claim that this orange is fitting to be the object of a pro-attitude. Conversely, if the orange is bad, it is fitting that it be the object of a contra-attitude. This implies, among other things, that someone who feels like eating an orange has, ceteris paribus, a good—although not necessarily overriding—reason to choose this particular orange.[29] In a similar vein, to claim that someone has capacity-responsibility, that is, that he can be held accountable for his behavior, is to claim that he has certain properties that can be tested empirically. On top of that, however, we also claim that he is in principle fitting to be subject to the attitude of asking for a justification for his behavior.[30] For example, if his behavior is such that he steps on our toes, our reactive attitude of asking him for his reasons for doing so would be fitting.

What does the foregoing imply about Dawkins' claim? According to the view I have just elaborated, even if Dawkins could, from a theoretical point of view, show that all of our actions, beliefs, desires, and decisions are causally determined, this proof would not say anything about the tenability of the concepts of free will and capacity-responsibility that accountability practices presuppose. For even if Dawkins is right in defending the truth of determinism, this truth in itself does not preclude the possibility of asking for and offering justifications for our behavior. The theoretical assumption that normal adult human beings are capable of practical reasoning and of acting upon their reasons does not conflict with determinism. We only need to presuppose the truth of this particular theoretical claim to be able to defend the claim, within the realm of practical reason, that they are therefore free, accountable, and sometimes liable and punishable for their behavior. Therewith I have rebutted claim 1a. The assumption that people can be free, accountable, liable, and punishable is not refuted by any of the scientific findings that Dawkins mentions.

However, Dawkins not only claims that human beings are not free and responsible in a theoretically interesting sense, he also seems to claim that practices of holding each other accountable are pointless (1b). So, even if I have convincingly argued that freedom and responsibility are terms within the perspective of practical reason that neurosciences cannot eliminate from that realm, I still have to show that Dawkins cannot consistently defend his claim that these practical perspectives are pointless. In section 3.4, I argue that practices of holding oneself accountable are not pointless. In section 3.5, I argue that practices of holding others accountable are not pointless either.

First- and Third-person Points of View

Remember that the central question of the practical point of view is: "What should I do?" For the sake of the argument, let us assume it is theoretically possible to take a purely theoretical perspective toward others and stop treating them as subjects who must justify their behavior.[31] In this section, I restrict myself to the question of whether it is possible to stick to the theoretical perspective in dealing with oneself.

Let me first stress that when one has to make a hard (moral, legal, or prudential) decision, it is not merely a possibility, but it is often useful to look at oneself as if one was an object of scientific research, subject to all kinds of causal influences.[32] Scientific research can teach us interesting things about the kinds of preferences we have—about the question of to what extent our beliefs, desires, and actions are influenced by our upbringing, by not having slept last night, by the use of drugs or alcohol—and it can also tell some interesting things about the questions of whether and when the reasons I make up are the reasons that truly guide my behavior.[33] Science can help to answer these and many other theoretical questions. We can take these answers into account when we make practical deliberations, and they might help us to find new reasons, to weigh existing reasons differently, and in the end to make a different decision about what we should do.

The fact that I take scientific information into account and the way in which I do so can in turn be described, explained, and predicted from a theoretical perspective, and these descriptions can in turn be taken into account as an argument within the practical perspective, and so on. In theory, this regress in turn-taking between posing and answering theoretical and practical questions can go on endlessly. It could even result in one's death before having made a final decision, let alone having acted on that decision.[34]

If Dawkins denies the tenability and the usefulness of practical perspectives, he places himself in such a situation. Scientific descriptions and explanations might help us to predict what others believe, aim at, and do and thus to anticipate their actions. However, when we look at our own reasons and our own behavior from the first-person point of view, all the descriptions and explanations in the world can never tell us what we are going to do.

A, if not *the*, most fundamental difference between the first- and other-person points of view is that whereas theoretical reason can tell others what I will do without affecting my own state of mind, by definition, it cannot tell me. We cannot predict our own behavior unless practical reason would be useless in any case or when our knowledge of the prediction would not affect our deliberations (Bok 1998:89).[35] As soon as I have been told what I am going to do, I can take this information into account, and this can change my reasons, my decision, and my action. From the first-person point of view, we can never bypass the problem of self-denying and self-fulfilling prophecies. As soon as we get new information, we must determine whether and how to decide and act on it.[36] In this sense, my future actions—that is, the things that do not just happen to me, but the things I will do—are not just metaphorically but truly open to me.[37]

Therefore, Dawkins cannot take a purely scientific view toward himself, unless he stops wondering what to do and reduces himself to an object that does not act upon reasons but that is only acted upon. Thus, the practice of holding oneself accountable is neither senseless nor pointless. On the contrary, it is inevitable.

Science, Too, Has an Ineliminable Practical Point of View

Although most philosophers speak about the practical point of view in singular terms and often implicitly equate it with the moral point of view, it is important to emphasize the fact that there exists a plurality of practical points of view. All practical points of view are alike in that they aim at answering the question of what I should do, but they use different criteria to answer that question. Moral points of view differ in this respect from prudential and legal points of view.[38] The practical point of view is not restricted to matters of morality, law, or prudence, however. For the purpose of this chapter, it is important to note that although the primary aim of most[39] sciences is to answer theoretical questions, all of them also are characterized by practices of holding each other accountable. So, if Dawkins wants to eliminate moral and legal practices, it seems he should also give up scientific practices.

Dawkins could argue that scientific accountability practices are unlike moral and legal practices in that they do not aim at determining what we should do but rather at what we should believe to be true, and that therefore scientific accountability practices are altogether different enterprises. This argument fails, however, if only because scientists do not only hold each other accountable for what they believe to be true or false; they also hold each other accountable for the things they *do* to test and confirm their beliefs. They hold each other accountable, among other things, for the setup of experiments and for testing the reliability of their sources and hypotheses. Scientific practices are not merely a matter of asking what beliefs we are justified to hold true but also a matter of asking what we should do to make our beliefs in the truth of our beliefs justified in the first place.

These remarks about the practical dimension of science may sound like trivial truths, but they are relevant to rebutting Dawkins' claim. It is important to note that Dawkins, as a scientist who criticizes the tenability and usefulness of moral and criminal accountability practices, necessarily takes part in scientific accountability practices within which scientists address each other as subjects who must offer proper reasons in order to justify their own theories and to criticize those of others. Although both the criteria and the ultimate "pointe" of scientific practical practices are different from those of moral and legal practices, nevertheless the direct aim of these scientific practices, too, is to hold oneself and others accountable, both for the scientific claims one makes and for the actions on the basis of which one makes those claims.

If Dawkins truly believes that the fact that our thinking and acting is causally determined implies that normal adult human beings do not have the freedom of will necessary for accountability, and if he truly believes that therefore practices within which we hold each other morally and criminally accountable have therefore been superseded, then he should also accept the view that any other practice within which we hold ourselves and others accountable is just as meaningless and useless as moral and legal practices. For obviously, our scientific beliefs and the things we do to establish their truth are no less "determined by circumstances that are predisposed by the physiology, the hereditary and the environment of the scientist" (Dawkins 2006:19) than our moral and legal beliefs and behavior. To eliminate the concept of accountability is to eliminate the possibility to pose the question of whether we have done the proper thing to establish the truth of our scientific beliefs.[40]

Thus, it turns out that Dawkins cannot make the claim that accountability practices have been superseded without contradicting himself. The only option available to him is to show his view by no longer making any claims, that is, by no longer participating in *any* accountability practice. However, if he does so, he not only withdraws from moral practices, he can also no longer partake in scientific practices. Therefore, I conclude that not only practices of holding oneself accountable, but also practices of holding others accountable, are neither senseless nor pointless.

Instrumental and Intrinsic Reasons for Holding Others Accountable

Let me briefly summarize my argument. I first argued that if we take freedom of will and capacity responsibility to be concepts of practical reason, the assumption of freedom of will does not conflict with a deterministic worldview. Thereby I have shown that, to this extent at least, accountability practices do not rest on untenable presuppositions. Next, I argued that one cannot propose to eliminate accountability practices without falling prey to contradictions. Finally, I have shown that partaking in accountability practices can contribute to realizing the "pointe" of these practices.

Although I argued in section 3.4 that we cannot take a purely instrumental view when it comes to holding ourselves accountable, readers might wonder whether we have more than instrumental reasons for holding others accountable. Stated differently, the question is whether we only have reason to take others as objects that are causally influenced by reasons and moral emotions, or also to view others as subjects who are truly guided by reasons.

From the findings of the last section, it follows that we should take others not merely as objects whose behavior is causally influenced by motives but also as subjects who are guided by reasons. First, whereas the point of theoretical reason is to arrive at truth and the point of practical reason is to arrive at right action, both aims are achieved by means of reason. Second, if we accept the Popperian view that both the theoretical and the practical points of view imply that we must subject ourselves to criticism, this seems to imply that we must regard other human beings as subjects who can truly argue with us, that is, as subjects who are truly guided by reason.

If we take the practical point of view seriously, it seems insufficient to merely envisage others as a special category of objects whose behavior we can influence, and whose behavior can influence our behavior through making conversation with each other. Such a perspective does not allow us to make a distinction between reasons that merely persuade us to hold certain beliefs or to do certain things on the one hand and well-founded reasons and actions on the other. It only allows us to distinguish between reasons that are factually influential and those that are noninfluential.

If we want to retain the normative distinction between good and bad reasons, it seems that treating others as reason-guided subjects is not merely instrumental to the fulfillment of the (moral, legal, or scientific) goals we have when we hold each other liable and punishable, but rather an ineliminable precondition of the viewpoints of *both* practical and theoretical reason. In section 3.5, I argued that we have to make use of practical reason in order to answer the question of theoretical reason. Therefore, treating others as reason guided is a precondition not only of practical reason but also of theoretical reason. Thus, to the extent that we commit ourselves to theoretical and practical reasoning, we have not merely an instrumental but also an intrinsic reason for holding others accountable (Bok 1998:193).[41]

CAN NEUROSCIENTIFIC FINDINGS CHANGE THE CONTENT AND THE APPLICATION OF OUR CONCEPTION OF CAPACITY-RESPONSIBILITY?

In section 3, I argued that, pace Dawkins, developments in the (neuro)sciences will not result in the elimination of our concepts of responsibility or of the accountability practices that contain these concepts. In this section, I go deeper into the question of whether neurosciences can play a more modest role in accountability (and in liability) practices. The focus of this section is on the question of whether neuroscientific views can result in a change of the content of the conception of capacity-responsibility (2a) or in a change of the application of the conception (2b).

One might wonder how I can argue for the view that they can, given what I said in the previous section. In the last section, I stated that the concept of capacity-responsibility belongs to the practical realm, whereas neurosciences belong to the theoretical realm. How could neurosciences have anything to say about a realm they do not take part in? In sections 4.1 to 4.4, I argue that neuroscientific findings can play a role in the assessment of (possibly conflicting) findings about human thinking and acting that other sciences come up with. Neurosciences can therewith contribute to a change of the conception of capacity-responsibility. In section 4.5, I first state that although no constitutive relation exists between neurological properties and capacity-responsibility, this does not preclude the possibility of an epistemological relation between them. If there exists such an epistemological relation in a concrete case, neuroscientific findings can change the application of the conception of capacity-responsibility. Thus, I argue that Gazzaniga's negative answer (2006:141) fails. I conclude that neurosciences might contribute to an improvement in both the content and the application of the conception of capacity-responsibility.

Triangulation as Cross-Examination

There are two reasons that neurosciences could contribute to our conception of capacity-responsibility. In the first place, as I argued in section 3.3, the property of capacity-responsibility supervenes upon theoretical properties that are established within the theoretical domain. Second, it is unclear and contested exactly what are the theoretical properties that capacity-responsibility supervenes upon. Thus, it seems that the conception of capacity-responsibility can do with a lot of *theoretical* specification and clarification of the underlying properties. By offering information about the brain mechanisms that underlie our capacity to act on reasons, neurosciences might very well contribute to this project.

Against this optimistic view, one might object that none of the criteria of application of capacity-responsibility refers to neurological properties. These criteria refer only to mental and behavioral properties. Because there is no type identity between mental states and brain states, let alone between brain states and behavior, it seems that neuroscientific findings cannot say anything about our conception of capacity-responsibility. Thus, it seems that although sciences might contribute to the clarification of the conception, only psychologists and psychiatrists, not neuroscientists, seem to be able to do so.

Here we should distinguish the question of whether neurosciences can change the conception of responsibility on their own from the question of whether they can adduce arguments to change that conception. There are two reasons that neurosciences can never alter our conceptions on their own. The first reason is that the final say about these conceptions lies in the practical realm. Even if neurosciences, or any other science, would adduce arguments for other factual criteria of application, the final decision to accept those criteria would belong to the domain of practical reason. The second reason, as was stated above, is that our present conception consists of mental and behavioral criteria only.

However, this does not imply that neurosciences cannot adduce any arguments at all. Most important, neuroscientific findings can be used in the *triangulation* of findings from psychology, psychiatry, and self-reports. If neuroscientific findings concur with psychiatric findings, psychological tests, and self-reports, this will reinforce the evidence from these sciences and self-reports. If neuroscientific findings conflict with aforementioned kinds of findings, this will be reason to ask further questions and possibly even to reject psychiatric findings, psychological tests, or self-reports. Stated differently, neuroscientific findings can be used to *cross-examine* psychiatric findings, psychological tests, and self-reports. In this manner, neurosciences might be able to contribute to the clarification of and the choice between conflicting conceptions.

To illustrate my claim, I discuss two conceptions of capacity-responsibility and show how neurosciences might help to clarify them and to choose between them.

M'Naghten Rule or "Substantial Loss of Volition"?

In many American states, the M'Naghten rule is the core of the insanity defense. It states that perpetrators are liable for their behavior only if they knew what they did and if they knew that what they did was wrong. The rule is contested, however. It is claimed to be too narrow in that it focuses on cognitive disorders only and not on volitional disorders. Critics of the M'Naghten rule argue that irresistible impulse or substantial loss of volition can also reduce or take away the liability of the offender (Sapolsky 2004:1790). Thus, in criminal law, we are confronted with contested and conflicting conceptions of capacity-responsibility.

The practical advantage of the M'Naghten rule is that most of the time it is possible, also in retrospect, to determine with a satisfactory degree of certainty whether someone knew what he did and whether he knew that what he did was wrong. Persons with mental retardation and psychotic delusions are good examples of suspects who are most likely to fail the M'Naghten rule. A problem of the criterion of irresistible impulse and loss of volition is that it is much more difficult, if not impossible, to ascertain whether someone suffered from lack of freedom of will that exculpates him, or whether it was merely a case of (normal and culpable) weakness of will. Moreover, some critics of the criteria of irresistible impulse and substantial loss of volition argue that the fact that someone was incapable of controlling himself should not be explicated in terms of a defect of will but in terms of a rationality defect (Morse 2002, also see Morse 2006:38–39). According to the latter view, a substantial loss of volition or an irresistible impulse takes away the accountability because the capacity to think was affected to the extent that the perpetrator did not know what he did or that what he did was wrong.

The debate about cognitive versus volitional defects seems particularly relevant when dealing with sociopaths, who do not seem to suffer from cognitive disorders. Accordingly, there is disagreement about whether and to what extent they can be held accountable for their criminal behavior. The question is whether neurosciences can help to improve the content of the conception of capacity-responsibility and whether they can help to improve the application of the conception to sociopaths.

Neuroscientific Findings

During the past few decades, a lot of research has been done with respect to the role of the prefrontal cortex in normative judgment and in the inhibition of behavior. Although positron emission tomography and functional magnetic resonance imaging still have their limitations (Goodenough and Prehn 2004:1715–1716), these techniques have helped to establish that in particular the ventromedial prefrontal cortex, orbitofrontal cortex, posterior cingulate cortex, and posterior-superior temporal sulcus play a central role in moral judgment (Goodenough and Prehn 2004:1718, Sapolsky 2004:1790).

Roskies (2003 and 2006) discusses research with respect to persons who have a damaged ventromedial cortex.[42] Despite the damage, their capacity for moral *judgment* does not seem to differ (much) from the capacity of subjects without a damaged ventromedial cortex. They do show deficiencies, however, when it comes to *acting* morally. For one thing, they are not motivated to act on their moral convictions.

In other words, they seem to know what they do, they seem to know that what they do is wrong, and they do not seem to suffer from substantial loss of volition or from an irresistible impulse. Instead, there seems to be a lack of (moral) emotions and motivation to do the morally proper thing. These subjects report that they experience no or at least less emotions in morally charged situations than "normal" subjects. These self-reports are corroborated by the fact that they do not show a normal physiological skin conductance response (Roskies 2006:19–20).[43] In morally charged situations in which they do show a normal physiological skin conductance response, they act like normal subjects (Roskies 2006:24).

Further psychological research has revealed that the moral reasoning of sociopaths is different in "up-close and personal" cases that evoke emotion and personal involvement in controls. One such example is the trolley case. In this hypothetical case, people are asked to decide what to do when a trolley is about to run into five people and kill them. In one version of the case, they are asked whether they would reverse a switch causing the trolley to take another track where it would kill only one person instead of five. In another version they can achieve the same result, not by toggling a switch, but by pushing a man into the path of the oncoming trolley. Whereas most controls believe it is morally correct to turn the switch, they believe it is immoral to push the man. Sociopaths, on the other hand, argue that both cases are alike in that saving five is better than saving one. Roskies hypothesis is that this is because they do not *feel* the difference between the two cases. Thus, lack of emotions, which is confirmed by neurological findings, does not seem to result merely in differences in compliance with moral and legal rules but also in differences in moral *judgments* after all.

Conception

The question, now, is what defects in the ventromedial cortex and the absence of skin conductance imply for our conception of capacity-responsibility. The results of brain scans, physiological tests, psychological tests, behavioral findings, and self-reports can be put together to achieve a consistent, although far from complete, picture of the subjects under investigation.

In this case, neuroscientific findings seem to concur with psychological and everyday findings that there is something different about sociopaths. The findings seem to conflict, however, with the M'Nagthen rule or at least with a restricted reading of this rule. First, neurosciences show that sociopaths are a category of persons who have defects in the ventromedial cortex, which is widely believed to play a crucial role in moral reasoning. Thus, it seems unlikely that such damage would not show up at all at the behavioral and psychological level. Therefore, these findings offer reason to surmise that despite the first appearance, their moral reasoning might differ from controls. This hypothesis is confirmed on the psychological level, among others, by the trolley case. Second, neurosciences show that the defects have to do with our emotion-regulation system, so there is reason to surmise that their emotions are "abnormal." The latter hypothesis, too, is confirmed on a psychological level.

These findings together are reason to reconsider the M'Naghten rule. For one thing, the combined neurological and psychological findings suggest that the claim that emotions only lead our judgments astray, as some ethical views want us to believe, is false. Rather (as 18th-century ethical theories have claimed), they seem to play a crucial role in arriving at proper moral beliefs (that there is a morally relevant difference in the trolley case between reversing a switch and pushing another human being[44]) and proper moral decisions (that the proper moral choice is not necessarily to save the highest number of people). Stated differently, emotions seem to play an important role in knowing what you do and in knowing that what you do is wrong.

This does not necessarily imply that we must reject the M'Naghten rule altogether. The findings do give reason, however, to reconsider the conceptions of understanding and reasoning that play a central role in the rule. Perhaps we should conclude on the basis of aforementioned combined psychological, neurological, and self-report findings that the capacity to understand and reason is not a sufficient condition to pass the M'Naghten test. It is interesting to note that similar problems, relating to the conceptions of understanding and reasoning and the role of emotions, play a role in the assessment of informed consent to and refusal of medical treatment of specific categories of psychiatric patients. The MacArthur Competence Assessment Tool-Treatment (MacCAT-T) is claimed to be the most fully developed standardized method of assessing competence in making decisions about medical treatment (Tan et al. 2003:698). It examines four abilities: to understand relevant information, to reason about risks and benefits of potential options, to appreciate the nature of one's situation and the consequences of one's choices, and to express a choice (Grisso et al. 1997:1415). The MacCAT-T appears to work well in assessing the competence of patients with a schizophrenic or schizoaffective disorder (Grisso et al. 1997:1415).

The Mac-CAT-T is a more sophisticated criterion than the M'Naghten rule. However, just like the M'Naghten rule, it seems to fail as a generally applicable conception of capacity-responsibility. Tan and colleagues (2003:704) claim that the performance on the MacCAT-T of a small group of patients with anorexia nervosa was comparable to that of a healthy population control group. Nevertheless, the investigators had serious doubts about the patients' competence in making a decision about their medical treatment. Tan and colleagues (2003) argue that the importance of the anorexia nervosa to the patients' sense of personal identity, and therewith the relative unimportance of death and disability compared with anorexia nervosa, might explain their ambivalence about and their refusal of treatment.[45] This evaluation of the relative weight of anorexia nervosa and death is not, however, picked up by the MacCAT-T as indicating an impairment of their capacity.

Here, too, the question seems to be whether we should regard patients with anorexia nervosa as fully competent or whether we should reconsider the adequacy of tests that assess their competence. As with sociopaths, lack of "proper" emotions seems to play a crucial role in the choices of patients. And given the conflicting scientific findings, it seems a viable approach to investigate to what extent neurological findings support or rather conflict with findings on the level of psychological tests and self-reports.

Application

Triangulation of different scientific findings and self-reports is not only helpful in the evaluation of conflicting conceptions of capacity-responsibility but can also be used to assess the application of conceptions. Let us assume that the M'Naghten rule is extended to include affective criteria. Let us also assume that the new conception still does not refer to any neurological condition, but rather is formulated in strictly behavioral and psychological terms. Nevertheless, or so I will argue, neurosciences can help to properly apply the conception. First note that, in many individual cases, empirical evidence will not be straightforward but rather will be indeterminate and sometimes conflicting. In these cases, triangulation of self-reports and behavioral, psychological, and neuroscientific findings can help to determine whether a particular agent fits the criteria. Thus, also with respect to application, neuroscientific findings can help both to strengthen and to put doubts on other findings.

I even want to go one step further and claim that although the property of capacity-responsibility supervenes on psychological and behavioral properties, this does not imply that it is in principle impossible that neuroscientific tests will be developed that assess capacity-responsibility without taking mental and behavioral findings into account. This claim might seem to conflict with my earlier statement that capacity-responsibility is defined in terms of acting on reasons. However, we should distinguish between, on the one hand, the *constitutive* relation that exists between theoretical properties and capacity-responsibility and, on the other hand, the *epistemological* relation that exists between them.

Although the constitutive relation is on the level of mental states and behavior, it is not impossible in principle that we will sometimes give priority to evidence from the neurological level. This only seems a plausible and potentially useful approach when the neurological findings are quite stable with respect to the category of persons under investigation, while at the same time the mental and behavioral findings with respect to a particular person are imprecise, inconclusive, or contradictory, or no such findings are available from the time of the act. This can especially become relevant in criminal law suits in which a defendant is not cooperative or a long time has passed since the crime was committed.[46] Thus, I conclude that neurosciences can play a role in the proper application of the conception of capacity-responsibility.

CONCLUSION

I have argued that neurosciences are not a threat to the continued existence of our accountability practices. They do not show these practices to be untenable, and they do not show them to be useless either. I have also argued that neuroscientific findings cannot dictate the content or the application of our conception of capacity-responsibility because decisions about the conception and the application are made in the practical realm. Nevertheless, or so I have argued, neurosciences can suggest new ways of looking at both the content and the application of our conceptions.

Neuroscientific findings can help, for example, to shape our thoughts about the question of whether the conception only consists of cognitive criteria, or whether and how volitional or affective criteria should also play a role. Moreover, neuroscientific findings can help to make those criteria more clear and precise. Neurosciences can also play a role in determining whether particular (categories of) persons actually fit the criteria. Although conceptions exclusively refer to mental and behavioral properties, it is not impossible in principle that, for some cases at least, neuroscientific tests will be developed and used next to or even instead of psychological tests. Thus, although capacity-responsibility supervenes on (verbal) behavioral properties, the decision about whether someone has those properties might in principle be partly or even largely based on neuroscientific evidence.

Some final questions remain. Although neuroscientific findings turn out not to be a threat to the existence of our accountability and liability practices, they might suggest that we should hold fewer (categories of) human beings accountable or liable than we do now. What should follow from the fact that we learn more about human behavior, more in particular about disorders and risk factors? Should the consequence be that only the state becomes more accountable and liable for the imposition of proper legal measures with respect to these persons? Or should we argue that at least some of these persons acquire a new responsibility, that is, to do something about their "disorder" and to avoid risky situations? Should knowing more about one's own disorders and risk factors result in more *culpa in causa*?

Will the growth of scientific knowledge result in an increase in state responsibility and a decrease in personal responsibility, or will it rather result in an inflation of both state and personal responsibility? From the perspective of theoretical reason, we can try to predict what will actually happen, but we will still have to decide, from a practical point of view, how much responsibility we should strive for.

NOTES

1. This paper is a fully revised version of Mackor (2008). I am grateful to Nicole Vincent and Vincent Geeraets for their helpful comments and to Stephen Morse for referring me to Bok (1998), who has defended a view very similar to mine. Her book enabled me to refine my own view on several points.
2. In this chapter, I deal only with the point of holding each other responsible. I do not deal with the further question of whether punishing each other will gain us anything.
3. Someone who, like Dawkins, believes that there are convincing arguments for the view that a particular practice is not useful or tenable offers an external type of critique. He could nevertheless also look into the practice and investigate whether the central concepts and conceptions of this practice are internally coherent. He would then pose the second kind of question and offer an internal type of critique, although merely for the sake of the argument.
4. These are the two most fundamental questions. One might ask other internal questions, such as whether neuroscientific findings can improve the treatment of some perpetrators and whether neuroscientific findings can contribute to a better judgment of the risk for recidivism of specific (categories of) persons. These questions will not be discussed in this chapter because they do not question the content of the conception of responsibility, nor its application. If my analysis of the internal questions is correct, these two questions should be answered affirmatively.
5. I have taken justice as an example, among others, because Rawls (1971:5) uses the distinction between concepts and conceptions with respect to justice.
6. The boundary between internal and external critique is not sharp, however, because a radical change of a conception could result in a fundamental change of the related concept, and if this concept is as central to a practice as "responsibility," it can therewith change the nature of the practice as a whole.
7. Unfortunately, it is not quite clear what Gazzaniga's claim is. He also makes the less controversial claim that "Neuroscience simply does not have as much to say on the issue of personal responsibility or free will as people would think or hope. Cognitive neuroscience is identifying mechanisms that help us to understand how changes in the brain create changes in the mind" (2006:145). This is not very helpful for understanding what neurosciences can and cannot do, however.
8. The use of the term *responsibility* in its causal sense might be a relic of past times in which causal relations and relations of imputation were not yet distinguished. Compare Kelsen (1960:86–89), who argues that that the concept of causality has

its roots in the concept of retribution, instead of the other way around. Dawkins (2006:19) offers the example of King Xerxes, who sentenced, in 480 B.C., the rough sea to 300 lashes because it had destroyed his bridge of ships. According to Dawkins, this example is just as hilarious as a judge who punishes a criminal.

9. Note that we are accountable not only for our actions but also for our behavior. If someone bumped into me because he was unexpectedly pushed by someone else, he did not act, but was acted upon. In that case, we cannot hold him accountable for an act or action, but we can still hold him accountable for his behavior.
10. Two other concepts of responsibility are often mentioned. One is role responsibility. To ascribe role responsibility is to claim that someone has special authority or duties that follow from his or her role, e.g., as parent, teacher, or policeman. The other is virtue responsibility. To call someone a responsible person is to say that this person is disposed to act responsible, i.e., to do the right thing, not to act immorally or carelessly.
11. Obviously, it is a normative question what counts as a *proper* role in the causal chain of events. The only claim being made here is that the action or the *inaction* of the parent plays a role in the *causal* chain and that therefore accountability presupposes some form of causal responsibility. See Quinn (1989a:292–294) on different conceptions of causality and the claim that both action and inaction can be the cause of a harmful event.
12. I agree with Bok (1998) that responsibility is not only about holding others accountable but also, first and foremost, about holding oneself responsible, i.e., about taking responsibility in the sense of accountability. Note that this claim does not conflict with the hypothesis that the capacity for taking accountability is not fully innate but has to be acquired, implying, among other things, that children first have to be held "as if" accountable by others before they truly are accountable because only then do they learn the "game" of taking accountability and holding accountable.
13. The "why" question is ambiguous. It can refer to a question for justification or a question for explanation. Here it is used in the sense of question for justification.
14. We must also understand what the "pointe" of holding each other accountable is. This question is dealt with in section 3.
15. Note that we can have all of these attitudes not only with respect to others but also with respect to ourselves.
16. Here I only discuss justification. Note, however, that when someone is held accountable, he can also offer an excuse. There is an important difference between justifying and excusing your behaviour. If I ask you, e.g., "Why did you take that book from the shop without paying?" you can justify your behaviour by answering, "You're wrong, I did pay for the book," implying that you did not do anything wrong and that, as far as you are concerned, the discussion should stop right there. If you excuse yourself, however, by stating, "I thought that you had paid for my book, but apparently I was wrong," you admit that you did something wrong, but you claim you are not (or at least that you are less) blameworthy for doing so. Thus, a justification differs from an excuse in that a justification prevents one from becoming liable, whereas a good excuse only prevents certain kinds of lia-

bility. For example, one might not be blameworthy, but one might still have the obligation to repair the damage done.

17. On liability and punishment, more in particular on the question of whether we can only punish for preventive or for truly retributive reasons, see Smart 1961, Carlsmith et al. 2002, Baird and Fugelsang (2004), Goodenough 2004, and Mackor 2008.
18. Dawkins (2006:19) states: "Each crime, however horrible it is, can in principle be *blamed* to the circumstances that are *predisposed* by physiology, heredity and the environment of the suspect" (my italicization).
19. Original text reads "...das [Interesse] des praktischen Gebrauchs [der Vernunft besteht] in der Bestimmung des Willens *in Ansehung des letzten und vollständigen Zwecks.*" (Kant, 1974:138[AA 120], my italicization).
20. Bok (1998, chapter 2, e.g., 63) defines the aim of theoretical reason in terms of description and causal explanation. This implies that mathematics, legal doctrine, and theology belong neither to theoretical nor to practical reason. I agree with Kelsen (1960), however, that legal doctrine belongs to the realm of theoretical reason in that it aims to describe and explain relations between events, even though it does not do so in terms of causal but rather in terms of a specific type of rational necessity, that is, imputation. See Mackor (2011), especially sections II and III.
21. Walter (2001:6) makes a now-famous distinction between three conceptions of freedom of will. To have free will implies that someone (1) could have acted otherwise, (2) is the source or initiator of his actions, and (3) acts for understandable (but not necessarily acceptable) reasons. Dawkins attacks the first distinction. Interestingly, however, the last claim, that we are capable of truly *acting on* reasons, might turn out to be a much more problematic assumption. Psychological investigations reveal that quite often we do not know our reasons for behavior, that we confabulate reasons, and that in fact our behavior is not guided by our conscious reasons, but that often we make them up afterward (see Wilson 2002 for an overview of some findings). However, we do not need to assume that we always act on the reasons we consciously hold, we only need to assume that we sometimes do. Moreover, although psychology has shown that we are less than perfect practical reasoners, it has also shown that we are less than perfect theoretical reasoners (compare, e.g., the famous studies of Kahneman and Tversky). The latter imperfections are never seen as an argument to stop doing theoretical reasoning, more specifically to stop doing science. So, again, there does not seem to be a fundamental difference between theoretical and practical reason in this respect.
22. The claim that one could have acted otherwise need not be explicated in terms of the claim that one can act contra-causally as Dawkins seems to presume. I will not go deeper into this. See Kelsen (1960:95–102) and Bok (1998, chapter 3) for a different account.
23. The capacity to think in hypothetical and counterfactual terms is the core not only of practical but also of theoretical reason. On a Popperian account, to be able to think in hypothetical terms is the precondition of the capacity to do science.

24. Adolescents are an interesting and widely discussed category of persons who are capable of hypothetical and counterfactual practical reasoning in laboratory conditions, but not always in complex everyday situations. It is argued that certain neurological immaturities play a crucial role in this real-life incapacity. See Mackor 2008, section 5.
25. I agree with Bok (1998, chapter 2) that this is the most fundamental distinction for a proper understanding of responsibility. For a similar view, see Kelsen (1960:95–102).
26. By "theoretical properties" I mean properties that belong to the theoretical realm. I do not mean to say that they are theoretical properties as opposed to empirical properties. On the contrary, at least part of these theoretical properties are empirically verifiable properties.
27. A problem is that it is often not quite clear exactly which (combination of) theoretical properties is prerequisite.
28. Supervenience and irreducibility as such are not distinctive characteristics of practical properties. Within the realm of theoretical reason, there also exist supervening properties, such as the property of water, which supervenes upon properties that collections of H_2O molecules have. Practical properties also do not differ from supervening theoretical properties in being irreducible. Many supervening theoretical properties can, unlike water, not (easily) be reduced to the underlying properties either. The fundamental difference is that practical properties make objects fitting to be the object of a certain attitude.
29. I will not go into the question of whether all evaluative properties are *eo ipso* also normative properties, i.e., whether they contain a norm that instructs people what to do. Accountability seems to be a normative property in that it implies that someone who is accountable is under the obligation to do something, that is, to offer a justification.
30. Compare section 2. More on the analogy between value and responsibility ascriptions in M. Zimmerman, *Responsibility, Reaction, and Value*, lecture Delft, August 2009.
31. Dawkins acknowledges that it is difficult if not impossible to do so in practice. "My dangerous thought is that we will outgrow all of this [i.e., the ascription of guilt and responsibility] and that we will even learn to laugh about it...*However, that state of enlightenment I will probably never reach myself*" (Dawkins 2006:19, my italicization). This paper deals only with the principled question about accountability. It neither deals with the practical question nor with the principled question about liability and punishment.
32. Introspection is not the only or even the best way to know oneself.
33. See Wilson (2002) for an overview of psychological research into the latter question.
34. This is what happened to Buridan's donkey. This strictly rational animal was placed between two identical haystacks. Because he did not have a rational argument to choose one stack or the other, he died in the midst of an abundance of food.
35. I can predict many things about myself, e.g., that I will grow older and eventually die. However, these are not things I do but merely events that happen to me.

36. "My struggles are not struggles against my fate... they *are* my fate" (Bok 1998:77).
37. For more on the sense in which we truly have alternatives, see Bok (1998, chapter 3).
38. The legal point of view itself is not monolithic either. In the first place, different countries use different conceptions of accountability and liability. Second, different conceptions are used within civil and criminal law.
39. There is some disagreement about the question of whether the purpose of theology (as opposed to empirical religious studies) and legal doctrine (as opposed to empirical legal sciences) is theoretical or practical in nature. Compare note 19. See Mackor (2011, section III.C).
40. Just like law and morals, sciences have not only accountability but also liability practices. The first, and sometimes the only, step is to hold someone accountable, not only for his scientific views but also for the actions through which those views are established. Only subsequently might the question arise of whether someone is liable and punishable for the views he has publicly defended. Thus, when someone makes a truly spectacular discovery, he will be praised and possibly rewarded, whereas he will be blamed and sometimes punished if his discovery turns out to be fraud or plagiarism. With respect to moral and legal as well as scientific liability practices, we can ask whether they contribute to the aim of the specific practice—in the case of science, to achieve better descriptions, explanations, or predictions of states of affairs and events in the world. I surmise that Dawkins will agree with me that not only accountability but also liability practices help to realize, though imperfectly, the very aim of scientific practices.
41. Note, first, that my argument differs from Kantian moral arguments that refer to the forms of respect that we owe to each other as *moral* beings to found our practices of holding each other accountable. Second, my claim has not been that I have herewith also shown that we have an intrinsic reason for *blaming* and *punishing* each other.
42. I have taken Roskies' paper as an example. I might just as well have discussed the work of Blair, Damasio, and Nichols, or others who stress the relevance of emotions for moral judgment and behaviour. Because I discuss sociopathy only to illustrate the way in which neurosciences might result in a change of the content and application of our conception, the details of different views about sociopathy are not relevant to my paper.
43. According to Roskies, the skin conductance response can be understood to be a physiological correlate of motivation because the reaction occurs independently of the object to which the motivation is related.
44. See, e.g., Quinn (1989b) for an argument that there is a morally relevant difference between such cases.
45. This finding gives rise to the question of why we look upon the lack of interest in the continuation of one's own life as "madness," whereas we are inclined to consider sociopathic lack of interest in the interests of others in the continuation of their life as "pure evil."
46. I do not mean to suggest that it should be possible to do neurological tests without the informed consent of the defendant. I primarily think of cases in which

a defendant has consented, but seems to be lying or cheating, or in which his self-reports do not seem dependable for other reasons (e.g., unreliable memory). In such cases, neurological tests might be more reliable than other evidence for the truth or falsity of claims about the mental capacity to deliberate and act on reasons.

REFERENCES

Baird, A. A., and J. A. Fugelsang (2004). "The emergence of consequential thought: Evidence from neuroscience." *Philosophical Transactions of the Royal Society of London, Series B* 359: 1797–1804.

Bok, H. (1998). *Freedom and Responsibility.* Princeton, NJ, Princeton University Press.

Carlsmith, K. M., J. M. Darley, and P. H. Robinson (2002). "Why do we punish? Deterrence and just deserts as motives for punishment." *Journal of Personality and Social Psychology* 83(2): 284–299.

Dawkins, R. (2006). "Straf is wetenschappelijk achterhaald." *NRC-Handelsblad* January 14: 19.

Dennett, D. C. (1984). *Elbow Room: The Varieties of Free Will Worth Wanting.* Cambridge MA: MIT Press.

Gazzaniga, M. S. (2006). "Facts, fictions and the future of neuroethics." In: *Neuroethics. Defining The Issues in Theory, Practice, and Policy.* J. Illes. Oxford, UK, Oxford University Press, pp. 141–148.

Goodenough, O. E. (2004). "Responsibility and punishment: whose mind? A response." *Philosophical Transactions of the Royal Society of London, Series B* 359: 1805–1809.

Goodenough, O. E., and K. Prehn (2004). "A neuroscientific approach to normative judgement in law and justice." *Philosophical Transactions of the Royal Society of London, Series B* 359: 1709–1726.

Greene, J., and J. Cohen (2004). "For the law, neuroscience changes nothing and everything." *Philosophical Transactions of the Royal Society of London, Series B* 359: 1775–1785.

Grisso, T., P. S. Appelbaum, and C. Hill-Fotouhi (1997). "The Mac-CAT-T: A clinical tool to assess patients' capacities to make treatment decisions." *Psychiatric Services* 48(11): 1415–1419.

Habermas, J. (2007). "The language game of responsible agency and the problem of free will: How can epistemic dualism be reconciled with ontological monism?" *Philosophical Explorations* 10(1): 13–50.

Hart, H. L. A. (1968). *Punishment and Responsibility.* Oxford, UK, Clarendon Press.

Kant, I. (1974) *Kritik der Praktischen Vernunft.* Herausgegeben von Karl Vorländer. Hamburg, Felix Meiner Verlag.

Kelsen, H. (1960). *Reine Rechtslehre.* Wien, Verlag Frans Deuticke.

Korsgaard, C. M. (1996). *The Sources of Normativity.* Cambridge, UK, Cambridge University Press.

Mackor, A. R. (2008). Wat hebben neurowetenschappen over verantwoordelijkheid te zeggen? In: *Kernproblemen van de Psychiatrie.* J. A. den Boer, G. Glas, and A. Mooij. Amsterdam, Boom, pp. 299–327.

Mackor, A. R. (2011). Explanatory non-normative legal doctrine. Taking the distinction between theoretical and practical reason seriously. In: *Methodologies of Legal Research. Which Kind of Method for What Kind of Discipline?* M. van Hoecke. Oxford, UK, and Portland, OR, Hart Publishing, pp. 45–70.
Morse, S. J. (2002). "Uncontrollable urges and irrational people." *Virginia Law Review* 88: 1025–1078.
Morse, S. J. (2006). Moral and legal responsibility and the new neuroscience. In: *Neuroethics. Defining the Issues in Theory, Practice, and Policy.* J. Illes. Oxford, UK, Oxford University Press, pp. 33–50.
Nichols, S. (2002). "Norms with feeling: towards a psychological account of moral judgement." *Cognition* 84: 221–236.
Quinn, W. S. (1989a). "Actions, intentions, and consequences: The doctrine of doing and allowing." *The Philosophical Review* 98(3): 287–312.
Quinn, W. S. (1989b). "Actions, intentions, and consequences: The doctrine of double effect." *Philosophy and Public Affairs* 18(4): 334–351.
Rawls, J. (1971). *A Theory of Justice.* Oxford, UK, Oxford University Press.
Roskies, A. (2003). "Are ethical judgements intrinsically motivational? Lessons from 'acquired sociopathy.'" *Philosophical Psychology* 16(1): 51–65.
Roskies, A. (2006). A case study of neuroethics: The nature of moral judgement. In: *Neuroethics. Defining the Issues in Theory, Practice, and Policy.* J. Illes. Oxford, UK, Oxford University Press, pp. 33–50.
Sapolsky, R. M. (2004). "The frontal cortex and the criminal justice system." *Philosophical Transactions of the Royal Society of London, Series B* 359: 1787–1796.
Smart, J. J. C. (1961). "Free will, praise and blame." *Mind* 70(279): 291–306.
Strawson, P. F. (1962). "Freedom and resentment." *Proceedings of the British Academy* 47. Reprinted in P. F. Strawson (1974). *Freedom and Resentment and Other Essays.* London: Methuen & Co, pp. 1–25.
Tan, J., T. Hope, and A. Stewart (2003). "Competence to refuse treatment in anorexia nervosa." *International Journal of Law and Psychiatry* 36: 697–707.
Walter, H. (2001). *Neurophilosophy of Free Will.* Cambridge, MA, MIT Press.
Wilson, T. D. (2002). *Strangers to Ourselves.* Cambridge, MA, Belknap Press of Harvard University Press.

4

Irrationality, Mental Capacities, and Neuroscience[1]

JILLIAN CRAIGIE AND ALICIA CORAM

The concept of mental capacity is of central importance in private law, where it is used to determine a person's ability to make certain personal decisions, including whether to consent to or refuse medical treatment. Questions about psychological capacities are also increasingly playing a role in criminal law, where culpability is being understood in terms of the ability for rational action. In both contexts, progress in the neurosciences has raised questions about how the science could be used to inform these legal judgments. In this chapter, we investigate this question by examining the relationship among judgments about capacities, the norms of rationality, and underlying psychological and neural mechanisms.

First, we aim to clarify the relationship between capacity and rationality as it relates to medical decision making in English law (law in England and Wales). The established position on this issue in England and Wales is that "the patient's right of choice exists whether the reasons for making that choice are rational, irrational, unknown or even non-existent."[2] We call into question the degree of independence between mental capacity and rationality that seems to be assumed here, and argue that judgments of incapacity are fundamentally made on grounds of procedural irrationality. The importance placed on distinguishing rationality and mental capacity in English law is, we claim, explained by concerns about judgments of irrationality made on substantive grounds, which would seem to threaten key political principles underlying this area of law.

This analysis of the relationship between capacity and rationality in the context of treatment decision making brings to light important differences when compared with the questions of rationality that are held relevant in assessments of criminal culpability. In particular, we focus on the role played by substantive norms in the criminal context. The essential role played by normative requirements in both kinds of determination, along with differences in the norms that are taken into account, is used to draw conclusions about what neuroscience can offer in these two legal contexts. We conclude that contemporary debates in moral psychology call into question key assumptions that underlie optimism about the potential role for neuroscience in assessments of culpability.

MENTAL CAPACITY

The concept of mental capacity (or, previously, competence) can be understood as a legal tool that is used to balance the sometimes competing values of self-determination and the protection of a patient's best interests (Buchanan 1990:29–41). A great deal of weight is placed on self-determination in liberal societies, though its importance is justified in a variety of ways. One strand of liberal thought on this issue is the idea that which treatment decision best promotes the patient's well-being depends on the person's commitments: their particular goals, projects, preferences, and values. The law, therefore, aims to preserve the patient's right to determine her own course of treatment, even in the face of dissenting medical advice. The legal limits of this respect for self-determination are decisions judged to have been made without mental capacity, in which case a surrogate decision maker is appointed.

Where the bar is set for capacity—how demanding the requirements for capacity are—is determined in significant part by the degree of importance society places on self-determination. The dilemma that arises when deciding what criteria to include in the assessment of capacity, and how stringently to interpret those criteria, is insightfully set out by Buchanan and Brock (1990:40–41). They note that legislators and courts must balance two errors. The first occurs when a patient is judged to have capacity when she is not in a position to be self-determining. The second occurs when self-determination is curtailed when the patient is in a position to make her own decision. Lowering the bar for capacity increases errors of the first kind but decreases the second kind, whereas raising the bar does the reverse.

The weight placed on respecting people's choices in the current political climate prioritizes, at least to some degree, avoiding the first kind of risk—one associated with traditional paternalistic practices in medicine. This is reflected in the legal presumption in favor of capacity. The Mental Capacity Act 2005 (MCA) is clear that judgments that a person lacks capacity should be made as a matter of last resort (section 1[1]). It is also an assumption that is often taken to be noncontroversial in the contemporary medical ethics literature; for example, Breden and Vollman write:

> [I]t is in moral and legal terms preferable that some patients, who are in fact incompetent, are misjudged to be competent. In this case the patient's right to free choice overrules the duty to help. (2004:279; see also Appelbaum 1998; Welie and Welie 2001)

Criteria for Capacity

Various legal standards have been used as the basis for judgments regarding mental capacity in different countries and jurisdictions. An influential investigation into notions of capacity in case law, drawn mostly from the United States, was carried out by Appelbaum and Grisso. They identified four legal standards that were variously used across jurisdictions. These referred to the ability to:

(a) "communicate a choice
(b) understand relevant information
(c) appreciate the situation and its likely consequences; and
(d) manipulate information rationally" (Appelbaum and Grisso 1995:109)

The ability to rationally manipulate information was later modified to the ability to "reason with relevant information so as to engage in a logical process of weighing treatment options" (Appelbaum and Grisso 1998:31).

In England and Wales the MCA was introduced to preserve best practice and common law principles in statutory law. According to the MCA, a person lacks capacity if, due to a dysfunction of mind or brain, "he is unable:

a) to understand the information relevant to the decision
b) to retain that information
c) to use or weigh that information as part of the process of making the decision
or
d) to communicate that decision" (section 3[1])

Although English law is clearly concerned with certain reasoning abilities as one element of capacity—as indicated by its reference to the weighing and using of information—its position on the relevance of judgments regarding the rationality of a decision for capacity is said to be "ambiguous" (Jackson 2010:230–233). For the most part, English law asserts the patient's right to make choices that are irrational without this having implications for judgments about their mental capacity, thus maintaining a significant degree of independence between incapacity and irrationality. However, as Owen and colleagues note, "despite denying that irrationality can on its own amount to incapacity, the courts are quite prepared to accept that it can be a symptom or evidence of incapacity" (Owen et al. 2009:1993), making it unclear what the relationship between capacity and rationality is in English law.

In what follows, we examine this position in the context of contemporary thought about rationality. We believe the tension identified by Owen and colleagues arises from the fact that English law currently fails to recognize that most judgments of incapacity are made on grounds that the person demonstrates irrationality. The apparent inconsistencies in the legal position and reluctance to explicitly bring rationality into questions of capacity are, we will argue, largely due to an underlying confusion regarding different norms of rationality.

RATIONALITY AND CAPACITY

The relevance of questions regarding rationality for the assessment of mental capacity has already received considerable attention in the medical ethics and law literature (for example, Buchanan and Brock 1990; Culver and Gert 2004; Pepper-Smith et al. 1996; Savulescu 1994; Savulescu and Momeyer 1997). However, these discussions reveal important differences in the way the terminology of rationality is used. In this section, we clarify these uses, focusing on the central distinction between assessing rationality on procedural versus substantive grounds, which we believe is essential for understanding the claims regarding irrationality and capacity in English law.

In the most general sense, theories of rationality are concerned with the norms that underlie the rational criticism of decision making. Theories of rationality are often divided into those concerned with the assessment of beliefs (theoretical rationality) and those concerned with action-guiding decisions (practical rationality). Although these are not neatly separable, it is the latter that are most relevant to capacity assessments, so our focus will be the norms of practical rationality.

Some Basic Norms of Practical Rationality

One way in which the rationality of a practical decision can be called into question is on grounds that it is based on a belief that is not appropriately sensitive to evidence (Pepper-Smith et al. 1996; Savulescu and Momeyer 1997). In a medical context, a person might express the belief that undergoing a particular medical procedure will almost certainly result in their death, when in fact it is a routine procedure with minimal risks. What might be called the "irrationality of their belief" is said to undermine the rationality of the associated decision.

Other ways in which the rationality of practical decisions is commonly criticized point to the deliberative process used in coming to a decision. So, for example, a decision can be judged irrational on grounds that the conclusion drawn from the relevant facts does not follow (Savulescu and Momeyer 1997). Alternatively, the decision maker might not take into consideration the full range of relevant information, or she might fail to recognize the implications that follow from a set of facts. The facts of a situation might not be properly integrated with other essential elements of the deliberative process, for example, the decision maker's projects, goals, or preferences. Thus, for example, a decision will be judged irrational if a person does not desire the means to a desired end (all other things being equal). If a person understands that having a particular procedure is the only way to save her life and wishes above all else to continue living, but refuses to consent to the procedure, then this provides good grounds to call the rationality of the decision into question.

Although this list is not exhaustive, these kinds of norms form the core of what practical rationality requires according to prominent contemporary theories. All of these grounds for rational criticism are directed at features of the procedure by which decisions are reached—either at beliefs on which the decision is based or features of the deliberative process—without taking into consideration the motivating states that ultimately guide the decision. We will use the term *ends* to refer to these motivating states, but this can be understood in terms of desires, goals, or values. Many contemporary theories adopt an instrumental approach to understanding rationality, according to which the norms of rationality are those constraints that further the goals of decision making. In the case of practical decisions, this most obviously (and least controversially) means enabling the person to effectively pursue her ends (Jones 2004). The kinds of procedural norms described previously further practical decision making by enabling the person to do what she wants to do. So, to describe a practical decision as rational is, in the most basic sense, to mean the decision is procedurally rational.

Criteria for capacity assess decision making on closely related grounds, examining a person's ability to make a treatment decision in accordance with her ends (Buchanan and Brock 1990:70). Under the MCA, the criteria for capacity (except the ability to communicate the decision) are all features that are relevant to the procedural rationality of a decision, such that close parallels can be drawn between the criteria for capacity and widely accepted norms of rationality in a procedural sense. So, for example, the requirement that a patient must understand the relevant information reflects the procedural norm that rational decision making requires having the relevant true beliefs. As a result of this connection, every judgment that a person lacks mental capacity as described in the MCA can be understood as a judgment of compromised procedural rationality (except in the case of an inability to communicate a decision).

In light of this, how can we make sense of claims in English law according to which the irrationality of a decision must not decide judgments regarding mental capacity? We will discuss two ways in which it might be appropriate to keep judgments of irrationality independent of capacity assessments. We believe it is the second of these that primarily underpins the position taken in English law.

Capacity and Procedural Rationality

First, there are good reasons to think that the criteria for full procedural rationality and the criteria for capacity will not be coextensive. Given that when a person is judged to lack capacity, this permits the state to limit the person's liberty with regard to the relevant decision, and given the value typically placed on self-determination in liberal societies, the bar for capacity must be low enough to ensure that most people in normal circumstances will be judged to have capacity. Therefore, the criteria for capacity will always be less stringent than the criteria for full procedural rationality. As a result, there are likely to be norms that provide good grounds for calling into question the rationality of a decision but that would not provide an appropriate basis for questioning a person's capacity, especially in the absence of any other reason to do so. It also seems that even in the case of norms that noncontroversially apply to both rationality and capacity—for example, in the case of the requirement that the decision maker understands the relevant information—such norms will need to be satisfied more easily in assessments of capacity. For example, a person may refuse conventional medical treatment for cancer and opt instead for a homeopathic remedy, citing the belief that this would be just as effective, despite strong evidence to the contrary. In this case, the apparent irrationality of the patient's belief would not necessarily mean her capacity was called into question (for discussion of a similar case see, Jackson 2010:231). In these respects, it will be true, perhaps in many cases, that a person can be judged to have capacity while making a procedurally irrational decision.

However, this way in which rationality assessments and capacity assessments come apart does not seem to explain the emphasis placed on keeping the two independent in English law. This explanation recognizes that the criteria for capacity essentially assess procedural rationality, albeit in a less stringent sense. This goes largely unacknowledged in English law, where it is explicitly claimed that the irrationality of a decision should not (on its own) be used to determine capacity. We therefore believe that the legal position is better explained by a more fundamental disagreement about the reach of rational criticism with regard to practical decisions.

Capacity and Substantive Rationality

The second way in which judgments of capacity and rationality come apart has to do with a central debate in the area of practical rationality, concerning the extent to which a person's action-guiding ends are themselves available to rational criticism. From a liberal perspective, it is often said that a person's ends are beyond the proper scope of criticism by the state in relation to personal decisions because to endorse any particular ends as superior runs contrary to value pluralism (e.g., Berlin 2002). Nevertheless, it is not uncommon for people to talk in a way that assumes that there are legitimate grounds on which ends can be criticized, for example, when a decision is called into question not because the person's ability to decide in accordance with what she wants is compromised, but because of what she wants—her decision-guiding ends (Jones 2004). For example, it might be said that someone with a long-standing desire to have a healthy limb amputated is irrational when she elects to have the limb surgically removed. However, it is often the case that this claim of irrationality cannot, at least in an obvious way, be made on procedural grounds because the person appears to be effectively pursuing what she wants. If such a criticism is justified, then it must be based on further norms of practical rationality.

Perhaps the most obvious candidates are norms according to which there are objective standards against which the content of ends can be assessed. For example, an Aristotelian-inspired account might argue that given the kind of creature we are, there are certain things that all humans are rationally bound to pursue, independent of their personal commitments. Philippa Foot has proposed that objective constraints of this kind might be derived from an understanding of how particular behaviors generally affect human flourishing (understood widely, e.g., to include growth, survival, reproduction, and emotional life) (Foot 2004). Such an approach goes beyond criticizing decisions on the basis of whether they are procedurally rational by offering what are often called "substantive" grounds to assess practical decisions. For example, someone might be judged irrational because her actions reveal a lack of concern for her own well-being. Substantive norms of rationality require that people should pursue certain things, regardless of what things they are actually motivated to pursue (on objective list theories see, Parfit 1984:493–502).

However, there are considerable difficulties associated with establishing that any such requirements are legitimate. Accounts of what is categorically worth pursuing are highly controversial, with some prominent schools of thought denying that there are any such norms (Williams 1981a). But these difficult conceptual issues can be largely sidestepped in the context of capacity assessments. Regardless of whether there are substantive norms of rationality, in the current liberal political climate there seems likely to be resistance to such norms being used to justify limiting self-determination by their inclusion in criteria for capacity.

It is this political commitment that we believe motivates the position in English law regarding capacity and irrationality. The motivating concern is that judgments regarding the substantive rationality of a decision must not decide a capacity assessment because it is not for the state to limit self-determination on grounds that people ought to live their life in accordance with certain interests or values. This commitment makes it inappropriate to let the content of a decision (the choice itself) determine the assessment of capacity. When a decision is said to be "irrational" in this part of English law, this is often taken to mean "odd," "bizarre," or "eccentric" (Jackson 2010:230–231)—terms that appear to be directed at the content of decisions rather than the processes by which they are reached. Therefore, "irrationality" in this context is understood to involve criticism of a decision on substantive grounds, which explains the importance placed on asserting the independence of capacity from irrationality.

The proposed underlying principle—that judgments regarding the substantive rationality of a decision should not be used to determine capacity—is explicitly endorsed in English common law, for example:

> [In assessments of capacity] it is most important that those considering the issue should not confuse the question of mental capacity with the nature of the decision made by the patient, however grave the consequences. The view of the patient may reflect a difference in values rather than an absence of competence and the assessment of capacity should be approached with this firmly in mind.[3]
>
> First, it is established that the principle of self-determination requires that respect must be given to the wishes of the patient... the doctors responsible for his care must give effect to his wishes, even though they do not consider it to be in his best interests to do so.[4]

The principle is also reflected in the wording of the MCA, which suggests that the content of a decision should not be used as the basis for judgments about capacity:

> (4) A person is not to be treated as unable to make a decision merely because he makes an unwise decision. (section 1[1])

The interpretation of the term *irrationality* as referring primarily, if not exclusively, to substantive norms appears to also occur in the US legal system, where it has influenced the criteria for capacity adopted by courts. In relation to Appelbaum and Grisso's "ability to rationally manipulate information" criterion, it has been observed that:

> Some courts have shied away from embracing this standard because of an apparent confusion regarding interpretation of the term *irrational.* A decision sometimes has been called irrational merely because the patient's choice was unconventional... In contrast, the "irrationality" to which this standard properly refers pertains to illogic in the processing of information, not the choice that eventually is made.(Appelbaum and Grisso 1995:110)

Capacity and the Consideration of Ends

Despite the position in English law regarding the assessment of capacity on substantive grounds, the powerful intuitive appeal of certain substantive principles has lead some commentators to propose that the criteria for capacity should include such considerations, for example, that capacity to consent should require a minimal concern for one's own well-being (Elliot 1997).

It is not easy to see how the tension between the principled rejection of substantive considerations in the assessment of capacity in English law and the intuitive appeal of certain substantive norms in this context can be resolved. However, there remains a further source of norms that may go some way to bridging this gap. Many contemporary accounts of rationality hold that, although the contents of ends are for the individual to determine, these are nonetheless available to rational criticism on grounds of internal inconsistency or coherence (see, e.g., Harman 1999; Shafir and LeBoeuf 2002). Such principles offer a way of taking a person's ends into consideration without appealing to the objective value of particular interests, by bringing the focus back to the individual and her particular ends in a global sense.

One such norm proposes that in practical decision making, people should aim to maximally advance their ends overall (Pettit 1984). According to this principle, the pursuit of a particular goal might be criticized on grounds that it would undermine other goals that the individual holds to be more important.

Taking on an increased workload in the pursuit of a promotion, for example, may frustrate the goal of living a healthier, less stressful lifestyle and could be criticized on these grounds if the individual places greater value on her health over career advancement.

Rational criticism based on a person's overall commitments is sometimes argued to include her commitments over time (e.g. Nagel 1970, chapters 5–8). According to this principle, acting rationally requires that a person give appropriate weight to her anticipated future desires. This norm is reflected in the commonplace criticism of practical decisions on the basis that they do not properly take into account the value that the decision maker is likely to place on being healthy in the future—for example, in relation to decisions to keep smoking or to not use safe sex practices.

Although these norms do not currently find formal expression in the MCA, references made to these kinds of considerations suggest that intuitively these kinds of principles are relevant to capacity. For example, Buchanan and Brock write that their account of competence to consent focuses on whether the patient's decision making is consistent with her "*underlying and enduring* aims and values" (Buchanan and Brock 1989:70).

Focusing on the coherence of a person's motivational states, Rudnick argues that a patient's decision-guiding ends should be considered in capacity assessments:

> [P]references that do not cohere considerably with most other preferences held by the person in the present or in the past are suspect… suggesting incompetence on the grounds of not being consistent with what that person would usually decide. Hence, coherence of preferences may be a neglected component of competence to consent to treatment, and if major depression without psychotic features considerably disrupts such coherence, then the refusal of psychiatric treatment by individuals afflicted with such major depression may not be arrived at competently. (Rudnick 2002:152)

It should be noted, however, that the inclusion of such considerations can nonetheless lead to strongly anti-intuitive conclusions about mental capacity. For example, it has been suggested that a person with a long-standing and strongly held commitment to having a healthy limb amputated may well be "globally rational" and could therefore be judged competent to consent to the surgery on these grounds (Bayne and Levy 2005). However, the anti-intuitive outcome in this case does not, in our view, undermine the legitimacy of such norms of rationality or their relevance to capacity. If mental capacity is compromised in such a case then this must be explained by an appeal to other norms. Alternatively, the anti-intuitive conclusion must be accepted. It is exactly the possibility that some choices that are difficult to understand will nonetheless be made with mental capacity, that the law seeks to protect.

Norms concerning a person's ends in a global sense—their internal consistency and coherence either at the time of a judgment or diachronically—offer a way of bringing rational criticism to bear on the ends that people pursue, which appears to be less politically and theoretically problematic than substantive approaches that directly criticise the content of those ends.[5] Like the basic norms of practical rationality, these norms are procedural, providing grounds for the criticism of decisions with reference to the decision maker's own motivational set (in some cases over time).

Capacity and Authenticity

These kinds of norms also seem to capture something that is essential to the ideal of self-determination, and this may help explain their intuitive relevance to capacity. The value currently placed on self-determination is based in the assumption that the person's decision-guiding ends, and resulting choices, are authentically her own (for example see the discussion of authenticity in Owen et al. (2009). In cases in which a person's choice is clearly inauthentic in the sense that it is imposed by another—for example, in cases of coercion—the value in respecting that choice would seem to lose its significance.

Other cases in which the authenticity of choices seems questionable, for example in some cases of mental disorder, the claim of inauthenticity is more difficult to explain. It has been argued that the authenticity of the value placed on thinness in anorexia nervosa might be called into question and that anorexic patients might lack the capacity to refuse treatment for the disorder on these grounds (Hope et al. 2011; Tan et al. 2006; for discussion see, Craigie 2011a). English law seems to recognize the relevance of this kind of consideration, stating that people with anorexia nervosa may lack capacity because "their compulsion not to eat might be too strong for them to ignore" (MCA Code of Practice 2005, 2007:49, section 4.22).

The issue of how authentic and inauthentic ends can be distinguished is highly contested, but appeals to global consistency represent one attempt at an answer (for discussion see, Hope 1994). Insofar as it may be appropriate to include norms that concern a person's ends in the criteria for capacity, it seems that considerations relevant to the authenticity of ends are consistent with the political commitments concerning substantive norms underlying this part of law. However, the inclusion of any new criteria in capacity assessments would nonetheless seem to raise the bar for mental capacity, which would constitute a shift in a paternalistic direction (for further discussion see, Appelbaum 2008; Spike 2008).

CULPABILITY, CAPACITY, AND NEUROSCIENCE

The foregoing discussion allows us to make some observations about the relationship between mental capacity and criminal responsibility and about the relevance of neuroscience to these two kinds of determination. The legal concepts of responsibility and capacity have, to a large extent, evolved independently. Questions of criminal responsibility have traditionally focused on issues such as causation, free will, and the moral justification of punishment, whereas questions of mental capacity have been concerned with the psychological functions necessary for autonomous decision making. However, a shift in debates about culpability from questions about causation to a focus on rationality—and in particular how this can be compromised in the context of mental disorder (Morse 1999)—suggests that the two traditions may be grappling with fundamentally the same issues. According to Stephen Morse, "[r]ationality is the touchstone of responsibility" (Morse 2007:197).

Understood in this way, questions of criminal responsibility can be taken to concern the functioning of certain psychological capacities that are necessary for rational action, and this has been identified as an important connection between the concepts of culpability and mental capacity (Beaumont and Carney 2003; Elliot 1997; Meynen 2009; Vincent 2010; Wilson and Adshead 2004). As the previous discussion has highlighted, questions about the rationality of practical decisions sometimes turn on whether a particular behavior satisfies particular stipulated norms. However, such questions can also concern which norms are genuine norms of rationality, or which norms apply in a particular situation—matters of the standards to which behavior should be held. We can therefore break down the question of what relevance neuroscience has for decisions about capacity or culpability in the following way: What could neuroscience tell us about the appropriateness of particular norms for the assessment of capacity or culpability? And how could neuroscience inform an assessment of a particular person's decision making in relation to the agreed-upon norms?

The Appropriateness of Norms

The question of whether any proposed norm is actually a requirement of rationality in either of these contexts is a conceptual issue—a matter of what we mean by "rational." Attempts to resolve these questions take place in philosophy using the tools of conceptual analysis, and in psychology through the study of decision making. However, it is difficult to see how facts about the brain could be used to resolve questions about, for example, whether a person demonstrates irrationality when her preferences are inconsistent over time. The point can be illustrated in the context of questions about whether decision making should accord with particular moral norms. Research has found that damage to the ventromedial prefrontal cortex is associated with more utilitarian decision making in response to certain moral questions (Koenigs 2007). However, the issue of whether this brain abnormality diminishes or augments moral decision making rests on whether utilitarianism is a good guide to what one ought to do in the relevant situation.[6] In the context of assessing either mental capacity or culpability, as long as there is disagreement about the norms with which behavior should accord, there will be disagreement about what relevance any particular piece of neuroscience has for these assessments.

Nevertheless, neuroscience might be employed to argue that certain norms do not constitute realistic constraints for human decision making. For example, there might be limits to cognitive processing that would make certain norms very difficult for most people to adhere to, and this could be offered as grounds for rejecting those norms as standards that can be reasonably used to criticize decision making.[7] This kind of consideration seems particularly relevant with regard to the criteria for capacity, given that the bar needs to be low enough that people are generally judged to be in a position to make their own treatment decisions, and in the case of the requirements for culpability, which must enable the law to hold most people, under normal circumstances, responsible. However, this sense in which neuroscience might be used to answer questions about the appropriateness of norms is pragmatic rather than conceptual. In the context of capacity assessments, we have argued that political considerations are used to limit the appropriateness of applying certain norms. Neuroscience can be thought of as informing this kind of decision about the norms that are suitable for inclusion in the criteria for mental capacity or in assessments of criminal responsibility.

Individual Assessments in Relation to Agreed-upon Norms

The task of assessing an individual's rational capacities with reference to particular norms for the purpose of determining culpability is an area in which commentators have seen a greater role for neuroscience, particularly in anticipation of future advances in this field (Vincent 2011). However, as

Stephen Morse has pointed out, given that assessments of rationality are judgments about a person's ability to act in accordance with certain behavioral norms, the person's behavior will always be the ultimate reference point for such assessments (Morse 2007). The same point applies to capacity assessments because of their grounding in norms of rationality. Judgments regarding rationality cannot be read directly from the neural level because the relevant normative questions are only properly directed to mental level explanations of behavior—explanations that refer to mental states such as beliefs and desires.[8] Although it may turn out that certain neural level features are closely associated with irrationality of a certain kind, the use of neuroscientific evidence to judge that a person lacks capacity or should not be held culpable will always rest on the assumption that the person's rationality is relevantly impaired. This means that regardless of how tightly correlated the neural and mental levels turn out to be, any contribution made by the neural level to assessments of capacity or culpability will always be open to revision based on the observation and interpretation of the person's behavior. As Morse argues in the case of criminal responsibility, "If the person's rational capacities, which we infer from her behavior, seem unimpaired, she will be held responsible, whatever the neuroscience might show, and vice versa"(Morse 2007:200).

In assessments of mental capacity, this dependence on mental level interpretations acts as a safeguard against unwarranted paternalism. This is because it is foreseeable that neuroscience might in some cases be used, however unintentionally, to legitimize what is effectively a substantive judgment about the patient's treatment choice. For example, in the context of a life-threatening treatment refusal, a doctor might point to a feature of the patient's brain (e.g., neural plaques associated with dementia) to call into question the patient's decision making ability, when it is in fact the likely consequences of the choice that are motivating the doctor's concern. In such a case, the neuroscience is being used to validate the doctor's underlying substantive judgment that the patient should want to live. However, the proposal that the patient may lack capacity can be answered by her demonstrating that she understands, retains, and can reason with the relevant information, despite the anomalous features of her brain. A patient's demonstrated ability to satisfy a normative requirement despite the presence of an anomalous neural feature that is often associated with a failure to meet that requirement could be explained by reference to idiosyncratic compensating functions, or a more general claim about the multiple realizability of psychological functions at the neural level (Papineau 1995). Regardless of the underlying explanation in any particular case, in the context of assessing mental capacity, the authority given to the ability to exercise a capacity over brain level observations helps preserve patient autonomy.

In contrast, a dependence on behavioral level explanations is often considered a weakness in assessments of culpability, and neuroscience has been proposed as a means to address such concerns. In particular, these concerns arise because in the context of assessing criminal responsibility, there are other viable explanations besides incapacity for the defendant not demonstrating the relevant capacities, and behavioral evidence alone may not provide the means to distinguish among such possibilities (McSherry 1997; Vincent 2011). The defendant may consider it in her interests not to do so, for example, if it's possible that the successful demonstration of a relevant incapacity would result in a supervision order or absolute discharge. Neuroscience is proposed to offer a solution by identifying neural features that are necessary for the relevant capacities, thereby offering a way of adjudicating between these two possible conclusions by reference to the presence or absence of abnormalities in these features.

However, the possibility of multiple realizability at the neural level or idiosyncratic compensating functions is also present in this context. Especially when the relevant capacities are high-level cognitive functions, there is the potential that they could be realized through a range of different psychological and neural level profiles. As other critics have noted, this limits the usefulness of neuroscience in addressing the perceived problems with behavioral methods of determining culpability. However, in this context, the danger of misapplying neuroscience is not unjustified deprivation of liberty in relation to medical care, but rather that blame and punishment by the state will be withheld (at least to some degree) in cases in which it is fully justified.

Substantive Norms and Criminal Responsibility

The different purposes of judgments about capacity and culpability also makes it likely that there will be differences in the norms that are relevant in these two contexts, and this points to further differences in what neuroscience can contribute. One salient difference between these two kinds of judgment is that the concept of criminal responsibility operates in the public sphere, concerning acts that affect others, whereas the concept of mental capacity is deployed in private law, where it concerns the decisions people make for themselves, which in most cases don't directly affect others. Although it is generally accepted that no single principle can be used to determine the boundaries of criminal conduct, substantive norms prohibiting harms to others, in particular direct harms, play a central role in theories regarding the criminalizing of acts (Feinberg 1984). Therefore, although the ability to act in accordance with substantive norms is not considered appropriate for inclusion in the assessment of mental capacity, it would seem to be a key component of being a competent agent in the context of a criminal act. This means that from the perspective of ascribing culpability, the competent agent has capacities over and above those that are currently considered relevant in the context of the right to make one's own treatment decisions. One interpretation of the role played by this further capacity is that in order to be held culpable, a person must understand that her conduct was wrong, not merely in the sense that she violated a legal rule, but in the sense that what she did was a moral transgression (McSherry 1997). For example, to be held responsible for a murder, the person must understand that the reasons for refraining are not merely instrumental reasons such as wanting to avoid incarceration. On this interpretation, the culpable agent must have the capacity for certain kinds of moral knowledge.[9]

The Capacity for Moral Knowledge

The proposed relevance of the ability to recognize certain moral norms in relation to culpability invites the conclusion that there will be psychological and neural functions that are necessary for culpability beyond those that are necessary for mental capacity. Understanding that one ought to refrain from certain actions for moral reasons, so the thought goes, requires more than the mechanisms necessary for procedural rationality as captured by capacity tests. However, this conclusion does not necessarily follow. One of the central debates in philosophical moral psychology concerns the scope of the role played by reasoning in reaching justified moral judgments. Broadly speaking, moral rationalists hold that moral knowledge is gained through procedural rationality alone, whereas moral sentimentalists understand moral evaluation necessarily in terms of human emotional response (D'Arms and Jacobson 2006). These two views, therefore, offer substantially different accounts of how it is that agents come to understand what morality requires, which suggest different conclusions about the psychological and neural functions that play a necessary role in moral agency.

The simplest forms of sentimentalism attribute reasoning the most minimal role in the acquisition of moral understanding, holding that "to judge that something is wrong (right) is to have a sentiment of disapprobation (approbation) towards it" (Jones 2006:45). According to this view, reasoning plays only an instrumental role in guiding action, according to which "reason's only task is to show how the pursuit of the passions—of subjective ends—is to be organized effectively and efficiently" (O'Neill 2005:833). More sophisticated forms of sentimentalism assign reasoning a more substantial role in moral agency. Sophisticated sentimentalists attribute critical reflection a key role in competent moral deliberation, incorporating the idea of "higher-order criticism and endorsement of our sentiments" in their accounts of moral agency (D'Arms and Jacobson 2006:216).[10] However, all forms of sentimentalism are committed to the view that reasoning is not, on its own, motivating. Accordingly, reasoning cannot be the ultimate source of reasons for action or action-guiding claims about what one ought to do. Moral knowledge on the sentimentalist view, therefore, seems to require psychological functions beyond those that underpin procedural rationality—moral judgment is fundamentally grounded in action-guiding emotional responses.[11]

In contrast, rationalist theories accord reasoning a more direct connection to action and moral evaluation, so that moral agency requires nothing more than procedural rationality. According to Michael Smith's account, for example, it is an essential part of our task as rational agents to engage in deliberative processes that could result in a systematically justified set of desires: "the same set of desires as our fellow rational creatures would come up with if they set themselves the same task" (Smith 2004:27). In this way, all fully rational creatures will converge upon moral requirements on action, although the degree to which a particular individual achieves moral understanding will depend on the extent to which she is a rational deliberator.[12]

The question of what psychological (and neural) features are required to enable an agent to be guided by substantive norms, which is often considered a relevant capacity in relation to criminal responsibility, is therefore one that can be answered in a multitude of ways. Moreover, philosophical thinking about the requirements for moral agency—and the scope of the role played by reasoning in action—leaves open the possibility that procedural rationality is all that is needed for the capacity to recognize moral norms. Versions of rationalism are compatible with a picture of the competent moral agent that does not require any capacities beyond those necessary for procedural rationality. It remains a further question whether the features of procedural rationality under examination in capacity tests such as the MCA, and the standards to which these features are held, would be sufficient for the evaluative capacity that is proposed to be necessary for responsibility in the criminal law. Just as the bar for mental capacity must be low enough to enable most people under normal circumstances to be judged to have capacity in the context of treatment decision making, arguably the requirements for criminal responsibility must be such that most people under normal circumstances can be judged blameworthy for their criminal acts. But given the unresolved nature of the previous meta-ethical issues, the question of what psychological (and neural) functions are necessary to meet this standard is at present wide open. Being a competent moral agent in the eyes of the criminal law will not necessarily require functioning of a kind or degree beyond that necessary for mental capacity.

Empathy and Moral Agency

The previous point is borne out in debates about the role played by empathic mechanisms in moral agency, and in this final section of the chapter, we make some observations about these debates in light of our discussion. In the context of psychopathy, diminished empathic responsiveness and associated neural features have been used to explain the violent antisocial behavior exhibited by psychopaths. Drawing on this research, it has been argued that because of this abnormality, psychopaths may lack the capacity for moral understanding and may therefore not meet the requirements for criminal responsibility (Fine and Kennett 2004). Yet it has been observed that autism is also characterized by serious problems in empathic understanding, but that moral agency seems to be preserved in this context, at least insofar as autism is not characterized by the criminal behaviors associated with psychopathy (Kennett 2002).[13] Taken together, this work raises the questions: What does the autistic person's sense of morality suggest about the role played by empathic functioning in moral agency? And does it follow that deficits in empathic functioning, including those present in psychopathy, should not be exculpating?

On the one hand, this set of observations has been used, in line with moral rationalism, to argue that moral agency can be achieved through reasoning alone—that the autistic person's recognition of moral norms is due to the capacity to be "moved directly by the thought that some consideration constitutes a reason for action" (Kennett 2002:357). Jeanette Kennett has argued that the apparent moral deficit in psychopathy is better explained by problems in reasoning than a lack of empathetic ability and therefore that "it is the capacity to take some considerations as normative, as providing reasons that are independent of our desires of the moment" that is at the core of moral agency (Kennett 2008:259).[14] However, alternative interpretations of the evidence have explained moral agency in autism with reference to other intact affective mechanisms (McGeer 2008), or have denied that moral agency is in fact preserved in autism (de Vignemont and Frith 2008). Such interpretations support a sentimentalist perspective on the research that assigns an essential role to capacities other than reasoning in moral agency.

There is, however, a third kind of interpretation that suggests there may be varieties of moral agency underpinned by a range of different psychological mechanisms. Victoria McGeer has proposed that there may be modestly different varieties that emerge in normal human development as the result of different cultural contexts, as well as more radically different varieties that are seen, for example, in the context of autism. In accordance with a sentimentalist perspective, McGeer holds that the diverse ways in which human agency can be realized are all grounded in an affect-laden valuing of particular ends.

However, one might reject the assumption that "valuing certain ends is fundamentally rooted in... one's affective life" (2008:247). Doing so envisages a more radical version of McGeer's proposal—one that rejects the dichotomy between rationalism and sentimentalism and instead proposes that the varieties of moral agency could include varieties grounded in responsiveness to the requirements of rational deliberation, as well as varieties grounded in human emotion (see Kennett 2008 for a defense of the first variety). Both versions of this position point to the existence—to different degrees—of a significant multiple realizability in how moral agency is achieved, whereby the requirements of morality can be met by individuals with radically different psychological and neurological profiles.[15]

Although some authors see the obstacles to discovering "precisely which capacities play an indispensible role in competent moral agency" (Vincent 2011:38) as being largely pragmatic, pluralistic accounts of moral psychology suggest that the search for essential capacities is misguided. McGeer suggests that such approaches—as exemplified in the literature on neuroscience and criminal responsibility—come from a "persistent tendency in philosophy and other academic inquiries to try to locate the essence of our moral nature in a single cognitive capacity or affective disposition" (McGeer 2008:229). It is true that pluralistic accounts of moral agency are compatible with the discovery of a collection of psychological profiles, at least one of which is necessary for moral agency. However, the complexity of such a story cautions against optimism that neuroscience will offer clear grounds for settling questions of criminal responsibility in relation to the ability to recognize moral norms, and reinforces the need to be attuned to the moral psychological assumptions that are involved in such a task.

CONCLUSION

In this chapter, we have focused on comparing the conceptions of capacity operating in mental capacity law and in ascriptions of criminal responsibility in order to explore the relevance of neuroscience in these different areas. We have argued that the connection between mental capacity and judgments regarding rationality is much closer than is currently acknowledged in English law, and that the criteria that have been developed in the MCA are essentially concerned with a person's ability to meet certain requirements of procedural rationality. In our view, the aversion to using the terminology of rationality in the assessment of mental capacity is explained primarily by the underlying political principle that capacity assessments should not be based on substantive norms of rationality, which require that people pursue certain ends. We propose that a patient's ends might nonetheless come into consideration within the current capacity framework, on grounds of procedural constraints concerning the coherence and endurance of the patient's motivating commitments.

Questions of culpability are also increasingly being understood in terms of rational capacities. This shared focus on rationality has the consequence that the usefulness of neuroscience will be similarly limited in these two areas of law. First, neuroscience has little to say on the question of the standards to which behavior should be held. Second, on questions of the ability to meet normative requirements, behavior will always be the ultimate point of reference.

However, the different legal contexts in which these two kinds of judgment are made—and differences in the norms that are held relevant—mean that these limitations will have different consequences for the application of neuroscience in assessments of mental capacity and criminal culpability. Although the motivation to rely on neuroscientific evidence will be greater in the criminal context, its use will be more theoretically problematic because of the relevance of substantive norms. We have argued that unresolved questions regarding the nature of moral agency make it particularly unclear what neuroscientific findings would be relevant in this context. Contemporary debates in philosophical moral psychology highlight key assumptions about the connection between rationality and morality that often go unnoticed in discussions of how neuroscience might inform ascriptions of criminal responsibility. This is illustrated by the fact that questions about whether a specific neuropsychological finding is relevant are often framed in terms of whether the finding affects the individual's capacity to act rationally *or* morally, which assumes that these are distinct questions.[16] However, according to a rationalist understanding of moral agency, the question of an individual's ability to act rationally already addresses the question of her ability to act morally. These considerations, and the further possibility that there are radically different varieties of moral agency, call for particular caution in the use of neuroscience as evidence regarding culpable agency in criminal law.

NOTES

1. Thanks go to Lisa Bortolotti, Fabian Freyenhagen, Genevra Richardson, Nicole Vincent and members of the King's College London KJuris seminar group for their very helpful comments on earlier drafts of this chapter. This work was done with the support of the Wellcome Trust [094910].
2. Lord Donaldson in re T (*Adult: Refusal of Treatment*) [1992] 4 All E.R. 649, at 653.
3. NHS Trust v T (*Adult Patient: Refusal of Medical Treatment*) [2004] EWHC 1279 (Fam).
4. Airedale NHS Trust v Bland [1993] AC 789.

5. We note that according to some versions of this kind of view, full practical rationality involves the systematic justification of desires—a process that guarantees convergence among rational agents on what it is desirable to do in any given situation (Smith 2004). Therefore, taken to their limits, procedural accounts look to become more substantive in nature, but it is not this kind of view that is our focus in this discussion. We will return to this account in our discussion of substantive norms and criminal responsibility.
6. Joshua Greene has proposed that neuroscience, in combination with evolutionary psychology, could help resolve such normative questions. This further issue goes beyond the scope of our discussion; however, we believe there is reason to be skeptical about the moral claims in support of utilitarianism that Greene has drawn from his empirical work (for further discussion see, Craigie 2011b:61–22; Greene et al. 2001; Greene 2008)
7. A parallel criticism has been made of normative constraints in moral theory (see, e.g., Flanagan 1996; Williams 1996.)
8. Take the requirement that a patient must understand the facts relevant to his medical condition. In order to find out whether a person has the ability to satisfy this requirement, the capacity assessment will investigate the person's relevant belief states. This appraisal will hinge on, among other things, the question of whether the patient's relevant beliefs are warranted given the available information. Beliefs are an appropriate target for this kind of normative question. However, it makes no sense to ask whether a certain pattern of neural activity is warranted, without reference to the associated effects on the agent in terms of beliefs, desires, or other mental states. There is nothing in the neural level description alone that could ground the claim that a particular state of affairs in the brain is irrational.
9. We note that this interpretation will not be appropriate in some jurisdictions; for example, in England and Wales, knowing that an act is wrong in the legal sense is generally considered sufficient for meeting this criterion for responsibility (Peay 2011).
10. This kind of theory is seen as an improvement on simple forms because it allows the sentimentalist to explain the observation that emotional responses can be unreliable guides to value. We can feel anger in response to someone's actions, while at the same time recognize that the evaluation associated with that emotion is not justified.
11. For the sentimentalist, these emotional responses are not making their contribution to moral decision making by facilitating procedural rationality.
12. Occupying a middle ground between sentimentalist and rationalist theories are accounts that hold that action can be guided by the norms constitutive of a particular society, but according to which those norms form the bedrock for practical reasoning—the norms themselves are not justifiable beyond being the norms constitutive of the relevant society. Such positions hold that there is only a limited sense in which there are reasons for action beyond instrumental reasons—the substantive norms in question are merely societal norms, not ultimately justifiable moral norms. According to such views, reasoning plays an instrumental role in guiding action and is employed in analyzing the connections between norms, but it is not engaged in the ultimate justification or revision of substantive norms

because no such justification is available (e.g., Williams 1981b). This conception of moral psychology might come into play in questions of culpability if it is assumed that a person knowing that what she did was wrong requires only that she understands that what she did was wrong in a merely legal or socially contingent sense (see for discussion, O'Neill, 2005).

13. However, see de Vignemont and Frith 2008 for concerns about this interpretation of the empirical evidence.
14. The kinds of rational deficiency that Kennett argues is exhibited by psychopaths might be understood in terms of constraints on global procedural rationality discussed previously.
15. A similar possibility has been proposed in relation to psychopathy. As Nadelhoffer and Sinnot-Armstrong note, "future neuroscience might teach us that psychopathy is really a bunch of separate mental illnesses... each with its own distinctive neural basis" but united by a common psychological profile (Nadelhoffer and Sinnot-Armstrong, 2012:XXX).
16. For example see, Broome, Bortolotti, and Mameli 2010; Vincent 2011.

REFERENCES

Airedale NHS Trust v. Bland [1993] AC 789.

Appelbaum, P. S. (1998). "Ought we to require emotional capacity as part of decisional competence?" *Kennedy Institute of Ethics Journal* 8(4): 377–387.

Appelbaum, P. S. (2008). "Response on Spike's comment: Patient's competence to consent to treatment." *New England Journal of Medicine* 358(6): 664.

Appelbaum, P. S., and T. Grisso. (1995). "The MacArthur Treatment Competence Study I: Mental illness and competence to consent to treatment." *Law and Human Behaviour* 19(2):105–126.

Appelbaum, P. S., and T. Grisso. (1998). *Assessing Competence to Consent to Treatment.* New York, Oxford University Press.

Bayne, T., and N. Levy. (2005). "Amputees by choice: Body integrity identity disorder and the ethics of amputation." *Journal of Applied Philosophy* 22(1): 75–86.

Berlin, I. (2002). *Liberty.* H. Hardy. Oxford, Oxford University Press.

Beumont, P., and T. Carney (2003). "Conceptual issues in theorising anorexia nervosa: Mere matters of semantics?" *International Journal of Law and Psychiatry* 26: 585–598.

Breden, T., and A. Vollman. (2004). "The cognitive based approach of capacity assessment in psychiatry: A philosophical critique of the MacCAT-T." *Healthcare Analysis* 12(4): 273–283.

Broome, M. R., L. Bortolotti, and M. Mameli. (2010). "Moral responsibility and mental illness: A case study." *Cambridge Quarterly of Healthcare Ethics*,19:179–187.

Buchanan, A., and D. Brock. (1990). *Deciding for Others: The Ethics of Surrogate Decision Making.* Cambridge, Cambridge University Press.

Craigie, J. (2011a). "Competence, practical rationality and what a patient values." *Bioethics* 25(6): 326–333.

Craigie, J. (2011b). "Thinking and feeling: Moral deliberation in a dual process framework." *Philosophical Psychology* 24(1): 53–71.

Culver, C. M., and B. Gert. (2004). Competence. In: *The Philosophy of Psychiatry*. J. Radden. Oxford, Oxford University Press, pp. 258–270.

D'Arms, J., and D. Jacobson. (2006). Sensibility theory and projectivism. In: *Oxford Handbook of Ethical Theory*. D. Copp. Oxford, Oxford University Press, pp. 186–218.

de Vignemont, F., and U. Frith. (2008). Autism, morality and empathy, In: *Moral Psychology, The Neuroscience of Morality: Emotion, Brain Disorders and Development*. W. Sinnott-Armstrong. Cambridge, MA, MIT Press, pp. 273–280.

Elliott, C. (1997). "Caring about risks." *Archives of General Psychiatry* 54: 113–116.

Feinberg, J. (1984). *Harm to Others: The Moral Limits of the Criminal Law*. Oxford, Oxford University Press.

Fine, C., and J. Kennett. (2004). "Mental impairment, moral understanding and criminal responsibility: Psychopathy and the purposes of punishment." *International Journal of Law and Psychiatry* 27(5): 425–443.

Flanagan, O. (1996). Ethics naturalized: Ethics as human ecology. In *Mind and Morals: Essays on cognitive science and ethics*. L. May, M. Friedman and A. Clark. Cambridge, MA, MIT Press, pp. 19–44.

Foot, P. (2004). Rationality and goodness. In: *Modern Moral Philosophy: Royal Institute of Philosophy Supplement 54*. A. O'Hear. Cambridge, Cambridge University Press, pp. 1–13.

Glannon, W. (2005). "Neurobiology, neuroimaging, and free will." *Midwest Studies in Philosophy* 29: 68–82.

Greene, J. D., R. B. Sommerville, L. E. Nystrom, J. M. Darley, and J. D. Cohen. (2001). "An fMRI investigation of emotional engagement in moral judgment." *Science* 293(5537): 2105–2108.

Greene, J. (2008). The secret joke of Kant's soul. In: *Moral Psychology Volume 3: The Neuroscience of Morality: Emotion, Brain Disorders, and Development*. W. Sinnott-Armstrong. Cambridge, MA, MIT Press, pp. 35–79.

Harman, G. (1999). *Reasoning, Meaning and Mind*. Oxford, Clarendon Press.

Hope, T. (1994). "Personal identity and psychiatric illness." *Philosophy* 37:131–143.

Hope, T., J. Tan, A. Stewart, and R. Fitzpatrick (2011). "Anorexia nervosa and the language of authenticity" *Hastings Center Report* 41(6): 19–29.

Jackson, E. (2010). *Medical Law: Text, Cases and Materials*. New York, Oxford University Press.

Jones, K. (2004). Emotional Rationality as Practical Rationality. In: *Setting the Moral Compass: Essays by Women Philosophers*. C. Calhoun. New York, Oxford University Press, pp. 333–352.

Jones, K. (2006). "Metaethics and emotions research—a response to Prinz." *Philosophical Explorations* 9(1): 45–53.

Kennett, J. (2002). "Autism, empathy and moral agency." *Philosophical Quarterly* 52(208): 340–357.

Kennett, J. (2008). Reasons, reverence and value. In: *Moral Psychology Volume 3: The Neuroscience of Morality: Emotion, Brain Disorders, and Development*. W. Sinnott-Armstrong. Cambridge, MA, MIT Press, pp. 259–264.

Koenigs, M., et al. (2007). "Damage to the prefrontal cortex increases utilitarian moral judgements." *Nature* 446: 908–911.

Lord Donaldson in Re T (*Adult: Refusal of Treatment*) [1992] 4 All E.R. 649

McGeer, V. (2008). Varieties of moral agency: Lessons from autism (and psychopathy). In: *Moral Psychology Volume 3: The Neuroscience of Morality: Emotion, Brain Disorders, and Development*. W. Sinnott-Armstrong. Cambridge, MA, MIT Press, pp. 227–257.

McSherry, B. (1997). "The reformulated defence of insanity in the Australian Criminal Code Act 1995 (Cth)." *International Journal of Law and Psychiatry* 23:135–144.

Mental Capacity Act (2005).

Mental Capacity Act 2005, Code of Practice (2007).

Meynen, G. (2009). "Exploring the similarities and differences between medical assessments of competence and criminal responsibility." *Medicine, Health Care & Philosophy* 12: 443–451.

Morse, S. J. (1999). "Craziness and criminal responsibility." *Behavioural Sciences and the Law* 17:147–164.

Morse, S. J. (2007). New neuroscience, old problems: Legal implications of brain science. In: *Defining Right and Wrong in Brain Science*. W. Glannon. Washington, DC, Dana Press, pp. 195–205.

Nagel, T. (1970). *The Possibility of Altruism*. Princeton, NJ, Princeton University Press.

NHS Trust v T (*Adult Patient: Refusal of Medical Treatment*) [2004] EWHC 1279 (Fam).

O'Neill, O. (2005). Practical reason and ethics. In: *The Routledge Shorter Encyclopedia of Philosophy*. E. Craig.London, Routledge, pp. 832–837.

Owen, G., F. Freyenhagen, G. Richardson, and M. Hotopf. (2009). "Mental capacity and decisional autonomy: An interdisciplinary challenge." *Inquiry* 52(1): 79–107.

Papineau, D. (1995). Mental disorder, illness and biological function. In: *Philosophy, Psychiatry and Psychology*. A. Phillips Griffiths. Cambridge, MA, Cambridge University Press, pp. 73–82.

Parfit, D. (1984). *Reasons and Persons*. Oxford, Oxford University Press.

Peay, J. (2011). "Personality disorder and the law: Some awkward questions." *Philosophy, Psychiatry & Psychology* 18(3): 231–244.

Pepper-Smith, R., W. R. C. Harvey, and M. Silberfeld. (1996). "Competency and practical judgment." *Theoretical Medicine*, 17:135–150.

Pettit, P. (1984). "Satisficing consequentialism." *Proceedings of the Aristotelian Society* 58 (Supplementary Volume): 165–176.

Rudnick, R. (2002). "Depression and competence to refuse psychiatric treatment." *Journal of Medical Ethics* 28:151–155.

Savulescu, J. (1994). "Rational desires and the limitations of life-sustaining treatment." *Bioethics* 8(3): 191–222.

Savulescu, J., and R. W. Momeyer. (1997). "Should informed consent be based on rational beliefs?" *Journal of Medical Ethics* 23: 282–288.

Shafir, E., and R.A. LeBoeuf. (2002). "Rationality." *Annual Review of Psychology* 53:491–517.

Sinnot-Armstrong, W., and T. Nadelhoffer. (20XX) Is psychopathy a disease? In: Neuroscience and legal responsibility. N. Vincent. Oxford, Oxford University Press: pp. XXX-XXX

Smith, M. (2004). Internal reasons. In: *Ethics and the a priori.* M. Smith. Oxford, Oxford University Press, pp. 17–42.

Spike, J. P. (2008). "Patient's competence to consent to treatment (comment on Appelbaum 2007)."*New England Journal of Medicine* 358(6): 664.

Tan, J., T. Hope, A. Stewart, and R. Fitzpatrick. (2006). "Competence to make treatment decisions in anorexia nervosa: Thinking processes and values." *Philosophy, Psychiatry and Psychology* 13: 267–282.

Vincent, N. (2010). "On the relevance of neuroscience to criminal responsibility." *Criminal Law and Philosophy* 4: 77–98.

Vincent, N. (2011)."Neuroimaging and responsibility assessments." *Neuroethics* 4: 35–49.

Welie, J. V. M., and S. P. K. Welie (2001). "Patient decision making competence: Outlines of a conceptual analysis." *Medicine, Healthcare and Philosophy* 4: 127–138.

Williams, B. (1981a). Internal and external reasons. In: *Moral Luck.* Cambridge, MA, Cambridge University Press, pp. 101–113.

Williams, B. (1981b). The truth in relativism. In: *Moral Luck.* Cambridge, UK, Cambridge University Press, pp. 132–143.

Williams, B. (1996). Toleration: An impossible virtue? In: *Toleration: An Exclusive Virtue.* D. Heyd. Princeton, NJ, Princeton University Press.

Wilson, S., and G. Adshead. (2004). Criminal responsibility. In: *The Philosophy of Psychiatry.* J. Radden. Oxford, Oxford University Press, pp. 296–311.

PART 2

Reappraising Agency

5

Skepticism Concerning Human Agency

Sciences of the Self Versus "Voluntariness" in the Law

PAUL SHELDON DAVIES

The findings of neuroscience cast grave doubts on the view of human agency implicit in the law. They do this by forcing us toward a form of skepticism concerning our capacities as agents. That is the thesis of this chapter.

The findings of neuroscience do not cast doubt in isolation. They do so when combined with findings in cognitive and social psychology and findings in evolutionary theory and primate cognition, and when integrated into a large-canvas view that sometimes results from informed philosophical reflection. This is a powerful methodological directive demonstrated throughout *On the Origin of Species*, a directive that ought to be adopted in the study of the self as much as in the study of life.

The logic of Darwin's (1859) argument is a sequence of abductive arguments concerning a broad range of distinct biological, geological, and geographical phenomena. None of his arguments is decisive taken alone, and some are stronger than others, but their combined power comes from a weighty convergence upon a single, unifying view of life, drawn from an accumulation of inferences to the best explanation concerning several distinct phenomena. This strategy, so potent in the study of life, is our best bet in the study of capacities that animate living things. Questions about human agency, for instance, including the viability of concepts of legal responsibility, cannot be settled with a small set of experiments or a localized hypothesis. What we need is a large-canvas view that integrates knowledge from the relevant sciences. And once we formulate such a view, we find that at present we do not know what kind of agent we are. An informed skepticism best describes where we are today.

The shape of my argument is as follows. Section I introduces the main target of my discussion: a concept of voluntariness that appears essential to a concept of criminal responsibility. I focus on "voluntariness" for the sake of concreteness. Once we appreciate the converging doubts against this concept, doubts concerning other concepts of agency naturally arise. Section II is a brief summary of my very general grounds for thinking that the methods with which we study the human self are in need of reform. The following two sections offer a more specific defense of this call for reform, as well as a few preliminary reformative steps, what I call *directives for inquiry*. I propose one directive in section III that is ameliorative or curative in nature and three additional directives in section IV that are exploratory rather than curative. Then, in section V, on the basis of my proposed directives, I defend my skepticism regarding human agency. I conclude by drawing out the implications of this skepticism for the specified notion of criminal responsibility.

"VOLUNTARINESS" AND CRIMINAL RESPONSIBILITY

For the sake of concreteness, I focus on the partial characterization of criminal guilt from section 2.01 of the Model Penal Code, which states that an agent is criminally guilty for a given action only if it was voluntary, where "voluntary action" is characterized in conditions (a) to (d):

> A person is not guilty of an offense unless his liability is based on conduct that includes a voluntary act or the omission to perform an act of which he is physically capable. The following are not voluntary acts within the meaning of this Section: (a) a reflex or convulsion; (b) a bodily movement during unconsciousness or sleep; (c) conduct during hypnosis or resulting from hypnotic suggestion; (d) a bodily movement that otherwise is not a product of the effort or determination of the actor, either conscious or habitual.

The core assertion is simple, at least on the surface: the attribution of guilt for an action is justified only if the agent's conduct was a bodily movement produced by "the effort or determination of the actor, either conscious or habitual."[1]

It may appear, however, that a person can be guilty in a quite different way, by failing to perform some action despite being physically capable. An agent may be guilty not by virtue of actions that result from effort or determination but simply by virtue of omissions, in which case this section of the Model Penal Code may be interpreted as articulating two distinct concepts of legal responsibility, only one of which employs "voluntariness." Although I am skeptical of any such "two concepts" interpretation, I shall, for the sake of this discussion, restrict my argument to acts of commission and silently pass over the question of whether criminal acts of omission rest upon a prior "voluntary" act that the agent performed or reasonably should have performed.[2] After all, if the single notion of legal responsibility applied to acts of commission falls to my skepticism, that is enough to show that the concept "voluntariness" in the Code is deeply problematic.

Note, then, that the characterization of "voluntariness" in (a) to (d) is remarkably uninformative. The first three conditions are entirely negative: bodily movements not "determined" by the agent—reflexive movements, sleepwalking, for example—are not voluntary. What, then, are the distinguishing properties of movements that are voluntary? We are not told. Condition (d) merely generalizes from the negative characterization in (a) to (c): only movements produced by the effort or determination of the actor are voluntary.

That this section of the Code is nonspecific is not an automatic indictment, however. Laws are tools designed to fulfill certain functions, and some functions can be executed with relatively blunt instruments. This, I surmise, is true of the above characterization of voluntariness. The relevant conditions are sparsely specified on the assumption that there is enough shared cultural knowledge concerning the causes of human conduct to fill the gaps. The lack of specificity in the law is tolerable, perhaps preferable, because our shared cultural knowledge enables us—lawyers, judges, and jurors—to apply the law in light of the particulars of each case. This may provide a degree of flexibility that a fuller specification of "voluntariness" may rule out.

If the above characterization of voluntary action is deliberately generic in this way, then the crucial assumption must be something like this: most adult citizens (including those likely to serve as jurors) know that we are agents who sometimes "determine" their actions and also know when, under what conditions, our actions are in fact the results of our "determinations." If this crucial assumption is false, then the law cannot fulfill its function. If there is not shared knowledge in areas where the law has jurisdiction—if lawyers, judges, and jurors do not know enough to reliably discern actions genuinely determined by the actor from those determined by other factors—then the law is defective.

The question, then, is whether this crucial assumption is true. Do most adult citizens know that we are agents who sometimes determine their actions? Do most know when, under what conditions, our actions result from such determinations? The question is not whether most citizens believe that they have such knowledge, but whether they in fact have it. To answer this question, we have no recourse but to turn to our best developed scientific theories of the self and, on the model of Darwin in the *Origin*, paint in vivid colors our most informed large-canvas view of our capacities as agents. Once we do that, we will see that the answer to this question, in light of current knowledge, is a decidedly negative one; the concept of voluntariness in the above notion of legal responsibility is at odds with what we know about ourselves.

THE AIMS AND STRATEGIES OF CONTEMPORARY THEORIES OF THE SELF

I turn to the general aims and strategies of contemporary theories of the self. I do not claim that these aims and strategies are explicitly endorsed by contemporary theorists in philosophy, psychology, or legal studies. I claim only that they accurately reflect the overarching commitments and methods of many theorists in those areas. As we will see, these aims and strategies tend to diminish rather than enhance our chances of discovering the truth about ourselves. The methods with which we study ourselves are in need of reform.[3]

At a high level of abstraction, the methods with which we study ourselves are either *conceptually conservative* or *conceptually imperialistic*. Neither conservatism nor imperialism by itself is objectionable but, when applied to *dubious concepts*, both methods retard our efforts at discovering the truth. The call for reform, in consequence, is a call for directives that cure us of conservatism and imperialism, as well as directives that guide us in our efforts to discover the truth.

Conceptual conservatism is a strategic orientation toward inquiry, affecting the way we frame our questions and answers. The overarching goal is to conserve or save as far as possible concepts of apparent importance, concepts that appear salient in our more general worldview. The preferred strategy for saving apparently important concepts is to "locate" them amid the concepts and claims of some preferred base theory. For naturalists, the preferred base is usually a well-developed scientific theory; for non-naturalists, it is some well-entrenched part of our inherited worldview. Ruse (2003) presents a book-length exercise in conceptual conservatism, aspiring to save a concept of normative functions in evolutionary biology even at the cost of resuscitating Kant's (1790) theory of natural purposes.

Conceptual imperialism is more ambitious than conservatism. The overarching goal is not to save apparently important concepts as far as possible but to force the rest of our conceptual scheme to accommodate certain concepts at any cost. Certain concepts, it is assumed, have dominion over other concepts and over methods of inquiry. These, according to the imperialist, are concepts without which we would be unable coherently to think or articulate a view of the relevant phenomena. Chisholm (1964) was an imperialist regarding the human self—he insisted that the libertarian concept of free will had to be retained even at the cost of accepting that every conscious, rational person is a little Thomistic god.

Conservatism and imperialism may be appropriate in some contexts but not when applied to dubious concepts. A concept is dubious when there are justified grounds for excluding it from our theorizing. Such concepts fall into two general groups. Some are *dubious by descent*. These are categories that descend to us from a worldview we no longer regard as true or promising, that have not been vindicated in any well-confirmed theories, but that nonetheless tend to influence the way we frame our inquiries. The concept of a nonphysical soul is illustrative. So far as we can surmise, the neural processes implementing human thought and action operate under the principle of causal closure. Our best evidence for this is the utter lack of experiments in which the best explanation of observed phenomena requires the postulation of a nonphysical cause.

Some concepts are *dubious by psychological role*. These are categories controlled by conceptualizing capacities prone to abundant false positives or false negatives. Consider by analogy visual illusions. Under a range of conditions, our visual capacities produce systemic errors. Similarly, under a range of conditions, our cognitive and affective capacities produce systemic errors. But there is a crucial disanalogy. Most visual illusions are easily identified and compensated for, whereas most conceptualizing illusions occur without the agent's notice. Some conceptualizing illusions are so much a part of our deliberative field that it never occurs to us to be troubled by them—not, at any rate, until we meet with an ingenious experiment that reveals the systemic error.

The much-discussed theory of Daniel Wegner (2002) is a case in point. The human mind, according to Wegner, comprises a system that generates the felt experience of consciously willing and thereby consciously controlling our actions. Not all of our actions, of course, because many of our actions are relatively thoughtless—just those we regard as the products of our will. What is provocative, and what reveals a tendency towards systemic error, is evidence adduced by Wegner that this system of conscious willing operates independently of the low level, nonconscious mechanisms that actually cause us to act.

The mechanisms that cause our actions, it appears, are not the mechanisms that give us certain beliefs and feelings about the causes of our actions. We are led astray by the very constitution of our psychology.

Wegner's theory, when integrated with theories from distinct areas of inquiry, wields considerable power. Indeed, Wegner combines experiments from his lab with evidence (some explicated later) concerning a broad range of affective and cognitive phenomena. The strength of his theory rests upon the integrated view of the self that emerges from these diverse phenomena, especially the view of our apparent capacity to control our actions by consciously willing. It is the breadth of converging evidence that makes it rational to hold that the concept of conscious willing, because it generates an abundance of false positives and negatives that are difficult to detect, is dubious by psychological role.

I will explicate Wegner's theory in due course, but I wish to highlight a general feature of our inclination toward conservatism and imperialism, namely, that we tend to be most conservative or imperialistic with respect to concepts most dubious. The greater the staying power of a conceptual category, the greater our tendency to try to save it, perhaps because we feel confident that long-lived concepts must be tracking something real and important. Indeed, the mechanisms that give concepts their staying power are among the very mechanisms that render some concepts so dubious. Some are preserved by culturally instituted mechanisms of transmission; some by the architecture of our psychology (mechanisms that produce persistent errors we tend not to notice); and some, no doubt, are preserved both by cultural and psychological factors.[4] Concepts preserved by any of these mechanisms are going to recur in our deliberative activities; they are going to appear important precisely because they are so tenacious.

Thus, we must be especially cautious in the study of the human self because the concepts with which we understand our capacities as agents are among the most dubious. Concepts such as "free will" and "moral responsibility" are clearly dubious by descent, thanks to our largely theological ancestry, and many of the concepts with which we understand our capacities as agents are dubious by psychological role, as we are about to see.

CURATIVE DIRECTIVES

If the above line of reasoning is correct, if the need for reform in our methods is real, we must diminish the retarding effects of our conservatism and imperialism regarding dubious concepts. To that end, I propose we adopt directives for inquiry formulated in light of our best theories of the very mechanisms that lead us astray. I begin with a directive designed to diminish the ill effects of concepts dubious by psychological role:

DP: For any concept dubious by psychological role, do not make it a condition of adequacy on our theories that we "save" or otherwise preserve that concept; rather, identify the conditions (if any) under which the concept is correctly applied and withhold antecedent authority from that concept under all other conditions.

Withholding antecedent authority from a concept comes to this: we frame our inquiries without that concept. This is not to adopt eliminativism concerning dubious concepts. The directive is to withhold dubious concepts from inquiry until we have reasonable knowledge of the conditions, if any, under which they can correctly be applied. A further aim is to cultivate intellectual creativity. The aim is not merely to avoid concepts that demonstrably lead us astray but also to put ourselves under pressure to create alternative categories with which to explain and predict the phenomena.[5]

When our knowledge of the mechanisms involved in the application of a concept gives rise to such doubts concerning that concept, the directive in DP is essential. We need a reliable process with which one part of our psychology can mitigate the ill effects of another part. To illustrate, consider a few details of our apparent capacity for consciously willing our actions. Wegner's proposed system is triggered when we consciously perceive instances of the following pair:

A thought about or the intention to perform action A
&
THE perception or recollection of oneself performing action A

You think about taking another sip of wine and then perceive yourself sipping. These conscious inputs trigger an interpretive system in your psychology, the function of which is to render your actions intelligible. It achieves this in two steps. It first produces a causal hypothesis to the effect that you, by virtue of your prior thought or intention, caused yourself to perform the action. The hypothesis is that your conscious thought is the means by which you controlled the production of your action. Then the system produces an accompanying affect, a felt sense of achievement, what Wegner calls the emotion of authorship.[6] This interpretive system does all this even though your action was caused by a separate set of mechanisms. This is Wegner's theory of *apparent mental causation.*

The power of this theory derives from the breadth of additional theories with which it integrates. Consider, for instance, the *forward model* of motor control.[7] Suppose I ask you to perform a simple intentional action. I ask you to touch the tip of your nose with your right pointer finger. As you move your right arm, your brain generates a continuous stream of predictions about where your arm ought to be at the next instant, relative to the goal of reaching the tip of your nose. That is why it is a "forward" model: the system generates predictions concerning the ideal future location of your arm. These predictions are useful because they are compared to the continuous proprioceptive feedback regarding the actual position and trajectory of your arm. Any mismatch between predicted and actual position is then used to update the signals sent to your muscles, thereby correcting your action in real time. All of this happens with breathtaking speed at a level of processing inaccessible to conscious awareness.

What is intriguing is that, in the course of executing these anticipatory functions, your brain suppresses its own ability to fully process incoming sensory information. Put generally, the processing of sensory information is suppressed or at least attenuated whenever we act intentionally. This appears clear from experiments reported in Blakemore, Wolpert, and Frith (1999).[8] In one study, experimenters first asked subjects to touch the palm of their right hand using a device manipulated with their left hand; subjects were asked, that is, to perform a simple intentional action. They then asked subjects to allow them, the experimenters, to touch subjects' right hand by manipulating the intervening device. The results were striking. When the experimenter initiated the action—when the act of touching was not intended by the subjects—subjects rated the sensation in their right palm as intense and tickly. When, however, subjects initiated the action themselves, they rated the sensation as less intense and tickly. Our motor control capacities, in the course of executing intended actions, suppress the processing of incoming sensory information.

The same attenuation occurred in a second study. Subjects once again were asked to touch the palms of their right hands by manipulating an intervening device with their left hands. What subjects did not know was that the experimenters were introducing very short delays in the operation of the device. With each trial the motion of the intervening device and the subsequent sensory perception were delayed relative to the subjects' initiation of the intended action. The results were striking. When the delay was short, subjects reported that the sensation was neither intense nor tickly. The brain, while executing the intended action, suppressed the processing of sensory input. But as the delay grew longer, as sensation became increasingly distant from action initiation, the sensation of being tickled increased. The sensation was increasingly processed as coming from something external to the self.[9]

This attenuation of sensory information is important in two ways. First, our motor capacities suppress an enormous quantity of sensory information whenever we act intentionally. One part of our psychology (motor control) conceals from another part (conscious awareness) a large set of causal information. And this contributes to what we might call a kind of *phenomenological quiet*, a degree of subjective silence against which the things that do come to conscious awareness—including the conscious inputs that trigger Wegner's interpretive system—appear salient in our conscious, deliberative fields. The factors that come to conscious awareness, against this backdrop of quiet, are bound to strike us as causally efficacious, especially when processed by an interpretive system dedicated to causal intelligibility. Second, the misleading effects produced by this phenomenological quiet arise from the very architecture of our psychology. This is no small point. Some of Wegner's critics try to dismiss his view by insisting that he is concerned with oddball illusions or marginal mistakes that our otherwise veridical capacities do not suffer. But integrating Wegner's view in this way shows that these critics are mistaken. We are seduced not at the margins but by capacities at our agential core.

Now consider the theory of *naïve realism*, which also integrates with the theory of apparent mental causation. The main elements of naïve realism are three: (1) We tend to assume that we see things in an unmediated and objective manner. (2) We tend to assume that other rational persons will see things as we do. (3) We tend to dismiss those who disagree as ignorant, slothful, irrational, or biased. The background suggestion is that, because each of us approaches a situation, especially situations involving other persons, with limited knowledge and extensive ignorance of what is going on, we must solve what researchers in Artificial Intelligence call the "frame" problem by quickly constructing an operable construal of the situation. We do this by imagining or filling in details that help us decide what is most significant about the situation we face.[10] The origin of our naïve realism, then, is that we construct a construal of the situation in the absence of a much-needed check. There is no check on the confidence that our construal is correct and, in consequence, no check on our confidence that our construal will be adopted by others.

Why this absence? Why are we devoid of a mechanism to remind us that our construal is gleaned from a particular perspective and that people with other perspectives will likely construe the situation differently? We do not know. If, however, our tendency toward naïve realism is manifest in social situations, we might do well to conjoin it with the *theory of mind* theory.[11] On this view, our construal of the desires or intentions that motivate the behavior of other agents may strike us with such force that we are affectively inclined to trust it as accurate. A tendency to respond in this way may have provided anticipatory advantages during our evolutionary history; unbridled confidence in one's construal may have had greater selective value than epistemic caution. Even today, the feeling that one is right in one's assessments of others may conduce to decisive action, better learning, greater career prospects, enhanced interpersonal relations, and increased survival. Arrogance concerning one's self may be less costly than accuracy, especially in social interactions.

The crucial upshot is that, in addition to the phenomenological quiet that accompanies our intentional actions, we naïvely overestimate the accuracy of our conscious assessments of the causes of our actions. We are seduced into thinking and feeling that the causal hypotheses that rise to conscious awareness are correct. And like the ill effects of phenomenological quiet, our naïve realism results from the constitution of our psychology. The former are by-products of a central system (motor control), whereas the latter result from an absence in the architecture of our psychology. Either way, these deficits are the direct effects of the system's normal operations; mistakes at the margins are not the issue.

The power of the directive in DP thus derives from the convergence of a range of theories, including those described here. The theory of apparent mental causation is confirmed in part by the extent to which it integrates with the forward model of motor control, the theory of naïve realism, and the theory of mind theory.[12] It thus is rational to conclude that the concept of "consciously willing" is dubious by psychological role and subject to the directive in DP. This is important for assessing the relevance of contemporary science to the view of agency presupposed in the law.

EXPLORATORY DIRECTIVES

The directive in DP is curative; it aims to cure us of conservatism and imperialism regarding dubious concepts. But it is limited. It helps us avoid what ought to be avoided without recommending an alternative strategy. The purpose of this section is to sketch a few components of an alternative that is *progressive* rather than conservative or imperialistic, and that is *exploratory* in the way that naturalists of the 19th century were explorers. Among the progressive's directives are the following:

> EH: For any capacity of the self we wish to understand, require that we frame our inquiry and our theory in terms of what is known concerning our evolutionary history.
>
> A: For any capacity of the self we wish to understand, assume that, as a consequence of our evolutionary history, it is endowed with the systemic function of anticipating objects or events relevant to organismic equilibrium, to the satisfaction of ecological demands, or to both.
>
> NC: For any conscious capacity of the human mind, expect that we will understand this capacity only after we discover the nonconscious, low-level, anticipatory mechanisms implementing that capacity.

Although EH and A appear banal, their effect on our inquiries can be substantial, altering the way we conceptualize the very capacities we wish to study. To illustrate, consider the hypothesis that human intelligence is best conceptualized as *social intelligence*. The hypothesis can be articulated in many ways, but the basic claim is that many of our affective and cognitive capacities evolved as tools enabling us to engage in myriad social relations. This is no mere speculation. It is based in part on knowledge concerning extant primate species. We know that, in one form or another, all primates are social animals,[13] capable of identifying con-specifics, recognizing social relations between con-specifics, recognizing one's own relations with others, responding appropriately to changes in those relations, and so on. And it is easy to generate hypotheses concerning the anticipatory functions of all these capacities.[14] We also know that *Homo sapiens* is the most social of primates. There is, for example, a clear difference in the breadth and depth of our cultural institutions, evidence that our capacities for social relations run wider and deeper. More specifically, human children by their fourth year clearly exercise the capacities posited in the theory of mind theory; 4 years is about the age at which most children begin to pass the false belief test. In addition, children as young as 9 months exhibit striking precursor capacities. They follow the gaze of adults, jointly focus on shared objects, imitate the behavior of others, and so on.[15] And ingenious, recent experiments suggest that human infants are reading minds, even attributing false beliefs, as early as 2 years of age.[16] By contrast, it is contentious whether other primate species possess the full suite of capacities posited in the theory of mind theory.[17]

There is also evidence of our social intelligence from neuroscience. Human infants attend preferentially to other humans. They attend to human faces more than any other visual stimuli and to human speech more than any other auditory stimuli. Infants as young as 2 days exhibit a distinctive cerebral blood flow when they hear a normal sentence but not when the sentence is played backward. And so on.[18] There is also the intriguing hypothesis that among the emotional systems implemented in the mammalian brain is what Jaak Panksepp (1998) dubs the PANIC system. This system functions to generate behavioral routines to free the organism from life-threatening situations. Effects of this system are evident in the distress calls of infants when separated from their mother, which are accompanied by physiological processes exhibited when an organism is suffocating, when it cannot catch a breath. This powerful response to separation is implemented in distinct neural structures and chemical processes identified by Panksepp. And the very same structures and processes that constitute the PANIC system also implement the reaction that adult mammals have to loss. Human grief is implemented in the brain's PANIC system.

Panksepp's hypothesis provides a striking account of our social emotions. If the basic function of the PANIC system is to generate behaviors to free the organism from threats to its life, a closely related function is to empower the organism to avoid or alleviate the experience of loss by establishing social attachments. The hypothesis is that the neural system that causes us to panic in response to loss is the very system that moves us to seek emotional attachments with others. Indeed, the PANIC system comprises neural structures known to implement certain forms of physical pain, suggesting that separation distress and grief are, literally, a form of pain and that the compulsion toward social relatedness is an anticipatory strategy for keeping some forms of pain at bay.[19]

This brief survey of the social intelligence hypothesis illustrates the power of EH and A. Notice, in particular, that our initial understanding of a capacity is altered by the application of these directives. Wegner's interpretive system, once again, is a case in point. What is the evolved, anticipatory function of a system that, by hypothesis, causes us to falsely believe that our conscious intentions cause our actions? Wegner (2007) offers several speculations, each keyed to an anticipatory, social function, and when we conceptualize our capacity for conscious willing in this way, as dedicated to some social function, our understanding is indeed altered. Instead of conceptualizing the feeling of conscious willing as evidence of a remarkable form of freedom, we conceptualize it in terms of our evolutionary history and the ways in which it prepares us for what is likely to occur next. We see the capacity as, for example, a mechanism that inclines us to inform one another about actions we are likely to perform, or a mechanism that causes us to feel a sense of obligation toward one another, and so on. And because there appears to be nothing parochial about conscious willing in this regard, the point here can be generalized: we should expect that, as our knowledge of the self progresses, our understanding of the phenomena we are trying to explain will shift in significant ways.

This shift also illustrates the power of the directive in NC. The feeling of willing, for instance, occurs at the level of conscious awareness; we are introspectively aware of some features of the process. And we tend to feel confident that the way things appear to us concerning the causes of our actions is an accurate reflection of the actual causes. Our confidence, however, is misplaced. Once we ask about the anticipatory function of any conscious capacity, we will likely discover mechanisms operating below conscious awareness that force us to revise our initial understanding. In general, it is rational to expect that our capacities for conscious experience will not be adequately understood until we discover the evolved, anticipatory functions of nonconscious mechanisms that implement those capacities.

If these exploratory directives are defensible, we must do more than withhold antecedent authority from dubious concepts. We must also fill the gaps in our conceptual repertoire left by the application of DP. We may begin by framing our inquiries with relevant knowledge from evolutionary biology and searching for the anticipatory functions of mechanisms that operate beyond the reach of conscious awareness. These are strategies informed by our best sciences of the self. And when we apply these strategies, we begin to appreciate how little of our capacities as agents we presently understand.

SKEPTICISM CONCERNING HUMAN AGENCY

If, then, the concept "conscious willing" is dubious by psychological role, and if the exploratory directives reveal that our former understanding of "conscious willing" is best replaced by a more informed understanding of the relevant capacity, then it is no longer rational to frame our inquiries in terms of this concept. It is no longer rational to assume that our alleged capacity for conscious willing is what we formerly took it to be. In particular, we cannot take it as given that our apparent capacity to consciously will our own actions reflects an actual capacity to control our actions. This is the basis for my skepticism concerning human agency.

The skeptical thesis is best formulated as an epistemic defeater: for any action we perform, we cannot justifiably claim to know from the first-person point of view the actual causes of our action. The claim is not that we never have true beliefs about the causes of our actions, but rather that we cannot reliably discriminate from the first-person point of view between cases in which our beliefs about our actions are true and cases in which they are false. This defeats the possibility of justifying, at least from the first-person perspective, beliefs about the causes of our actions. That is the lesson on which the theories described above appear to converge.

Still, this defeater appears to conflict with any number of ordinary cases in which we intuitively take ourselves to know the causes of our actions. Suppose it is Monday afternoon and you see me walking across the parking lot and entering my daughter's school. You ask me what I am doing. I tell you I am picking up Cassie and taking her to her piano lesson. Being nosy, or perhaps being an inquisitive social psychologist, you ask me why I am doing this. Being a congenial philosopher, I tell you that several weeks ago my wife and I agreed on a weekly schedule. I am taking Cassie to her lesson because that is what I agreed to do.[20]

Such intuitions, especially about cases in which the relevant action has been planned in advance, may appear to challenge the epistemic defeater.[21] But in fact they do not. In responding to your question, it is plausible to suppose that my episodic memory quickly recollects a relevant event. It retrieves an agreement I made several weeks ago, the content of which is that every Monday afternoon during the semester I will take Cassie to her lesson. My memory did this, presumably, because the social situation demanded a timely response to your question. But the mere fact that my memory retrieved this recollection does not entail or even suggest that the content of this recollection is the actual cause of my action. My episodic memory, in recalling my earlier agreement, appears nicely attuned to considerations of relevance, but relevance requires nothing stronger than association.

Moreover, the actual causes of my action, whatever they might be, are factors that caused me to leave the house and drive to Cassie's school, and for all I can tell from the first-person point of view, the agreement I made weeks ago is causally unrelated to those factors. Of course, it certainly feels to me that the content of my recollection is causally related, but the reliability of this feeling is undermined by the theories surveyed above.[22] I also grant that I may truly believe that my recollection is causally relevant, but true belief does not suffice for justification. The theories surveyed above show that, from the first-person perspective, we cannot reliably discriminate cases in which our prior thoughts cause our actions from cases in which they do not. That is the basis of my defeater.

We might vary the locus of the objection. Instead of fixing on my present recollection of an agreement made several weeks ago, fix on the conscious thoughts that occurred just before the action. Suppose I was engrossed in work all afternoon until I happened to glance at my watch. "Oh," I exclaimed, "time to collect Cassie's music and get to her school!" Suppose I even exhorted myself: "I cannot renege on my agreement with Ann!" Suppose, finally, that upon reaching the school and hearing your question, I consciously recall and report to you the exclamations and exhortation that occurred just before initiating my action. Does this show that I know the actual causes of my action?

Not at all. It is true that my agreement with my wife was recalled to conscious awareness just before I began my Monday afternoon routine. But, again, it does not follow that this recollection is part of the actual causes of the action. All manner of nonconscious processes were no doubt occurring in me as I realized it was time to stop working, and I have conscious access to virtually none of them. And we know from the theories of Wegner and Wolpert and Blakemore and others that the things which do rise to conscious awareness often seduce us toward causal beliefs that are demonstrably false. That, to repeat, is the upshot of the theories canvased above: we know that in many cases the correlations we observe among our conscious perceptions or recollections concerning our actions are unreliable indicators of genuine causal connections, and nothing available from the first-person perspective enables us to discriminate the causes from mere correlations.

In general, it makes no difference where in the sequence we fix our attention. Even when the relevant action was planned weeks in advance, the actor is faced with an open question concerning the actual causes that finally move him to act. This is the basis of my skepticism. To refute it, we need an alternative theory of the self based upon a convergence of evidence of equal or greater strength. Short of that, no matter how unintuitive or unsettling it may be, skepticism is the rational position to adopt.

There is, moreover, an analogue to the above defeater that applies from the third-person perspective. When we claim that some other agent acted voluntarily, we posit a causal process that includes what we traditionally describe as "conscious willing." Yet, as we have seen, because the concept "conscious willing" is dubious by psychological role, it ought to be factored out of our inquiries and replaced by considerations from our evolutionary and social history. If that is right, then we cannot justifiably frame our inquiries in terms of a concept so deeply dubious. Precisely that is the basis for an additional defeater: for any action performed by another agent, we cannot justifiably claim to know that the agent acted voluntarily by consciously willing it. It must be emphasized that this defeater rests upon the above directives. The curative directive directs us to withhold antecedent authority, and the exploratory directives direct us to conceptualize the relevant capacity in terms of our history. The reason, therefore, that we cannot justifiably claim to know whether another person consciously willed her action is that, in light of our best sciences of the self, the central conceptual category has no legitimate role in contemporary inquiry.

You might worry that my skepticism refutes itself by rendering impossible all forms of rational debate. If we are indeed faced with my defeaters concerning our reasons for acting, and if adducing evidential or logical relations for a scientific or philosophical thesis qualifies as an action, it appears we can never know the reasons why anyone ever accepts one theory over another, which would undermine the very possibility of rational debate. It would seem to show, in particular, that I cannot give any reasons for skepticism concerning human agency.[23]

This worry is motivated by the apparent phenomenology of actual intellectual discussions. When in conversation you challenge some part of my view, I focus my attention on specific features of the world. When, for instance, you ask me why I hold a given thesis, I appeal to features of the world I judge to be evidentially potent. What seems crucial is that the features to which I appeal are consciously accessible to me. How could it be otherwise? How could I appeal in conversation to considerations that do not come to conscious awareness? The worry, then, is that my defeaters conflict with this bit of phenomenology. My first defeater seems to entail that I cannot justifiably claim to know my reasons for defending the relevant thesis, which seems to preclude the possibility of my rationally defending my view. My second defeater seems to entail that you cannot justifiably claim to know my reasons for defending the thesis, which appears to make reasoned exchange between us utterly impossible.

There are several reasons why this objection is wide of the mark. I will mention just two. First, the phenomenology of our rational discussions is concerned with a relatively narrow notion of "reasons." My reasons for accepting a given thesis are patterns of evidential relations between facts in the world and the contents of the thesis, or logical relations between the thesis and other theses. As such, these sorts of reasons are limited. Even if they reveal substantive relations between the thesis and certain facts or certain other theses, they fail to explain why I endorse the thesis. It is naïve to assume that I endorse any thesis simply because of the substantive relations it bears to certain facts or to other theses. This is not to confess a foible or infirmity unique to myself. It is true of any intelligent organism whose capacities for acting are a mix of cognitive and affective capacities that operate mostly beyond the reach of conscious awareness. The evidential or logical relations I consciously acknowledge as my reasons are surely supplemented and in some instances supplanted by a host of nonconscious processes. That, at any rate, is the upshot of the scientific theories surveyed above. And that means there is a much fatter notion of "reasons" relevant to all forms of human action. This fatter notion is surely applicable to the actions we perform in the course of intellectual debates, but it is even more pertinent to actions that fall under concepts of legal responsibility. Indeed, this relatively fat notion of "reasons" is at the heart of the concept of criminal responsibility described in section I.

My second response is that it is false that my view rules out the possibility of knowing our reasons for acting. It rules out the possibility of knowing our reasons in certain ways, including ways assumed by many philosophers, legal theorists, and laypersons, but it is compatible with knowledge acquired in other ways. So long as my first defeater stands, we cannot justifiably claim to know our reasons from a first-person point of view, but that leaves open the possibility of subjecting our agential capacities, including our reason-giving capacities, to scientific investigation. My second defeater, moreover, suggests we cannot justifiably claim to know that another agent acted voluntarily by virtue of conscious willing. But that is compatible with the scientific study of our capacities as agents in terms of conceptual categories other than "conscious willing" and "voluntariness." Whether we can at present articulate an alternative concept is not to the point. It would be the most egregious form of conceptual imperialism to insist that we must preserve our traditional concepts of agency just because we have yet to formulate other concepts informed by what is actually known about the human self.[24]

"VOLUNTARINESS" AND LEGAL RESPONSIBILITY

I come at last to the troubling implications that my defeaters raise for "legal responsibility." In the Model Penal Code, a notion of criminal responsibility is explicated in part by appeal to "voluntariness," though the explication given, as we saw in section I, is remarkably uninformative. This is so, I surmise, on the assumption that there exists sufficient shared cultural knowledge to fill the gaps in any given case. The crucial assumption, then, is that most citizens know that we are agents who sometimes determine their actions and also know when, under what conditions, our actions result from such determinations. And this crucial assumption is precisely where we meet the troubling implications of my skepticism, for this assumption is indeed false.

It should be clear by now that we do not possess shared cultural knowledge concerning the nature of human agency because we are burdened with the previously described epistemic defeaters. Thanks to progress in knowledge, we cannot justifiably claim to know from the first-person perspective the causes of our actions. Nor can we justifiably claim to know of some other person that he "determined" his own action because the central concept is so clearly dubious and our most fruitful methods direct us to conceptualize the relevant capacity in very different terms. In general, recent progress in the scientific study of the human self reveals that we do not know what kinds of agent we are. The characterization of criminal responsibility given in the Model Penal Code cannot serve its intended function.

Of course, the assumption that there exists shared knowledge of our capacities as agents runs deep and wide in our culture. That I do not deny. But the persistence and power of that assumption can be explained without granting its truth. It can be explained, in particular, by the persistence and power of concepts dubious by descent and by psychological role, and by our stubborn inclination toward conceptual conservatism and imperialism. It may be possible, moreover, to alter or eliminate this widespread assumption. If the expectations and intuitions of informed citizens are brought up to speed, if they come to reflect the larger implications of our best sciences, then appeals to commonsense may become increasingly impotent, even pathetic, when opposed to the findings of contemporary science, and our traditional concepts of the self may be revised or replaced. Until then, however, we remain burdened with laws that, for wont of knowledge of ourselves, cannot fulfill their functions.

These skeptical doubts concerning human agency are grave in two ways. They are grave because they are derived from demanding methods of inquiry. The argumentative strategy of Darwin's *Origin* is illustrative. Its power stems from the convergence upon a single view of life from a broad range of distinct phenomena concerning living things. The same holds for understanding core capacities of living things, including our capacities to deliberate, choose, and act. There is a growing convergence on the nature of the self across the relevant sciences. Not a fully articulated view of the self, to be sure, but enough to articulate some important claims: (1) A great deal of our mental lives is lived beneath the level of conscious awareness. (2) At least some of the phenomena comprising conscious awareness are partial, misleading, or illusory. (3) As a consequence, we are in a muddle about our capacities as agents; we know enough to appreciate how little we understand our experiences as selves. That is one reason why the doubts are grave: they emerge from a breadth and depth of current scientific knowledge that cannot rationally be ignored.

The doubts appear grave in another way, in terms of vital practical matters. The most pressing question is whether these doubts concerning human agency will take hold in the larger culture and, if they do, what effects they will likely provoke. One problematic effect will be the lack of a clear alternative. We live in a period of profound uncertainty about the nature of our selves; we have no choice but to endure a great deal of confusion. Another problem is knowing when to trust the converging results of human inquiry. How integrated and mature must a set of scientific theories be for us rationally to use it as the basis for policies that affect social stability, fairness, and human well-being? This is a deeply vexing question, especially for organisms who need to anticipate and feel a sense of control. Finally, there is the question of whether we have the stomach for periods of conceptual confusion, for not knowing how to think or feel ourselves as agents, and whether we will respond with creativity to better conform our beliefs and practices to the way the world actually is, or whether we will panic and revert to conservatism and imperialism.[25]

NOTES

1. We tend to attribute guilt for what a person actually does, not for what a person merely thinks. Hence the focus here on bodily movement.
2. On this point, see chapters 6 and 7 of this text.
3. The discussion in this and the next two sections is a highly compressed version of portions of my recent book (Davies 2009). Compression of course tends to distort. I hope, however, there is intelligibility enough to recommend the fuller discussion in the book.

4. Cultural mechanisms of conceptual stasis are discussed in Norris and Inglehart 2004, Richerson and Boyd 2005, chapter 6 of Davies 2009, and elsewhere. Evidence for the efficacy of psychological mechanisms of stasis is described throughout this essay and in Davies 2009.
5. In Davies 2009, part 2, I defend an additional directive to diminish the ill effects of concepts dubious by descent.
6. Since "sense" and "emotion" hardly appear equivalent, you might worry that I am being conceptually flat-footed. Perhaps so. But conceptual fussiness is a virtue only when it makes a difference in substance. Wegner is interested in not one specific affective response but rather a whole cluster. He is interested in the full range of responses to our own actions that tempt us to feel that we are conscious controllers of those actions.
7. See Wolpert at al 1995. Wolpert 1997 is an accessible overview.
8. See also Blakemore, Frith, and Wolpert 2000.
9. Choudhury and Blakemore 2006 provide a recent overview.
10. For experimental evidence, see Ross and Ward 1996, Pronin et al. 2002, Pronin et al. 2004, and Pronin 2007. This last paper surveys recent studies of the biasing effects of naïve realism.
11. Tomasello 1999 and Leslie 2000 are good overviews.
12. The full integrative picture is much broader than I can depict here. Also relevant is John Bargh's work on automaticity (e.g., Bargh et al. 2001; Dijksterhuis and Bargh 2001), Timothy Wilson's on knowing our reasons for acting (e.g., Nisbett and Wilson 1977a and 1977b; Wilson 2002), Martin Conway's on autobiographical memory (e.g., Conway and Pleydell-Pearce 2000; Conway 2003), and so on.
13. See Smuts et al. 1987 for the remarkable range of social structures among primate species.
14. Indeed, the challenge is generating experimental evidence with which to discriminate among all the possible hypotheses.
15. Tomasello 1995 and 1999.
16. Baillargeon et al. 2010.
17. Tomasello and Call 1997.
18. These references come from Cheney and Seyfarth's marvelous 2007 book on social intelligence in baboons and humans (p. 6). Cheney and Seyfarth cite Dehaene-Lambertz et al. 2002 and Peña et al. 2003.
19. See chapter 14 of Panksepp 1998. The implications of this view for the so-called "reactive attitudes" are considerable. Or so I think. See chapter 9 of Davies 2009.
20. This oversimplifies, of course. There are several reasons why I take Cassie to her lessons, including the pleasure of being with her.
21. I am grateful to Walter Sinnott-Armstrong for raising this challenge.
22. The reliability of this feeling may vary with certain features of the action and the situation in which it occurs. This appears to be an implication of the theory defended in Wilson 2002. When action and context are relatively simple and unambiguous, reliability may be greater; the interpretive system that helps us make sense of our actions may be less prone to error in uncluttered contexts. (I discuss this point in Davies 2009:146ff.) This is not, however, a difference we

can read off our feelings at the time, and it leaves ample room for error in the complex and ambiguous real-life cases in which knowledge of our reasons matters most.

23. I am grateful to Nicole Vincent for pressing this worry.
24. These two replies merely sketch the direction that a fuller response would likely take.
25. This essay descends from a presentation given at Delft Technical University in August 2009 on the occasion of an interdisciplinary conference, *Moral Responsibility: Neuroscience, Organization, and Engineering*, organized by Neelke Doorn, Jessica Nihlen Fahlquist, and Nicole Vincent. I am grateful to the organizers for a setting in which scholars from a wide range of fields engaged constructively. My travels to the Netherlands were supported by the Wendy and Emery Reves Center for International Studies at the College of William and Mary and by Silvia Tandeciaraz, Dean of Educational Policy at the College. I am grateful for both sources of support. I also received help from several good thinkers. Thanks to Stephen Morse and Walter Sinnott-Armstrong for much-needed advice; to Walter, George Harris, and Nicole Vincent for probing comments on an earlier draft; to an anonymous referee for OUP Press for constructive resistance; and to Nicole for thoughtful advice throughout.

REFERENCES

Baillargeon, R., R. M. Scott, and Z. He (2010). "False-belief understanding in infants." *Trends in Cognitive Sciences* 14(3): 110–118.

Bargh, J, P. Gollwitzer, A. Lee-Chai, K. Barndollar, and R. Troetschel (2001). "The automated will: Nonconscious activation and pursuit of behavioral goals." *Journal of Personality and Social Psychology* 81: 1014–1027.

Blakemore, S. J., C. Frith, and D. Wolpert (1999). "Spatiotemporal prediction modulates the perception of self-produced stimuli." *Journal of Cognitive Neuroscience* 11: 551–559.

Blakemore, S. J., D. Wolpert, and C. Frith (2000). "Why can't you tickle yourself?" *NeuroReport* 11(11): R11–R16.

Cheney, D., and R. Seyfarth (2007). *Baboon Metaphysics: The Evolution of Social Intelligence*. Chicago, IL, University of Chicago Press.

Chisholm, R. (1964). "Human freedom and the self." The Lindley Lecture, University of Kansas.

Choudhury, S., and S. J. Blakemore (2006). Intentions, actions, and the self. In: *Does Consciousness Cause Behavior?* S. Pockett, W. Banks, and S. Gallagher. Cambridge, MA, MIT Press.

Conway, M. (2003). "Cognitive-affective mechanisms and processes in autobiographical memory." *Memory* 11(2): 217–224.

Conway, M., and C. Pleydell-Pearce (2000). "The construction of autobiographical memories in the memory system." *Psychological Review* 107(2), 261–288.

Darwin, C. (1859). *On the Origin of Species by Means of Natural Selection, or the Preservation of Favoured Races in the Struggle for Life*. (A facsimile of the first edition with an introduction by Ernst Mayr, 1964.) Cambridge, MA, Harvard University Press.

Davies, P. S. (2009). *Subjects of the World: Darwin's Rhetoric and the Study of Agency in Nature*. Chicago, IL, University of Chicago Press.

Dehaene-Lambertz, G., S. Dehaene, and L. Hertz-Pannier (2002). "Functional-neuroimaging of speech perception in infants." *Science* 298: 2013–2015.

Dijksterhuis, A., and J. Bargh (2001). The perception-behavior expressway: Automatic effects of social perception on social behavior. In: *Advances in Experimental Social Psychology, Volume 33*. M. P. Zanna. San Diego, CA, Academic Press, pp. 1–40.

Kant, I. (1790). *Critique of Judgment*. (Translated by J.H. Bernard, 1951, Hafner Press.)

Leslie, A. (2000). "Theory of mind" as a mechanism of selective attention. In: *The New Cognitive Neurosciences*, 2nd edition. M. Gazzaniga. Cambridge, MA, MIT Press, pp. 1235–1247.

Nisbett, R., and T. Wilson (1977a). "Telling more than we can know: Verbal reports on mental processes." *Psychological Review* 84: 231–259.

Nisbett, R., and T. Wilson (1977b). "The halo effect: Evidence for unconscious alteration of judgments." *Journal of Personality and Social Psychology* 35: 250–256.

Norris, P., and R. Inglehart (2004). *Sacred and Secular*. New York, Cambridge University Press.

Panksepp, J. (1998). *Affective Neuroscience*. New York, Oxford University Press.

Peña, M., A. Maki, D. Kovacic, G. Dehaene-Lambertz, H. Koizumi, F. Bouquet, and J. Mehler (2003). "Sounds and silence: An optical topography study of language recognition at birth." *Proceeding of the National Academy of Science* 100: 11702–11705.

Pronin, E. (2007). "Perception and misperception of bias in human judgment." *Trends in Cognitive Science* 11: 37–43.

Pronin, E., D. Lin, and L. Ross (2002). "The bias blind spot: Perceptions of bias in self versus others." *Personality and Social Psychology Bulletin* 28: 369–381.

Pronin, E., T. Gilovich, and L. Ross (2004). "Objectivity in the eye of the beholder: Divergent perceptions of bias in self versus others." *Psychological Review* 111: 781–799.

Richerson, P., and R. Boyd (2005). *Not by Genes Alone: How Culture Transformed Human Evolution*. Chicago, IL, University of Chicago Press.

Ross, L., and A. Ward (1996). Naïve realism in everyday life: Implications for social conflict and misunderstanding. In: *Values and Knowledge*. T. Brown, E. Reed, and E. Turiel. Hillsdale, NJ, Erlbaum, 103–135.

Ruse, M. (2003). *Darwin and Design: Does Evolution Have a Purpose?* Cambridge, MA, Harvard University Press.

Smuts, B. B., D. Cheney, R. Seyfarth, R. Wrangham, and T. Struhsaker (1987). *Primate Societies*. Chicago, IL, University of Chicago Press.

Tomasello, M. (1995). Joint attention as social cognition. In: *Joint Attention: Its Origin and Role in Development*. C. Moore and P. Dunham. Hillsdale, NJ, Erlbaum.

Tomasello, M. (1999). *The Cultural Origins of Human Cognition*. Cambridge, MA, Harvard University Press.

Tomasello, M., and J. Call (1997). *Primate Cognition*. New York, Oxford University Press.

Wegner, D. (2002). *The Illusion of Conscious Will.* Cambridge, MA, MIT Press.

Wegner, D. (2007). Self is magic. In: *Psychology and Free Will.* J. Baer, J. C. Kaufman, and R. F. Baumeister. New York, Oxford University Press.

Wilson, T. (2002). *Strangers to ourselves: Discovering the adaptive unconscious.* Cambridge, MA, Harvard University Press.

Wolpert, D. (1997). "Computational approaches to motor control." *Trends in Cognitive Science* 1(6): 209–216.

Wolpert, D., Z. Ghahramani, and M. Jordan (1995). "An internal model for sensorimotor integration." *Science* 269: 1880–1882.

6

The Implications of Heuristics and Biases Research on Moral and Legal Responsibility

A Case Against the Reasonable Person Standard

LEORA DAHAN-KATZ[1]

The heuristics and biases tradition of cognitive psychology, hereinafter "Bias Research," demonstrates that human reasoning processes are often non-normative; that people are not perfectly rational but rather exhibit "bounded" rationality. Specifically, such research has demonstrated that human reasoning processes often rely upon inaccurate rules of thumb, or heuristics, and are subject to a wide variety of identifiable biases, often as a result of reliance upon such heuristics. These, in turn, lead to errors in judgment that may impact human behavior. This chapter argues that this impact cannot be ignored when dealing with questions of moral and legal responsibility. It argues that when heuristic reasoning has an impact upon human judgment and decision-making, this fact can ultimately negate moral culpability. Furthermore, it argues that where legal responsibility is justified based on moral considerations (such as notions of desert in the criminal law), the findings of Bias Research must also inform and limit the imposition of legal responsibility. Specifically, it will be argued that in light of Bias Research, the current reasonable person standard employed in negligence offense at common law is unjustifiable and must be replaced with a standard of responsibility that better reflects individual culpability.

INTRODUCTION: THE MORAL, THE LEGAL, AND THE COGNITIVE

Persons are generally regarded as responsible agents.[2] Nevertheless, it is generally recognized that situations may arise in which a person will not bear moral responsibility because of the circumstances under which he or she acted. This is because moral responsibility is dependent upon what may be termed *necessary conditions of responsibility.* Moral responsibility has been so construed since Aristotle, who writes in both the Nicomachean Ethics and Eudemian Ethics that responsible action is only action that may be defined as "voluntary" (*hekousion*). Although Aristotle presents slightly different accounts of voluntary action in different works, he generally defines such action as that which is (1) "up to" the agent (the "principle" or the origin of the action is internal to the agent rather than external), and that which (2) is not performed in ignorance.[3]

A classic example of a situation that does not meet Aristotle's first condition is one involving force. Generally, causing physical harm to another is regarded as morally unacceptable. Thus, an agent who inflicts harm upon another is viewed as morally responsible for having done so. However, if the agent was physically forced to hurt another against his will and had no way of overcoming such force, in light of the first condition of moral responsibility noted by Aristotle, the agent is not morally responsible for his action or the consequences thereof. In such a case, although the agent performed an action that could be subject to moral responsibility (he has done "wrong"), he is not morally responsible for what he has done because he is not "culpable" — it is not his fault that he has done wrong.[4]

A classic example of a situation that does not meet Aristotle's second condition is one involving error. Consider, for instance, the case of a host who unknowingly pours his guest a shot of poison, causing the immediate death of his unfortunate friend. In this case, knowledge, or rather ignorance, is relevant to the moral status of the host. Had the host been fully aware of the lethal contents of the bottle, this would be a basis for properly judging him to be morally culpable for his action.[5] However, if the host honestly and faultlessly believed the bottle to contain nothing but whisky, this fact negates his moral responsibility for his action.[6]

This idea of necessary conditions of responsibility has since been adopted and developed by many moral philosophers. Aristotle's own conditions continue to serve as the basis for a major conceptual distinction between two different types of conditions: volitional (or having to do with one's powers or capacities) and cognitive (or having to do with one's mental state or level of awareness),[7] although theorists vary in their positions about which conditions should be included in each of these categories.

What Is the Relevance of Bias Research to Questions of Moral Responsibility?

Bias Research is a field of cognitive psychology that studies the cognitive limitations that affect human reasoning. At the center of this research is the claim that human mental capacities are limited; that there are certain imperfections which plague the human mind and render persons less rational than may initially have been believed — that render man "boundedly rational."[8] Bias Research exposes human reasoning to be riddled with flaws and imperfections, which manifest themselves in errors in judgment and decision making. It studies these errors and demonstrates when people do and can be expected to make judgmental errors — to be mistaken or ignorant. Bias Research is thus significant to moral theory because it is relevant to the evaluation of one of the necessary conditions of moral responsibility: the condition of knowledge. Bias Research reveals that on many occasions people can and do faultlessly err, and thus ought not to be held morally responsible for their actions. Consequently, it is crucial that the findings of Bias Research inform evaluations of moral responsibility to ensure that responsibility is not attributed excessively.

These moral implications have further legal implications: where legal responsibility is stipulated upon the moral responsibility of the agent, such as is classically considered to be the case in much of criminal law, Bias Research must also inform questions of legal responsibility. Otherwise, legal responsibility will be imposed excessively, upon those who are not at fault.

This chapter focuses on the implications of Bias Research for one important legal standard: the objective reasonable person standard (and its variants) imposed in negligence offenses at common law. Pursuant to this standard, courts will find a defendant guilty where it is deemed that he was "negligent" — where it is found that although he was ignorant with respect to a relevant element of his action, such as the dangers involved in such action (and thus does not meet standard mens rea requirements), it is deemed that he ought to have been aware and acted otherwise. Yet, employing the objective reasonable person standard, the court will not evaluate responsibility for negligence individually. Rather, it will only evaluate whether a hypothetical "reasonable person" would have been aware and would have acted otherwise under the circumstances. The court may thus impose liability upon a defendant even where, due to the heuristic and biased nature of her own reasoning processes, she herself could never have realized that her conduct was dangerous. Thus, although she acted faultlessly and should not be held legally responsible, the criminal law, employing an objective standard, will hold her criminally accountable.

Of course, the discrepancy between objective standards and moral responsibility precedes the findings of Bias Research. However, Bias Research significantly affects the question of the justification of such standards. Objective standards are generally viewed as problematic from a moral point of view because they do not evaluate responsibility individually, and thus allow the conviction of even the faultlessly ignorant. They have nonetheless been justified, even by culpability-oriented or fault-based theories of criminal law, based on pragmatic considerations relating to the realistic constraints of the legal institution, which allegedly cannot allow itself to evaluate responsibility individually. As will be demonstrated, however, Bias Research sheds new light on the question of the justification of the reasonable person standard, undermining the considerations raised against the individualization of responsibility.[9] Drawing from the implications of Bias Research, it will be argued that the current reasonable person standard is unjustifiable and must not be maintained. It will be argued that ideally, the standard should be removed from the criminal law and replaced with a subjective standard of responsibility that reflects individual fault. Alternatively, if it is found that institutional considerations nonetheless cannot allow for a fully subjective evaluation of responsibility in negligence offenses, the law must, at the very least, internalize the findings of Bias Research. It must do so to avoid the unnecessarily excessive imposition of criminal responsibility upon blameless persons, who merely acted based on erroneous judgments (without having been indifferent, lazy, careless, or any other disposition that could render them culpable). This may be achieved by employing a more accurate "boundedly" reasonable person standard that reflects the heuristic and biased nature of human reasoning and does not idealize away common, cognitive error.

BIAS RESEARCH

What Are Heuristics and Biases and How Do They Manifest Themselves in Human Judgment?

Heuristics are general, inaccurate principles used by the human mind to aid it in arriving at judgments and decisions. They are essentially mental shortcuts or cognitive tools used by the mind to circumvent the need to engage in the mental procedures necessary to arrive at precise results. Generally, these tools prove to be satisfactory. However, the use of heuristics is not without consequences. Research in the field of cognitive psychology has demonstrated that in certain, often predictable, cases, relying on heuristics will create cognitive biases that can lead to severe and systematic errors in judgment.

Daniel Kahneman and Amos Tversky, often credited as the fathers of Bias Research, focused their research on the heuristics and biases prevalent in situations of "judgment under uncertainty," that is, on the question of how people assess the probability of uncertain events and the value of uncertain qualities. In their research, Kahneman and Tversky found that when faced with such tasks, human reasoning relies upon three major heuristics: availability, representativeness, and 'anchoring and adjusting.'[10] With reference to the availability heuristic, for example, Kahneman and Tversky found that when estimating risks and probabilities, rather than rationally or normatively considering relevant information,[11] people base their assessments on the ease with which instances or occurrences can be brought to mind.[12] For example, when asked to assess the probability of heart failure, a person will often base his judgment upon the rate of heart attack among his friends and family rather than on base-rates that refer to the general population.[13] As a result of this reliance, events that are more easily recalled are judged to be more probable, whereas events that are less easily retrieved are judged to be less probable. Although this heuristic may be a helpful tool in many real-life situations, it often leads to errors in judgment. Kahneman and Tversky explain that this is the case because ease of retrievability is influenced by factors irrelevant to actual probabilities such as familiarity and salience: For example, events that one is familiar with will seem more probable, whereas less familiar events will seem less probable.[14] Thus, the fact that one has never come into contact with a cancer patient and is not familiar with the situation will influence one's assessment of the probability of contracting cancer, possibly causing one to underestimate such probabilities, though of course, this fact is irrelevant to the likelihood of becoming ill.[15] Similarly, vivid or memorable events will also be more readily retrievable because of their salience, causing one to overestimate the likelihood of their occurrence, despite the fact that there is little connection between salience and probability. After having witnessed a car accident, for example, a person's assessment of the probability that a car accident will occur generally rises.[16]

Reliance on the availability heuristic, though cognitive, is problematic because the errors in judgment it leads to may affect human behavior. For example, if a person must estimate the risks involved in a certain activity, such as the handling of tools in a dangerous manner, he may err because his judgment is distorted by elements such as his familiarity with and the salience of the dangers involved in the potential activity. This can lead the agent to seriously misassess how dangerous the activity is and to engage in the activity, though had the agent properly assessed the risks, he would have refrained from acting or taken additional precautions.

Other documented biases, relevant not only to situations of judgment under uncertainty, include biases such as the overoptimism bias. Overoptimism denotes the tendency people have to be overly optimistic with respect to themselves, creating naïve feelings such as "it would never happen to me."[17]

Dr. Neil Weinstein explains that "[s]uch ideas imply not merely a hopeful outlook on life, but an error in judgment . . . ".[18] A wide variety of studies demonstrate this bias and show that whenever assessing probabilities relating to their own lives, people cannot escape the subjective viewpoint through which they judge the world, and tend to overestimate the probability of encountering good outcomes and underestimate the probability of encountering bad outcomes.[19] Bias research has further found that the overoptimism bias persists even where people are fully aware of actual probability distributions relative to the general population.[20] The potential severity of this bias was demonstrated in a study conducted by Baker and Emry, who asked couples applying for marriage certificates about the probability that their union would end in divorce. Although most of the couples questioned were aware that the divorce rate in the United States was close to 50%, the modal response given by the couples in the study was 0![21]

Neuroscientific studies, which have recently begun to work toward understanding the neural correlatives of psychological biases, have begun to investigate the neurological underpinnings of this bias. A study conducted by Tali Sharon and colleagues found that signals in the amygdala, an emotion center of the brain, as well as a region toward the rostral (front) portion of the anterior cingulate (rACC) with which it is connected, were weaker when imagining negative future events than when imagining positive future events. The activities of the rACC and the amygdala were also found to be more strongly correlated with one another when imagining positive future events. Researchers speculate that the selective reduction of activity when imagining possible negative outcomes could be a neural mechanism for the optimism bias, which leads to errors in judgment.[22]

The distortion caused by this bias can result not only in bad judgment but also in bad choices of action. As argued in the following section, the fact that subsequent action is grounded in such cognitive error is morally significant and can serve to undermine the moral responsibility of an agent.

Yet another well-documented cognitive bias relates to the non-normative way in which people adjust their beliefs and judgments based on the introduction of new information. Studies have shown that, in their daily life, people act as lay scientists, constantly forming hypotheses about what has happened, what is happening, and what will happen.[23] These hypotheses are modified as new information becomes available. Logic dictates that when trustworthy evidence is brought that either supports, proves, destabilizes, or refutes a hypothesis, belief in its truth ought to be modified accordingly. Many people recognize this logical reason to modify their beliefs theoretically and when asked, generally agree, for example, that where one discovers that the basis of a belief is false, that belief should be abandoned — "you ought not to go on believing something you no longer have reason to believe."[24] Unfortunately, however, this logical understanding is not unequivocally applied in practical situations, and people exhibit what has been termed "belief perseverance." Studies have found that people generally seek out and exaggerate the significance of evidence that confirms their current hypotheses and beliefs, while disregarding the significance of evidence to the contrary.[25] This bias, however, is not necessarily conscious or intended. Rather, as people move through the world internalizing the information available around them, the effects of this bias are essentially to highlight evidence that stands in support of current beliefs and to downplay evidence that stands in contradiction to such beliefs. Thus, people become irrationally biased toward continuing to believe what they already believe, though evidence to the contrary may be standing right before them.[26] As a result of the effects of this bias, even where a person has access to all relevant information, information processing may be impaired and lead one to misjudge reality.

The effects of this bias can clearly have implications for questions of moral responsibility. In the context of judgment under uncertainty, for example, whenever a person is faced with the task of assessing the probability that a future event will occur, or the potential risks involved in some activity, the person may make an initial assessment based on the information that has been arbitrarily presented first. Often, new data will subsequently become available that ought to significantly change the person's original assessment. However, because of belief perseverance, such modification may never take place or may be seriously understated. The effects of this bias are disturbing because they may influence the subsequent choices and actions taken by an agent, including those relating to precautions taken against risk. As demonstrated later, if moral theory and criminal law ignore the cognitive source of such error and behavior, responsibility may be incorrectly attributed in many cases.

This may have been the case, for example, in *Commonwealth v. Keech*.[27] In *Keech*, the court found that the defendant and driver, Andrew Keech, had turned westbound onto the eastbound section of a highway. At the point where he had turned onto the highway, the eastbound portion of the highway was not visible through dense foliage, and Keech formed a mistaken hypothesis that he was driving on the correct side of the road and, therefore, that he was driving perfectly safely.[28] Keech continued to drive in the wrong direction on the highway for more than 8 miles, over the course of 5 to 7 minutes, during which he passed numerous reversed traffic signs, as well as drivers driving in the opposite direction in adjacent lanes who attempted to call his attention by flashing lights at him and waving. Yet Keech did not properly internalize this information or find that it contradicted his current belief, and responded by simply waving back.[29] Ultimately, Keech collided head on with an oncoming vehicle, not having attempted to slow down before the collision, causing the death of its driver. As a result of his dangerous driving, Keech was found guilty of involuntary manslaughter. Although the court, in all its instances, found that Keech was not aware of the risk his driving created to human life (did not realize he was driving at highway speeds in the wrong direction, creating an unreasonably high likelihood of causing an accident and subsequent injury), his conviction was upheld based on the conclusion that "he should have known of the risk his conduct created" and therefore such knowledge was "chargeable to him." "A callous act of indifference to the safety of others."[30] The court insisted that his conduct was "a callous act of indiff erence to the safety of others."[30] And reflected wanton disregard for life, among others, based on the length of time during which he continued to drive in the wrong direction and the numerous indications he "disregarded" or "ignored" that attested to this fact. Yet, as clarified by Bias Research, where a person is under the impression that a hypothesis is correct, indications to the contrary are not necessarily "rationally" considered — beliefs tend to persevere more than they ought. Thus, although Keech or anyone like him may have realized his mistake after seeing reversed traffic signs had he suspected he was in the wrong lane, such indications failed to make such an impression in a situation in which he had (in his mind) no reason to believe he was in the wrong. Evidence of his belief perseverance is furthered by the fact that upon noticing drivers waving at him to get his attention, he indeed responded, yet not by properly processing the calls for his attention as warnings of danger, but by waving back. His cognitive interpretation of the indications before him produced an erroneous picture of reality, within which his behavior would not have been considered wrong. Thus, owing to such error, Keech is simply not morally responsible for having caused the accident or the death of the other driver, unfortunate as these events may have been. Consequently, as shall be argued later, the law should not have imposed criminal responsibility upon such an agent.[31]

In yet another case in which liability seems to have been imposed excessively as a result of the failure of the law to consider the full implications of bounded rationality, a father of seven was found responsible for the death of his four younger children in a tragic traffic accident.[32] Nigel Gresham, a father and mechanic, had made several repairs to his vehicle, among them to the rear axle. According to his own statements, nothing was wrong with the vehicle, and it had no defects that might contribute to an accident.[33] Nevertheless, when mounting on a verge, the rear axle system fractured, and the car rolled down a bank into a river, causing the death of his four youngest children. The court found Gresham guilty on four counts of causing death by dangerous driving based on the allegations that the repairs made by Gresham were "ill judged" and on expert testimony that the car was left in "very poor condition."[34] The judge stated that, under the circumstances, it would have been "blindingly obvious" that driving the vehicle in its condition "would be dangerous."[35] Yet although this may have been "blindingly obvious" to the court judging the case after the fact or to the hypothetical reasonable person, it was not obvious to Gresham. Nor was any evidence brought alleging that Gresham bore any culpable state of mind: that he was careless, indifferent, or a daring driver. Rather, Gresham seems to have been subject to cognitive biases (namely, over-optimism and overconfidence), which led him to underestimate the likelihood that such a tragic accident could befall him, and overestimate his ability to safely repair the vehicle. These, coupled with the actual indications Gresham had that his vehicle was in fact safe (having passed a road safety test not 6 months before the accident), could have caused this belief to persevere despite any indications to the contrary. Although it is tragic that Gresham drove the car in its condition, he is thus not culpable for having done so and should not have been held criminally accountable for his unfortunately dangerous actions.

Dissenters may argue that people simply "ought not" to allow their current beliefs to persevere, "ought not" be overly optimistic. However, this underestimates the ingrained nature of many of the cognitive mechanisms being discussed. To illustrate, reliance on the availability heuristic creates yet another bias, known as "the effectiveness of a search set." In one of their more famous studies, Kahneman and Tversky asked a group of subjects whether it was more likely that the letter "r" would appear in the first or third position of a random English word. Most subjects replied that it was more likely that the "r" would appear in the first position, whereas in fact there are more English words with the letter "r" in the third position.[36] Kahneman and Tversky explained this discrepancy by theorizing that people approach such tasks by attempting to recall words that fall into each of the two suggested categories. Because it is easier to search for words by their first letter rather than by their third letter, words belonging in the first category will be more easily retrievable, and therefore the category will appear to be more numerous. Because of the availability heuristic, people will assume that more easily retrievable items or events are in fact more frequent or probable, an assumption that often leads them to assess frequency or probability incorrectly.[37] This bias is especially telling because it highlights the fact that the errors resulting from the heuristic and biased nature of human thought are not a matter of personal laziness or carelessness. Rather, the mechanisms of the human mind often prevent a person from judging reality properly, consequently affecting human behavior.

BIAS RESEARCH AND MORAL RESPONSIBILITY

As noted previously, Bias Research has serious implications for questions of moral responsibility. The details of when precisely responsibility will be undermined by cognitive error have been expounded elsewhere and are not the focus of this chapter.[38] However, briefly, it is here suggested that where the errors in judgment caused by heuristic and biased reasoning cause a person to be mistaken about morally significant particulars of her action, and such error leads her to believe she is engaging in morally acceptable behavior, whereas in fact her behavior is morally "wrong" (e.g., creates an unreasonable risk of harm to human life),[39] such person is not morally responsible for having engaged in the action she perceived to be acceptable.[40]

To illustrate, imagine a person who creates an unreasonable risk to human life by his driving and causes an accident and the subsequent death of a fellow driver. Now imagine that the driver had every intention of driving safely, yet because of the heuristic and biased nature of his thought, he judged that the manner in which he was driving to be reasonably safe. Note that his error and subsequent behavior did not arise from any negative moral attitude such as indifference toward human life or carelessness. Rather, had the driver been able to assess the situation correctly (had he had "full" mental rational capacities), he would have realized that such behavior was dangerous and morally unacceptable. Yet, because of the bounded nature of his rationality, he did not so conclude and caused the death of another driver. Such a person, it is contended, is *not* culpable for having driven as he did, nor for the consequences thereof; he is not morally responsible for his actions because the error caused by the bounded nature of his rationality prevented him from being aware of the morally unacceptable nature of his behavior (as was the case in the previously mentioned cases of Keech and Gresham).

Yet one may ask: what is the significance of heuristics and biases within these arguments? If ignorance generally undermines moral responsibility, what is the relevance of one's cognitive limitations where ignorance caused by any other factor will also result in the negation of an agent's moral responsibility?

To understand the significance of heuristics and biases in the context of the discussed conclusions, it must be clarified that ignorance of fact is not *always* exculpatory. Ignorance of the wrong-making features of an act *can* undermine moral responsibility. However, it will not do so in many situations because in many cases in which persons are unaware of the gravity of their actions, their ignorance is not "innocent." Rather, in many cases, though an agent "did not know" about a morally relevant circumstance of his action (such as the harm it might cause to others), it may nonetheless be the case that the person "ought to have known" and thus remains morally responsible for his action despite his ignorance.[41] This will be the case, for example, where it is either relatively simple to avoid ignorance (and thus moral obligations require one to overcome such ignorance) or, even if it is more difficult to do so, where a special duty exists that morally obligates the agent to at least attempt to avoid ignorance. An example of the first situation may be of a doctor who prescribes penicillin to a patient while being ignorant of the fact that the patient is allergic to penicillin. In such case, the doctor was in fact ignorant of a morally relevant circumstance relating to her action. However, if all the doctor had to do to become aware of the patient's allergy was to check the patient's chart, wherein the allergy was clearly noted, the doctor will bear moral responsibility despite her ignorance because in this case, she "ought to have known" about the allergy.[42] Further, even if avoiding ignorance is not quite so simple, where a special duty of care applies, for example, as a consequence of one's position or profession, an agent may still be morally required to avoid ignorance. For instance, even if a patient's chart does not specify any allergies, a doctor may be required to test for allergies before treating a patient to ensure that the chosen course of treatment does not harm her patient. In such cases, although the agent may be ignorant of a morally relevant fact, her ignorance is "morally tainted," or "culpable." Therefore, in none of these cases will her ignorance fully undermine her moral responsibility. (If it did, people could intentionally avoid awareness, thereby consistently avoiding moral responsibility for their actions). Thus, more accurately, only where a person is *faultlessly* ignorant will ignorance fully undermine responsibility.

What is significant about heuristics and biases, then, is that the errors caused by such cognitive features generate *faultless*, or nonculpable, ignorance and thus result in the negation of the agent's moral responsibility. Although a full justification of this argument has been discussed elsewhere and is beyond the scope of this chapter, it is important to at least briefly address this argument.[43] For one could object that agents who err as a result of heuristic and biased reasoning ought not to have erred. Unfortunately, however, the general implication of much of Bias Research is that, in many cases, people do not have the mental capacity necessary to overcome the effects of such cognitive features. The use of heuristics as reasoning tools is viewed as the result of a long evolutionary process of human mental development, wherein the human mind adopted such tools in order to cope with the tasks it was faced with on a daily basis. Yet despite their relative utility, these tools lead to judgmental errors, errors that many people are not equipped to overcome, because such tools are in fact the primary, and often the only, tools available to human reasoning.[44] Thus, heuristic thinking, biased reasoning, and the subsequent errors caused by these are often inescapable. This conclusion is supported by the work of many leading figures in the field of Bias Research. Slovic, Fischhoff, and Lichtenstein, all experts in the field, write, "[i]t appears that people lack the correct programs for many important judgmental tasks... We have not had the opportunity to evolve an intellect capable of dealing conceptually with uncertainty."[45]

This conclusion is further supported with regard to specific biases by a line of research that studies "de-biasing" practices. These studies test whether people can, through a variety of methods, overcome the distortions caused by heuristic and biased reasoning and arrive at more normative judgments. Many of these studies have found that people are generally resistant to de-biasing practices and continue to make faulty judgments about the world as predicted by Bias Research despite myriad attempts to aid them in overcoming their own biased thought processes.[46] For example, Hanson and Kysar conclude that the overoptimism bias "is an indiscriminate and indefatigable cognitive feature, causing individuals to underestimate the extent to which a threat applies to them even when they can recognize the severity it poses to others."[47]

Yet, even if people can overcome judgmental errors in certain cases, this is still not sufficient to argue that a person ought to have known or was faultlessly ignorant. To advance such an argument, it must also be argued that it was *wrong* of the agent not to have avoided ignorance. Morality does not require that we do all that is within our capacities; it only requires that we do that which it would be morally wrong not to do. Yet it is simply not the case that there is a standing moral obligation to do all that is necessary to avoid cognitive error. For example, it has been argued that education in the field of probability may help reduce cognitive error in many situations.[48] However, it seems clear that people are not under a standing moral obligation to educate themselves in statistics or probability theory such that they may generally avoid the errors that might be caused by heuristic and biased reasoning. To claim otherwise would be to imply that every person who has not majored in probability theory—which in and of itself is a developing science and may be faulty—or otherwise educated himself, is in violation of his moral duties.

In sum, it is not the case that a person acting "under the influence" of heuristics and biases *ought to* have known; rather, such agents are faultlessly ignorant and thus not morally responsible for their actions.

BIAS RESEARCH AND LEGAL RESPONSIBILITY—GENERAL

How Do the Above Moral Arguments Affect Questions of Legal Responsibility?

As noted earlier, where legal responsibility is stipulated upon the moral responsibility of the agent, legal responsibility must not be imposed upon agents who acted in faultless ignorance. Although not all agree that criminal responsibility and punishment are or should be based on notions of moral responsibility, this is the view of one major justificatory theory of the criminal law, the theory of culpability-based retributivism. Pursuant to this theory, criminal responsibility and punishment should be imposed based on considerations relating solely to the moral culpability of the agent. This argument is based on a Kantian-type maxim that holds that people should not be treated merely as means, but rather as ends in themselves. Thus, the only proper justification for punishment on this account is the agent's own desert — his own moral responsibility (and not, e.g., the question of deterrence or greater social welfare). It follows that, pursuant to this theory, if we have discovered a range of cases in which agents are not morally responsible for their actions, such as cases in which agents are rendered ignorant due to the heuristic and biased nature of their reasoning as discussed in the previous section, such agents should not be held criminally accountable for their consequent actions.

Yet, one need not accept retributivism as the underlying principle of criminal law to accept such a conclusion. Whereas culpability-based retributivism also implies, for example, that an agent's desert is a *sufficient* condition for responsibility, this premise is not necessary to arrive at the conclusion that where an agent is not morally responsible, she should not be held criminally accountable. To accept such a conclusion, one need only accept that notions of moral responsibility should act as the *limiting* principle of criminal law.[49]

Without venturing too far into the question of the justification of such a principle, it is here contended that the scope of criminal responsibility ought to be restricted in this way. Briefly, it may be noted that the criminal law is an especially injurious area of law, used to justify extreme violations of personal liberty, which are further aggravated by the social repercussions of being found criminally responsible. Therefore, only those persons who bear moral responsibility with respect to their crimes — only those who actually deserve such harsh punishments — should be subject to the harms brought on by criminal liability. Consequently, where the law imposes liability irrespective of the question of moral responsibility, it ought to be modified.

Is the Value of the Limiting Principle Recognized Only by Retributive or Desert Oriented Theories of the Criminal Law?

Even if one rejects the use of desert-based principles as guidelines for criminal law, a similar limitation of criminal liability and thus similar conclusions may be derived from utilitarian principles as well.[50] A simple utilitarian justification of the limiting principle argues that because the public views retribution as the general principle of criminal punishment — because much of the general population believes that the law ought to impose criminal liability based on the fault of the agent — not admitting such a limitation would cause greater harm than good.[51] Jeremy Bentham argued, on other utilitarian grounds, that moral-like excuses should be admitted into the criminal law, generally limiting the imposition of criminal responsibility. However, he diverged in arguing that the justification for entertaining such excuses (e.g., duress, insanity, mistake) was the fact that in such cases, agents cannot be deterred by the threat of future punishment and, therefore, there would be no utilitarian value in punishing them. Punishment in such cases, he writes, "must be inefficacious" because "it cannot act so as to prevent the mischief."[52] H.L.A. Hart disagrees with Bentham's argument,[53] yet he too offers a non-retributive justification of a similar limitation of criminal responsibility. According to Hart, considerations of fairness and justice to individuals demand that punishment be restricted only to those who had a normal capacity and fair opportunity to obey the law:[54]

> Thus a primary vindication of the principle of responsibility could rest on the simple idea that unless a man has the capacity and a fair opportunity or chance to adjust his behavior to the law its penalties ought not to be applied to him. Even if we punish men not as wicked but as nuisances, this is something we should still respect.[55]

Thus, the limitation of criminal responsibility and punishment to cases in which the agent may be said to be morally responsible or at fault is a principle justified by both retributive and utilitarian theories alike.

BIAS RESEARCH AND LEGAL RESPONSIBILITY—THE REASONABLE PERSON STANDARD

In view of the limiting principle, it is clear that persons who are faultlessly ignorant must not be subject to criminal responsibility. Following Bias Research, those acting "under the influence" of heuristics and biases must be recognized as faultlessly ignorant and therefore undeserving of criminal responsibility.[56] This understanding is not only generally significant to criminal law but has implications for specific legal doctrines as well. One such doctrine is the reasonable person standard imposed in negligence offense at common law. In negligence offenses, defendants are held accountable for substantially deviating from a reasonable standard of conduct, where although they were unaware of some relevant element or consequence of their activity, it is deemed that they acted negligently because they ought to have been aware and thus ought to have acted otherwise. Yet the question of their negligence (mens rea) is measured, not individually, but rather objectively, against a "reasonable person" standard.[57] The question posed by the court is never: "Could the defendant have known and should she have known?" Rather, the question posed is whether a reasonable person in the position of the defendant would have known, where the ability of the reasonable person to have known dictates the normative judgment that the defendant herself should have known and is thus responsible.

The use of objective standards of responsibility in criminal law is clearly problematic from the point of view of the limiting principle because they impose responsibility regardless of the question of individual fault. Nonetheless, arguments have been raised in support of objective standards. One strategy for the justification of objective standards argues that they improve individual behavior and thus benefit society. On this view, the reasonable person standard is *intended* to hold persons up to a desirable standard of conduct regardless of the question of their culpability so as to ensure that they do not fall beneath a certain level of conduct. Evaluated against the limiting principle, however, this "justification" must clearly be rejected. Sacrificing the individual for the good of society in such a way as to impose criminal sanctions even upon the morally innocent is simply unacceptable under this view.

This regulative argument, however, is not the only view that has been put forth in support of the reasonable person standard. Other views, including those that respect the limiting principle and would adhere to it as long as is realistically possible, argue that the reasonable person standard is nonetheless justifiable in light of pragmatic, institutional considerations that necessitate its utilization. According to these views, the objective reasonable person standard is imposed in the place of a more desirable subjective standard of responsibility in order to overcome practical difficulties. For example, the standard might be intended to allow for greater efficiency in an area where proof is difficult: because it is difficult to ascertain the precise knowledge and abilities of each individual agent, a figure is constructed who represents a standard member of society, who has the basic rational capacities of such a member and any knowledge that such a member of society can be expected to have. On this view, the standard is intended, not as a regulative construct, but rather as a generalization over society. Thus, the reasonable person standard is expected to produce generally satisfactory results even from a fault-based point of view.

Bias Research, however, sheds new light on these arguments. It exposes the justifications raised in support of the reasonable person standard to be unsubstantiated, and indicates that the sacrifice made by utilizing such objective standards is much higher than may previously have been believed. Thus, the limiting principle, taken together with the absence of overriding considerations, begs the conclusion that the current reasonable person standard employed in negligence offense cannot be maintained.

Undermining the Justifications Raised in Support of the Reasonable Person Standard

One often-raised objection to the individualization of culpability is that the law, as an institution, cannot afford to accommodate all of the minor, idiosyncratic impairments people suffer from when regarding questions of responsibility. Realistically, it is argued, if the law is to be efficient, it must generalize even if at the expense of an idiosyncratic minority whose unique qualities would render its members free from criminal responsibility had the law evaluated their culpability individually. And in much of the literature discussing objective standards, even in that which opposes such standards, those cases in which "nonstandard" individuals are sacrificed and punished despite their innocence (despite their "diminished capacity") are in fact viewed as "unusual cases," as described by George Fletcher,[58] which affect only "some unfortunate individuals," as described by Hart.[59]

Bias Research, however, clearly demonstrates otherwise. It establishes that the "diminished capacity" caused by such heuristics and biases is neither idiosyncratic nor the unfortunate calamity of a select few. These are common cognitive deficiencies that persistently plague human judgments and beliefs. Bias Research essentially reveals that, given the prominence of cognitive error among the general population, the sacrifice made by objective standards is much greater than previously appreciated. Thus, the claim that for the sake of institutional efficiency, the law is justified in ignoring the few, idiosyncratic cases of "diminished capacity" is exposed to be false.

A second objection voiced against the individualization of responsibility is that the law should only take gross incapacities and irrationalities into account, such as infancy or insanity, and should not relate to minor deviations from rationality.[60] Justice Holmes writes that standards of criminal law "take no account of incapacities, unless the weakness is so marked as to fall into well known exceptions such as infancy or madness"; unless "a man has a distinct defect of such a nature that all can recognize it as making certain precautions impossible."[61] If this line of argument is applied to heuristics and biases and the ignorance and mistakes caused by such cognitive features, regardless of the commonness of such errors, it would follow that they may be disregarded. However, it is unclear why this argument is justified. If the reason amounts to a question of proof — that it is easy to discover when a person suffers from full fledged insanity, whereas it is more difficult to demonstrate slighter deviations from the "norm" — this is a plausible argument, which will be discussed next. Otherwise, there is no underlying principle that justifies recognizing gross, though not less severe, deviations from "rationality." The fact that the common law, which once limited exculpatory ignorance to cases of infancy and insanity, has since abandoned this requirement in knowledge offense, allowing ignorance caused by less extreme sources of irrationality to undermine responsibility, indicates that this argument is in fact not accepted by modern legal systems. Yet, it is here suggested that the common law failed where it refrained from further abandoning such requirements with respect to negligence offenses as well.

A more serious objection to the individualization of criminal responsibility is that the reasonable person standard allows the law to overcome evidentiary difficulties. According to this argument, proving precisely what a defendant did or could have known is extremely difficult. Consequently, to simplify the judicial process and overcome the difficulties involved, an objective standard is employed. Under this standard, once it is established that a reasonable person would have and therefore could have known, it is not necessary to delve into the mental state of the defendant and prove that he could have and therefore should have known.[62]

This argument, however, is outdated and ignores the simple fact that the cognitive research discussed in this paper provides precisely the type of evidence that can aid in overcoming such difficulties of proof. One of the most important contributions of the research conducted by Kahneman and Tversky is that it clearly demonstrates that errors caused by heuristics and biases are both systematic and predictable.[63] Although admittedly, this research cannot prove completely that a specific individual was so affected in a specific situation, it can provide strong evidentiary support for such arguments, demonstrating that the average person would in fact not have known and establishing the types of situations in which distortions of judgment are likely to be exacerbated.[64]

Furthermore, the best indication that the problem of proof is not so severe as to be disabling is the current state of negligence offenses in much of continental criminal law. German law, for example, does not restrict exculpatory mistakes in negligence offenses to only those mistakes that are reasonable, and does evaluate culpability on an individual basis, considering the capacities and incapacities of the individual agent.[65] Yet, this task has yet to wreak havoc in continental criminal law.[66]

In sum, in light of Bias Research, arguments brought in support of the reasonable person standard imposed in negligence offenses prove to be groundless. Therefore, it is contended that the standard, violating the limiting principle, must not be maintained.

Rather, in light of the limiting principle, the objective reasonable person standard should be removed from the criminal law and replaced with an individualized standard of responsibility, which evaluates the subjective culpability of each agent. Under such a standard, the court would be asked to pose the question of whether the individual agent herself ought to have known, as opposed to its current objective evaluation of the question of whether a hypothetical construct would have known and acted otherwise. Such an evaluation would mimic the individualized evaluation of mens rea currently in place in knowledge offenses, wherein courts judge whether the *individual agent* was or was not aware of some fact under the circumstances, and not whether a reasonable person would have been aware (though in the past, knowledge offenses were also evaluated in this way).[67]

If it is demonstrated, however, that the law as an institution truly cannot accommodate a fully subjective evaluation of responsibility in negligence offense,[68] the reasonable person standard itself must *at least* be reconstructed to better reflect actual human capacities so as to minimize the violation of the limiting principle. A new, "boundedly reasonable person" must be employed, which incorporates the findings of Bias Research into evaluations of responsibility. Such a standard, though not ideal, would nonetheless result in far more realistic and "reasonable" expectations (of human knowledge, foresight, and behavior), thus calming much of the criticism directed against the current idealized standard.[69]

POTENTIAL OBJECTIONS

Reasonable Though Already Boundedly Rational?

The previous criticism of the reasonable person standard presumes that the current standard does not already take into account the bounded nature of human rationality. Yet, some may protest that perhaps the current reasonable person standard already accommodates the types of errors demonstrated by Bias Research. This argument, however, fails to take into account a number of crucial elements. First, in many cases, courts do not even intend to employ the reasonable person standard as a measure of "typical error," but rather employ it as a regulative tool that strictly requires a certain level of conduct (and thus knowledge or foresight) from every agent.[70] Second, even where the courts aim to employ a more descriptive version of the reasonable person standard, the character of the reasonable person is often viewed as that of an upstanding citizen, interested in the good of society and the welfare of others.[71] Thus, when questioning whether an "oversight" of a defendant was reasonable or not, the court will be asking whether the typical *upstanding* citizen would have made such an error, immediately excluding all error attributable to nondesirable attitudes (e.g., indifference or carelessness). Yet, as noted previously, Bias Research demonstrates that many of the cases that may have been previously assumed to involve fault-based error are actually cases of faultless error rooted in the bounded nature of human rationality.

Finally, even if a court *intends* and attempts to reflect common error, it may nonetheless overestimate human capabilities. This is because of the effects of a bias that distorts the reasoning processes of the *judges and juries* who evaluate agents' actions after the fact — the "hindsight bias." Studies demonstrating the hindsight bias illustrate that knowledge of the fact that an event has occurred causes estimations of the *ex-ante* likelihood that the event would come to pass to be higher than they would have been had such information not been available.[72] As articulated by Fischhoff, "in hindsight, people consistently exaggerate what could have been anticipated in foresight."[73] Consequently, the courts, who judge situations after the fact, knowing that the defendants' activities did in fact cause harm to others, will tend to incorrectly assume that many of the mistakes made by defendants would not have been made by the ordinary person. Thus, courts will tend to attribute the "ordinary person" with greater knowledge and foresight than such an agent would actually have had.[74] The reasonable person standard will therefore result in the excessive imposition of liability, despite the court's attempt to accurately estimate what an ordinary person would have known or foreseen. For all these reasons, the bounded nature of human rationality is unfortunately not incorporated "naturally" into the reasonable person standard.

The Moral Obligation to Take an Extra Measure of Care

Some might suggest that although heuristics and biases cause faultless ignorance, if people are aware that they are boundedly rational or are aware of the fact that they often err in their judgments and estimations, they are under a secondary moral obligation to "err on the safe side of caution" — to be more careful than their judgments suggest they should be. Even if an agent is unaware of all the risks involved in her activities, if she *knows* she may be unaware of such risks, she is morally obligated to take extra precautions so as to take account of such potential error. Thus, if she does not do so, she is culpable for violating this secondary moral obligation.

The foremost problem with this argument, however, is that it assumes that people are aware of their cognitive deficiencies or of their tendency to err. Many people, however, are not so aware and thus are not responsible for violating such a moral obligation. One may want to argue that even if people are not aware, perhaps they ought to be aware; however, it is not clear what this would mean. The Roman jurist Gauis once argued that an incompetent muleteer, who is unaware of his infirmity, is nonetheless (legally) responsible if he causes an accident because he ought to have known about his own incompetence.[75] In response, Tony Honoré writes:

> If the muleteer knew he was incompetent, fault was indeed, by our ordinary criteria of judgment, present... But if he did not appreciate his own

> incompetence, what is the force of the assertion that he "should have known" of it? Perhaps he was too obtuse to realize his own infirmity, and no one pointed it out. In his own eyes he was an ace muleteer. If he then has to pay for the accident he is paying for his defective makeup, not for fault.[76]

Bias Research itself indicates that ignorance of one's tendency to err may in fact be built in (and thus nonculpable). It demonstrates that another bias many people are subject to is the overconfidence bias, which leads agents to believe that they are more capable than they actually are. The overconfidence bias may even cause agents to "misremember" their own *ex ante* judgments and believe that they accurately predicted future outcomes, although in fact they did not.[77] Thus, many people are cognitively biased against recognizing their own deficiencies. It is therefore problematic to claim that they ought to have been aware of their tendency to err or that they are responsible for not having done so.[78]

The Need to Motivate Desirable Behavior and Reduce Social Harms

Moral considerations aside, imposing an idealized standard of responsibility may have certain positive consequences. For example, it may motivate persons to take more care than they otherwise would have in their daily activities: if people are aware that they will be held up to a standard that exceeds their capacities, and punished according to such a standard, this may cause them to take precautions that they otherwise would not have taken.[79] Clearly, such a regulative standard cannot be reconciled with the limiting principle. Yet it is difficult to ignore the need to motivate socially desirable behavior that is met by utilitarian conceptions of the criminal law.

Here again, Bias Research may be useful in finding a solution to the discussed problem. Bias Research not only illuminates that agents often faultlessly err, it can also *predict* that in certain situations, agents will err. With this information in hand, it is possible to efficiently encourage desirable behavior by appropriately formulating *ex-ante* regulation that requires specified, desirable behavior in those situations in which it is clear that left to their own estimations, people will tend to err. This is yet another legal implication of Bias Research: in the field of *ex-ante* legislation, it implies that specified regulation *should* be imposed where such problems are foreseen and that matters should not be left up to human discretion, as is done in the case of generalized statutes that require "reasonable" conduct.

In fact, legislators are not the only ones implicated. Any body that has the capacity to minimize social risks and harms should utilize Bias Research.

Thus, those who design the layout of the social environment (e.g., city designers, road builders) should apply these findings to optimally prevent common error.

In all these manners, the goal of promoting desirable behavior and reducing the undesirable risks and harms may be achieved without violating the limiting principle. In any event, the method of imposing criminal responsibility upon faultlessly mistaken agents so as to enhance socially desirable behavior is not an acceptable means by which to achieve these goals.

CONCLUSION

The findings of the heuristics and biases tradition of cognitive psychology demonstrate that people are not as "rational" as may have been assumed in many contexts. This chapter suggests that two such contexts, the moral and the legal, must incorporate new understandings about the bounded nature of human rationality. In the moral arena, it must be recognized that bounded rationality limits the possibility of moral responsibility and undermines the responsibility of agents who are rendered ignorant of morally relevant particulars by the heuristic and biased nature of their thought processes. In the legal arena, following the discoveries of Bias Research, Hanson and Kysar have suggested:

> Ultimately, any legal concept that relies in some sense on a notion of reasonableness or that is premised on the existence of a reasonable or rational decisionmaker will need to be reassessed in light of the mounting evidence that a human is a "reasoning rather than reasonable animal."[80]

The current reasonable person standard imposed in criminal offense is one such doctrine. Idealizing away common cognitive limitations, it is exposed to be an entirely unjustified standard of criminal responsibility that must be modified or removed from the criminal law.

The legal implications of heuristics and biases research are thus far and wide. They demand the rejection of objective standards of liability, such as the reasonable person standard of culpability imposed in negligence offenses. Mordechai Kremnitzer notes that adopting subjective standards may have further implications for criminal law: in the event that negligence offenses no longer disregard the fault of the agent, it will no longer be necessary to restrict the imposition of liability for negligence or to relate to it as an irregular form of criminal responsibility, allowing the scope of negligence offenses in the criminal law to be appropriately broadened.[81] In addition, Bias Research has further implications with respect to desirable legislative policy: in the ongoing debate regarding the question of whether legislation regulating conduct should come in the form of specified rules or generalized standards, such research provides additional support for imposing rules as opposed to standards (both in and out of criminal law), especially where legislation is aimed at preventing undesirable risks and harms.[82] Once again, however, where this is not possible or where this is not currently the state of the law, responsibility must be imposed only where the agent may be said to be morally responsible or at fault and, therefore, never in cases in which one's bounded rationality undermines such moral responsibility.

NOTES

1. JSD candidate at Yale Law School and graduate student at the Hebrew University of Jerusalem Philosophy Department. Special thanks to Alon Harel and David Enoch for the extensive feedback they offered on these ideas, and to Yaron Dahan, Michal Dahan and Aaron Katz for their helpful comments.
2. This is not always the case. For example, incompatibalists who accept determinism argue that people may not be subject to moral responsibility.
3. Aristotle, *Nicomachean Ethics* 1110a17–18, 1111a22–24, 1113b20–21; *Eudemian Ethics* 1224a 8–1225b 10. In the Eudemian Ethics, both force and compulsion are viewed as rendering an action involuntary. This is also the opinion expressed in the Magna Moralia (although this work is no longer unanimously attributed to Aristotle). In the Nicomachean Ethics, on the other hand, force is viewed as an element that renders actions involuntary, whereas compulsion is not. See Irwin 1980:120–122. Irwin argues that they are not posed as sufficient conditions based on the argument that these conditions alone would allow both animals and children to bear moral responsibility for their actions, a conclusion that Irwin is unwilling to attribute to Aristotle. *Ibid.*, 165, 178. This indicates to Irwin that Aristotle's discussion of voluntary action is actually concerned with virtue and may not be concerned with moral responsibility at all.
4. Note that a more contemporary analysis of the case of force would argue that where physical force is involved, the agent has in fact *not acted*. Thus, there is no action for which the agent may be said to be morally responsible (rather than positing that there is an action, yet one that was performed involuntarily). Both analyses, however, invite concluding absence of responsibility.
5. Subject to the nonexistence of other conditions of responsibility (e.g., not having a gun to his head perhaps changes things).
6. One could repaint the situation in such a way as to undermine this conclusion: perhaps the host had himself placed the poison in the whiskey bottle, but had not bothered to properly label it, thus himself creating the risk that the bottle may at some later date be mistaken for an innocent beverage. Perhaps he was negligent in his behavior. These scenarios would be ones in which although he did not "knowingly" pour his guest poison, it would still be appropriate to view him as culpable, although perhaps not to the same extent (or for the same action) as he would have been had he poured the poison intentionally.

7. Rosen 2008 discussing the distinction between cognitive and volitional conditions.
8. A term coined by Herbert Simon (Simon 1957).
9. The arguments of this paper are also generally applicable to objective standards such as the reasonable person standard imposed in the self-defense excuse.
10. Kahneman and Tversky 1982:3–20.
11. Rational or normative reasoning is that which accords with norms derivable from valid logical systems such as deductive logic and probability theory. Botterill and Carruthers, 1999:128.
12. Kahneman and Tversky, *supra* n. 10, 11.
13. Kahneman and Tversky, *Ibid.*, 51.
14. See Tversky and Kahneman 1973.
15. *Ibid.*
16. *Ibid.*
17. Korobkin and Ulen 2000; Hanson and Kysar 1990:656.
18. Weinstein 1980:806.
19. Hanson and Kysar, *supra* n. 17, 658.
20. Yates 1990:203.
21. Baker and Emery 1993:439–450.
22. Sharot et al. 2007:102–115.
23. Hanson and Kysar, *supra* n. 17, 647.
24. Botterill and Carruthers, *supra* n. 10, 108.
25. Hanson and Kysar, *supra* n. 17, 647.
26. In other words, "[f]irst impressions carry disproportionate weight." Hanson and Kysar, *supra* n. 17, 647.
27. 9 Va. App. 272, 1989.
28. *Ibid.* 274.
29. *Ibid.* 275.
30. *Ibid.* 282.
31. Irrespective of this fact, it may have been the case that the law required the imposition of such responsibility.
32. R. v. Gresham [2009] EWCA Crim 1005, par. 4, citing R. v. Gresham T2008 7130 Lincoln Crown Court, 18/12/2008 (Unpublished).
33. This statement was made in a pretrial interview. See "Land Rover Crash Dad's Trial," http://www.horncastlenews.co.uk/news/Land-Rover-crash-dads-trial.4774764.jp.
34. *Ibid.* pars. 7, 12.
35. *Ibid.* par. 12. Note that the objective standard employed in the offense of "causing death by dangerous driving" is the standard of the "careful and competent" driver. Nonetheless, the mens rea is at its base equivalent to the reasonable person standard (actually demanding a higher level of negligence), which is alternatively satisfied if the defendant was subjectively aware of the "dangeroursness." See Road Traffic Act 1991, s 2A(3).
36. Kahneman and Tversky, *supra* n. 10, 12; Yates, *supra* n. 20, 187.
37. *Ibid.*
38. See Dahan-Katz.

39. Of course, there will always be great debate as to which actions are morally wrong, for example, as to which risks are too great and which risks are morally reasonable under given circumstances. Regardless of the exact distinction between right and wrong, reasonable and unreasonable, however, what is significant to this paper is that such a distinction exists, and that its positioning dictates the moral status of a given action or agent.
40. An action may appear to be acceptable as a result of one of two different reasons: (1) incorrect assessment of actual facts or probabilities involved, as discussed previously; or (2) disagreement as to the nature of a given action—where others view it to be unacceptable, the agent's different moral sensibilities lead him to conclude that the very same action is acceptable. The paper deals with the first case—a case of ignorance of fact—but does not deal with or argue for a moral exemption in the second—a case of moral ignorance.
41. Although the agent may be less culpable than agents who engaged in the morally undesirable act knowingly or intentionaly (or culpable for a different action — his previous action of having knowingly entered into a situation of ignorance), he will nonetheless fail to fully escape moral responsibility owing to the culpable nature of his ignorance.
42. This example is borrowed from Rosen, *supra* n. 7, 593–594.
43. Dahan-Katz, *supra* n. 37.
44. Kahneman and Tversky make this claim with respect to the untutored mind relative to situations of judgment under uncertainty.
45. Slovic et al. 1976:174.
46. See, for example, Weinstein and Klein 1995:138; Rachlinski 1998:573.
47. Hanson and Kysar, *supra* n. 17, 658–659. In fact, based on additional studies, they further conclude that under certain circumstances, careful reasoning may even *foster* overconfidence. *Ibid.*
48. For other arguments, see Dahan-Katz, *supra* n. 38.
49. This view is otherwise known as "negative retributivism." See Duff 2009.
50. See Murphy and Coleman 1990:73, 122.
51. Robinson 2009 notes, however, that such a justification would result in the limitation of the law to social views of desert as opposed to "justice" or "desert" in the transcendent sense.
52. Bentham 1823:3, 5–10 (chap. XIII).
53. See Hart 1968:19.
54. *Ibid.* See 22–24, 181–182, 201.
55. *Ibid.* 181.
56. Elsewhere, I have raised the concern that in some cases, morally non-neutral characteristics may manifest themselves as biases, complicating the scope of this argument. Dahan-Katz, *supra* n. 38.
57. §2.02(d) of the Model Penal Code may possibly leave space for some individualization. See Hornle 2008:24.
58. Fletcher 2000:708.
59. Hart, *supra* n. 53, 154.
60. Hart, *Ibid.* 155.
61. Holmes 1991:51, 109, respectively.

62. Hart, *supra* n. 53, 33.
63. Kahneman and Tversky, *supra* n. 10.
64. Note that this can be the case with respect to offenses requiring any form of knowledge, wherein the defendant wishes to substantiate the existence of an honest (and faultless) mistake.
65. *See* Hornle, *supra* n. 57; Fletcher, *supra* n. 58.
66. Even if the problem of proof is substantial, at the very most, this justifies a *presumption* that the defendant has the capacities of a reasonable person. It does not justify barring the defendant's opportunity to present evidence to the contrary.
67. Perkins 1939:35.
68. I tend to doubt this argument, especially considering the fact that other legal systems, such as German law, do not employ the objective standard even in negligence offense.
69. Readjusting the reasonable person standard to reflect a less capable figure with cognitive deficiencies may be to set the standard too low, allowing persons with above-average cognitive capacities to benefit from the lesser capacities of others.
70. See, for example, Holmes' discussion of the reasonable person standard, which requires levels of conduct and awareness "at the peril" of the agent. Holmes *supra* n. 61, 109.
71. Among others, the reasonable person has been referred to as "just and reasonable," and "ordinarily prudent." See Dressler 1982; see also C.A. 5604/94 State of Israel v. Chemed P.D. 48(2), 498; Ripstein 2001:7.
72. Fischhoff 1975; Korobkin and Ulen, *supra* n. 17, 1095–1096.
73. Fischhoff 1982:341.
74. See Fischhoff, *supra* n. 72; Kamin and Rachlinski 1995.
75. Digest 9.2.8.1 (Gaius 7 ed. prov.). In: Honoré 1999:21.
76. Honoré, *Ibid.*
77. Fischhoff and Beyth 1975:1–16.
78. Unpredicted negative outcomes may not even indicate error; they may merely be the unfortunate manifestation of a perfectly reasonable risk.
79. Note that extra precautions are not always desirable even from a utilitarian point of view (overdeterrence).
80. Hanson and Kysar, *supra* n. 17, 635. Quoting Alexander Hamilton.
81. Kremnitzer 1994:94.
82. For a discussion of the advantages and disadvantages of rules vs. standards, see Mautner 1988:321–352.

REFERENCES

Aristotle (1992). *Eudemian Ethics: Books I, II, and VIII.* Michael Woods trans. New York, Oxford University Press.

Aristotle (1998). *The Nicomachean Ethics.* J. L. Akrill and J. O. Urmson eds., David Ross trans. New York, Oxford University Press.

Baker, L., and R. Emery (1993). "When every relationship is above average: Perceptions and expectations of divorce at the time of marriage." *Law and Human Behavior* 17: 439.

Bentham, J. (1823). *Introduction to the Principles of Morals and Legislation, Volume II.* London: W. Pickering.

Botterill, G., and P. Carruthers (1999). *The Philosophy of Psychology.* Cambridge, UK, Cambridge University Press.

C.A. 5604/94 State of Israel v. Chemed P.D. 48(2). [Israel Court of Appeals]

Commonwealth v. Keetch 9 Va. App. 272, 1989.

Dressler, J. (1982). "Rethinking heat of passion: A defense in search of a rationale" *Journal of Criminal Law & Criminology* 73: 421.

Duff, A. (2009). "Legal punishment." *Stanford Encyclopedia of Philosophy.* Retrieved December 6, 2009, from http://plato.stanford.edu/entries/legal-punishment.

Fischhoff, B. (1975). "Hindsight =/ foresight: The effect of outcome knowledge on judgment under uncertainty" *Journal of Experimental Psychology* 1: 288–299.

Fischhoff, B. (1982). "For those condemned to study the past: Heuristics and biases in hindsight." In: *Judgment Under Uncertainty.* D. Kahneman, P. Slovic, and A. Tversky. New York, Cambridge University Press, pp. 422–444.

Fischhoff, B. and R. Beyth (1975). "'I Knew It Would Happen:' Remembered Probabilities of Once-Future Things" *Organizational Behavior and Human Performance* 13: 1–16.

Fletcher, G. (2000). *Rethinking the Criminal Law.* Oxford, UK, Oxford University Press.

Hanson, J. D., and D. A. Kysar (1990). "Taking behaviorism seriously." *N.Y.U. Law Review* 74: 630.

Hart, H. L. A. (1968). *Punishment and Responsibility: Essays in the Philosophy of Law.* Oxford, UK, Clarendon Press.

Holmes, O. W. Jr. (1991). *The Common Law.* New York, Dover Publications.

Honoré, T. (1999). *Responsibility and Fault.* Oxford, UK, Hart Publishing.

Hornle, T. (2008). "Social expectations in the criminal law: The 'reasonable person' in a comparative perspective." *New Criminal Law Review* 1: 1.

Irwin, T. (1980). "Reason and responsibility in Aristotle." In: *Essays on Aristotle's Ethics.* A. O. Rorty. Berkeley, CA, University of California Press.

Kahneman, D., and A. Tversky (1982). "Judgment under uncertainty: Heuristics and biases." In: *Judgment Under Uncertainty: Heuristics and Biases.* D. Kahneman, P. Slovic, and A. Tversky. (Cambridge, UK, Cambridge University Press.

Kamin, K. A., and J. J. Rachlinski (1995). "Ex post? Ex ante: Determining liability in hindsight." *Law & Human Behavior* 19: 89.

Korobkin, R. B., and T. S. Ulen (2000). "Law and behavioral science: Removing the rationality assumption from law and economics." *California Law Review* 88: 1051.

Kremnitzer, M. (1994). "On criminal negligence: A mental element, a factual element or both?" *Mishpatim* 24: 88. [Hebrew Text].

"Land Rover Crash Dad's Trial." Available at: http://www.horncastlenews.co.uk/news/Land-Rover-crash-dads-trial.4774764.jp.

Leora Dahan-Katz, *The Implications of Heuristics and Biases Research on Questions of Moral and Legal Responsibility* [On Record with Author].

Mautner, M. (1988). "Rules and standards: Comments on the jurisprudence of Israel's new Civil Code." *Mishpatim* 17: 321. [Hebrew Text].

Murphy, J. G., and J. L. Coleman (1990). *Philosophy of Law: An Introduction to Jurisprudence*. Boulder, CO, Westview Press.

Perkins, R. M. (1939). "Ignorance and mistake in criminal law." *University of Pennsylvania Law Review* 88: 35.

R. v. Gresham [2009] EWCA Crim 1005.

R. v. Gresham T2008 7130 Lincoln Crown Court, 18/12/2008 (Unpublished).

Rachlinski, J. J. (1998). "A positive psychological theory of judging in hindsight," *University of Chicago Law Review* 65: 571–625.

Ripstein, A. (2001). *Equality, Responsibility and the Law.* Cambridge, UK, Cambridge University Press.

Robinson, P. H. (2009). "Empirical desert." In: *Criminal Law Conversations.* P. Robinson, S. Garvey, K. K. Ferzan. Oxford, UK, Oxford University Press.

Rosen, G. (2008). "Kleinbart the oblivious and other tales of ignorance and responsibility." *Journal of Philosophy* 105(10): 591–610.

Sharot, T., A. M. Riccardi, C. M.Raio, and E. A. Phelps (2007). "Neural mechanisms mediating optimism bias." *Nature* 450: 102–115.

Simon, H. A. (1957). *Models of Man: Social and Rational.* New York, John Wiley and Sons.

Slovic, P., B. Fischhoff, and S. Lichtenstein (1976). "Cognitive processes and societal risk taking." In: *Cognition and Social Behavior.* J. S. Carol and J. W. Payne. New York, Erlbaum.

Tversky, A., and D. Kahneman (1973). "Availability: A heuristic for judging frequency and probability." *Cognitive Psychology* 5: 207.

Weinstein, N. D. (1980). "Unrealistic optimism about future life events." *Journal of Personality & Social Psychology* 39: 806.

Weinstein, N. D. (1987). "Unrealistic optimism about susceptibility to health problems: Conclusions from a community-wide sample." *Journal of Behavioral Medicine* 10: 481.

Weinstein, N. D., and W. M. Klein (1995). "Resistance of personal risk perceptions to debiasing interventions." *Health Psychology* 14: 132–140.

Yates, J. F. (1990). *Judgment and Decision Making.* Englewood Cliffs, NJ, Prentice Hall.

7

Moral Responsibility and Consciousness

Two Challenges, One Solution

NEIL LEVY

Does moral responsibility require consciousness? Most philosophers have apparently assumed that it does, if only implicitly. In the wake of the automaticity revolution in social psychology, however, this is an assumption in need of defense. The automaticity revolution has been characterized by a growing sense among psychologists that conscious processes are epiphenomenal. As John Bargh, one of the leaders of this latest revolution, puts it, "there ultimately is no future for conscious processing in accounts of the mind, in the sense of free will and choice" (Bargh 1997:52). Bargh's conclusion is based on a substantial body of evidence showing that functions once thought to be the exclusive preserve of consciousness can in fact be accomplished by nonconscious processes. Similarly negative conclusions about consciousness have been reached by Benjamin Libet (1999) and Daniel Wegner (2002) on other grounds: Libet on the grounds that conscious processes are too slow to initiate action; Wegner on the grounds that the conscious experience of willing dissociates from actually acting.

Whether consciousness is epiphenomenal is ultimately an empirical question, not one that philosophers are especially well placed to answer. There is good reason to think, however, that the enthusiasts for epiphenomenalism have overplayed their hand. Even if—and this is a very big if—it can be established that consciousness is not directly involved in the initiation of action, there are other ways in which it might matter to free will and moral responsibility. In particular, there is extensive evidence that it is important for the integration of information encoded in different regions of the mind (Moresella 2005). Nothing in the evidence presented by Bargh, Libet, Wegner, and other proponents of epiphenomenalism conflicts with this theory of the role of consciousness (Baumeister and Masicampo 2010; Baumeister, Masicampo, and Vohs 2011).

It might be doubted, however, whether this integrative function is enough to establish that moral responsibility requires consciousness. In this chapter, I argue that it is: consciousness is (normally) crucially important to morally responsible action. In so doing, I will sketch a reply to a number of philosophers who, quite independently of the work of the psychologists and neuroscientists mentioned, have challenged the assumption that consciousness is necessary for moral responsibility. The challenge focuses on what we need to be conscious *of* in order to be morally responsible: the philosophers I have in mind deny that we need to be aware of the facts that make our actions morally significant in order to be morally responsible for them. I call the claim that we need to be aware of these facts in order to be morally responsible for our actions the *awareness thesis*; I argue that the awareness thesis is true. In doing so, I hope also to assemble the elements of a proper response to the empirically driven epiphenomenalists: by the end of this chapter, we shall, I hope, be able to see that their claims are unthreatening to moral responsibility even if they are true.

I present two separate, but related, defenses of the awareness thesis, one directed at each of the two major accounts of moral responsibility in the literature. The first, directed at the Strawsonian or reactive attitudes account of moral responsibility, is designed to show that actions caused by our unconscious attitudes do not express our good or ill will toward others, and that therefore we ought not to hold agents responsible for such actions if this account of moral responsibility is correct. The second, addressed to what I call the *evaluative account* of moral responsibility (according to which agents are morally responsible for actions that express their evaluative agency) is designed to show that actions caused by unconscious states do not express our evaluative agency.

CHALLENGES TO THE AWARENESS THESIS

Traditionally, philosophers have distinguished two conditions the satisfaction of which are jointly sufficient for moral responsibility: a *control* condition and an *epistemic* condition. Most discussion has centered on the control condition, as a consequence of the almost exclusive focus on the apparent threat to free will and moral responsibility from causal determinism. If causal determinism is true, some have worried, agents do not have the right kind of control over their actions. Responses to this worry, and sophisticated elaborations of it, have been so central to the free will debate that it is often just called the *free will and determinism debate*. In the meantime, the epistemic condition has been relatively neglected. Most philosophers have simply assumed that agents need to be aware of what they are doing and why in order to be morally responsible for their actions. To deal with obvious counterexamples—what, for instance, of the man who gets behind the wheel of a car while too drunk to appreciate what he is doing?—they introduce a tracing condition; agents are *directly* morally responsible for actions regarding which they knew what they were doing, and *indirectly* responsible for the foreseeable consequences of actions for which they were directly responsible.

The new attack on the awareness thesis is not so easily dealt with. Consider three representative examples from the recent philosophical literature:

1. Huck Finn, Nomy Arpaly (2002) argues, is not conscious of the facts that entail he ought to help Jim escape from slavery. Quite the contrary; he is conscious of what he takes to be the moral requirement that he turn Jim in. Yet he aids Jim's escape, and does so for the right reasons. He is morally praiseworthy for his action; therefore, we need not be conscious of the facts that make our actions right in order to be praiseworthy for them.
2. Agents who forget their friends' birthday and therefore omit to offer them good wishes cannot be conscious of the facts that make their actions wrong, as Angela Smith (2005) reminds us. Yet they can be blameworthy for their omissions. Hence, we need not be conscious of the facts that make our omissions wrong in order to be blameworthy for them.
3. Ryland, a character in one of George Sher's examples (Sher 2009:28), is too self-absorbed to notice that "her rambling anecdote about a childless couple, a handicapped person, and a financial failure is not well received by an audience that includes a childless couple, a handicapped person, and a financial failure." Obviously, Ryland was not aware of the facts that made her actions wrong, yet she is blameworthy for her actions. Hence, we need not be conscious of the facts that make our actions wrong in order to be blameworthy for them.

The challenge here focuses on the content of agents' awareness. In all three cases, the agent is not aware of the facts that make it the case that the action has the moral status it actually has (that it is an instance of aiding someone to escape terrible injustice; that it is a violation of the norms of friendship; that it is gratuitously hurtful). Yet in all three cases—the philosophers who urge them claim—the agent is morally responsible for the action. Moreover, in all three cases, the agent is *directly* morally responsible. There need be nothing in the chain of events that leads to their action which renders them responsible for it; nothing, that is, prior to their action. The awareness thesis is, therefore, too strong, they claim.

These philosophers offer different positive accounts of moral responsibility to explain why agents are responsible despite their failures of awareness. For Smith (2005), the epistemic condition ought to be replaced by a causation condition: agents are morally responsible for actions caused by their *judgment-sensitive* attitudes, where an attitude is judgment sensitive if in an ideally rational agent it alters as a function of the evidence available to the agent. For Arpaly (2002), the epistemic condition should be replaced by a responsiveness condition: was the agent responding to moral reasons? For Sher (2009), we can accept the epistemic condition as it is usually understood, but only with an important addendum: agents are not to be excused if they are unaware of the relevant information because of psychological states that are constitutive of them. I shall not attempt, here, to assess these proposals. Rather, I shall argue directly for the awareness thesis. Before I begin, however, it is necessary to say a few words about what kind of awareness is in question.

THE CONTENT OF AWARENESS

Philosophical debates about consciousness typically focus on *phenomenal* consciousness. They focus on the alleged explanatory or metaphysical gap between the neural states countenanced by physicalism and phenomenal states. Clearly, in denying that agents need be conscious of the facts that give their actions the moral significance they have, opponents of the awareness thesis are not taking a stand on debates over phenomenal consciousness. Their claim is orthogonal to that debate. They are concerned with denying that agents need access to certain *information*, not with the properties of states of consciousness. The awareness thesis is a thesis about *what* the agent need be aware of, not about phenomenal consciousness. In the terms made famous by Ned Block, it is claim about *access consciousness*, not phenomenal consciousness (Block 1995).

To set out the awareness thesis, we must answer two questions. What kind of awareness is in question? What content must this awareness have? There are different ways of being "aware" of something. We are aware of whatever is phenomenally conscious, but also of a great deal besides: we speak of awareness, for instance, when it is some dispositional state that we have in mind. When we look at what opponents of the awareness view are denying—that is, at what their (putatively) responsible agents lack—we see some of this variety. Smith's case involves an agent who forgets the information that makes her omission significant (that it is her friend's birthday). The information is dispositionally available to her—she has not had it removed from her brain by nefarious neuroscientists, for instance—but it is temporarily forgotten. Ryland may or may not have the relevant information dispositionally available to her; it depends on how the case is spelt out. In one variant, Ryland is a basically decent person, who, we might imagine, might later be aghast at her own behavior when she thinks back on it. In another variant, Ryland might be unrepentant once she comes to see that she has caused offense *because* she is morally flawed. In the first case, Ryland has dispositional knowledge of the facts that make her actions wrong, whereas in the second she lacks it. Finally, Huck, as Arpaly presents him, does not even need to have dispositional knowledge of the facts that make his action morally right in order to be morally responsible for them; "The belief that what he does is moral need not even appear in Huckleberry's unconscious" (2002: 77), Arpaly maintains. It is sufficient that he is responsive to these facts.

Let us say an agent is *strongly* aware of a fact at a time if she is introspectively aware of it at that time; that is, she is currently entertaining the thought of that fact. She is *weakly* aware of it if she is not strongly aware of it but is dispositionally aware of it, where an agent is dispositionally aware of a fact if she would be strongly aware of it given the right cues. Obviously, the denial that agents need to be strongly aware of the facts that make their actions wrong is a weaker claim than the denial that agents need to be either strongly or weakly aware of the same facts. Opponents of the awareness thesis mean to defend the stronger claim. Indeed, the weaker claim is uninteresting: no one thinks that agents must be strongly aware of the relevant information in order to be morally responsible for their actions (consider the murder plan that is so complicated that as the assassin completes it, she is totally absorbed in her calculations and therefore fails to be strongly aware of the goal toward which she works). The awareness thesis, then, the thesis I wish to defend against philosophers like Arpaly, Sher, and Smith, is that agents must be *dispositionally* aware of the facts that make their actions morally significant.

This is not yet precise enough, however. Information is dispositionally accessible, as we saw, if the agent would be introspectively aware of it given the right cues. But there is a continuum of states that fit this description. Some dispositionally available information is nevertheless exceedingly difficult to access and some easy: your middle name is probably exceedingly easy for you to access right now, whereas it would take much more effort for you to recall what you had for dinner a week ago last Tuesday. Indeed, attention to states like the latter might mislead us into *overestimating* the availability of the relevant information: in asking you to recall what you had for dinner a week ago last Tuesday, I give you the opportunity to begin searching for cues to give yourself. Knowing the kind of thing you are looking for, you can begin to try various search strategies, selecting from a relatively circumscribed space of possibilities. In the cases that the opponents of the awareness thesis present, coming to be aware of the relevant information is exponentially more difficult because the agents have neither a cue to begin a search, nor any cues as to how to conduct it. Now, it should be common ground that *some* dispositionally available information is not available enough; that is, that agents are not morally responsible for failing to guide their behavior in its light. Consider dementia patients: at least in the earlier stages of the disease, their memories remain dispositionally available. It is simply that the range of cues they would need to recall them is much narrower than is the case with normal agents, such that the information comes to mind far less frequently than previously.

As I will understand it, the awareness thesis holds that only some of the information on the dispositional availability continuum is available enough to ground moral responsibility. I will say that information must be *personally* available in order to be available enough to ground moral responsibility. Information is personally available if either the agent is strongly aware of it or it is dispositionally available for *easy and effortless* recall. Information that is not personally available includes information that is entirely unavailable—like the information of which Huck is unaware—in that it is not represented anywhere in the agent's mind, and information that is merely subpersonally available, as well as some that is not available for easy recall. Information that is subpersonally available may play a role in guiding behavior, but I argue that it is not available enough to ground moral responsibility. From now on, I use "aware" and "conscious" (and their cognates) to refer to personal availability.

So much for the kind of awareness at issue. A word, finally, on the content of the awareness. I have said that agents must be aware of the facts that make their actions morally significant. This is not quite right, for two reasons. First, a *belief*—true or false—that an act is morally significant—seems sufficient to ground moral responsibility. People are praiseworthy for actions that express good or ill will, or their evaluative agency, and we express these things by responding to the facts as we see them. We need not be correct in our beliefs. Second, we need not be aware of *all* the facts that make our actions morally significant. We are responsible for actions regarding which we are aware of some facts that make them morally significant. That is, it is sufficient that when I act I take the circumstances in which I act to have certain features that entail my action is morally significant (i.e., circumstances, including my mental states, are such that it would be appropriate for others to describe my action as expressing good or ill will). Our responsibility for the action is a function only of those facts; if the action is also morally significant in other ways, we are not (directly) responsible for the action *qua* significant in those ways. I may be morally responsible, for instance, for saying something offensive, given that I am aware that my words are likely to cause offense, but I am not morally responsible for the fact that the offensive words are especially hurtful to my victim if I am (nonculpably) unaware of the facts about their history that make this true.

THE SIGNIFICANCE OF THE AWARENESS THESIS

Why does it matter whether the awareness thesis is true? There are several reasons. One stems from the work constituting the automaticity revolution mentioned previously. This work has, for our purposes here, two central features: (1) it has produced an ever-growing body of research on the role that our unconscious attitudes play in causing behavior and (2) it has shown that there is often a divergence between the content of these attitudes and our explicit attitudes. Implicit attitudes are acquired by way of enculturation. We absorb them from the mores and patterns of responses of those around us. These stereotypes are acquired early in life, before we have the cognitive capacities to assess them; moreover, they are acquired by means that bypass our capacities to reflect on them (Devine 1989). We often come to acquire explicit beliefs that conflict with these implicit stereotypes, but the implicit attitudes remain ingrained in our patterns of response.

Our implicit attitudes can be revealed in many ways. One of the best-known methods is the Implicit Association Test (IAT). The IAT measures subjects' speed and accuracy in associating positive and negative terms with items that activate typically stereotyped concepts: for instance, with black and white faces, or male and female names. Most white Americans (for instance) show an implicit bias against blacks, women, and homosexuals; that is, they are slower to associate positive terms with black, female, or gay-associated concepts than with white, male, or heterosexual concepts, and make more mistakes in associating positive terms with these former three concepts than the latter (Dasgupta 2004). Because these attitudes are acquired in ways that bypass our capacities for critical reflection and are typically insensitive to the content of our explicit beliefs, we usually retain them, even though many of us go on to reject racist, sexist, and homophobic attitudes. A dramatic illustration of this fact is that though blacks show more variance in their implicit attitudes with regard to race than whites, they often absorb and retain the negative stereotype associated with blacks as well (Nosek, Banaji, and Greenwald 2002)

Implicit attitudes are not beliefs. Hence, it may be unfair to think worse of those who have them, inasmuch as they are not responsible for acquiring them and may have made wholehearted efforts to eliminate them. We might wonder, therefore, whether they matter at all. There are, I think, good reasons to be concerned about our implicit attitudes: under a variety of conditions, they cause behavior, sometimes very significant behavior. Scores on an implicit association test seem to be a better predictor of certain kinds of subtle racist behavior than are our conscious attitudes toward other races (McConnell and Leibold 2001). Moreover, implicit attitudes can explain certain kinds of lethal force. Priming with black faces raises the likelihood that agents will identify ambiguous stimuli or nongun tools as guns (Payne 2001); this fact may partially explain why police are more likely to use deadly force when confronted with black suspects. When we are under time pressure, stressed, tired or distracted, our actions will often reflect our implicit attitudes, even when our explicit attitudes conflict with them. Unfortunately, many of our most significant actions occur under these kinds of pressures. We do not have the luxury of calm reflection in life-or-death situations, or in those we take (wrongly) to be life or death.

When we act on implicit attitudes, we often represent the facts to ourselves in ways that diverge from reality. We are unaware, say, that the man in the doorway is holding his wallet in his hand; instead, we take him to be holding a gun. If we can only be responsible for actions that respond to the facts as we are aware of them, in situations like this we will fail to be (directly) responsible for our actions, when they are reasonable responses to what we *take* ourselves to see. Hence, the awareness thesis matters: on its truth depends whether we are responsible for some of our most consequential behaviors.

Obviously, the awareness thesis matters for the law, too. It matters because the law ought to be responsive to whether agents are *really* morally responsible for their actions. There is a case for saying that the law ought to be responsive to other facts, too—to facts about the consequences of legislation, for instance—but facts about moral responsibility ought to be given *some* weight in legal thought: to the extent to which doing so is compatible with the other goods the law seeks to protect, it ought to find people *legally* culpable only when they are *morally* responsible. The conflict between the law and the awareness thesis comes out clearly with regard to the question of negligence. Jurists hold people to be negligent when they fail to act in the way that a reasonable person would in their circumstances. A negligent person is *not* aware of the facts that make their actions morally significant (if they are aware of these facts, the act is reckless and not negligent). Typically, jurists hold that finding people to be negligent is morally justified, holding that the law may find agents to be responsible for their negligence because agents *are* responsible for their negligence. Hence, the awareness thesis matters for the law because on its truth depends whether negligence is justified.

DEFENDING AWARENESS

I believe that the awareness thesis is in fact true, and that agents are therefore morally responsible less often than its opponents claim. Indeed, as my reflections on negligence indicate, I think that because the awareness thesis is true, agents are less often responsible than has *generally* been thought. I shall advance two arguments for this claim. There are two major accounts of moral responsibility in the literature; the arguments will each be addressed to supporters of one of the accounts and will assume that that account is true.

The two accounts I have in mind are the reactive attitude account, associated with Peter Strawson (1962), which has subsequently become the dominant view among philosophers, and what I will call the evaluative view, associated with Scanlon (1998), which is rapidly growing in popularity. On the reactive attitude account, the attitudes we express in holding one another morally responsible are deeply rooted in our ordinary lives: they are attitudes that develop from and are best understood by reference to our interpersonal relationships. In blaming someone, or in being grateful to him, I implicitly express my appreciation of the significance of his actions for my actual or counterfactual relationship to him. Insofar as the attitude expressed is a properly reactive attitude, which is a response to an agent as a morally responsible actor, and not an objective attitude, which is a response to an agent as an object to be manipulated or otherwise dealt with, my attitude is a response to his good will, ill will, or indifference, as it was expressed in his action. Insofar as his action expressed ill will toward me or others, I respond with reactive attitudes in the negative spectrum (resentment, indignation, anger, and so on); insofar as it expresses good will, I respond with positive reactive attitudes (gratitude, esteem, respect, and so on). Indifference might also be met with negative attitudes, inasmuch as someone's indifference toward me fails to accord me the respect I deserve as a person.

On the evaluative view, in contrast, I am responsible for my actions insofar as they are expressions of my evaluative agency. That is, if my behavior is caused by an attitude of mine, and that attitude is of the kind that is *in principle* responsive to my rational judgment, then I am responsible for it. There are two reasons to think that the class of actions for which agents are responsible is circumscribed by our evaluative agency. First, in acting on such attitudes, we express ourselves as evaluating beings. These actions are expressions of us as engaged in the "activity of evaluative judgment" (Smith 2005:237). Second, insofar as we act on these attitudes, we take a stand on what counts as reasons for us. To have a pro-attitude toward something is to see it as reason giving for us (Scanlon 2002:177), or to judge it as "good in some way" (Smith 2005:270). Proponents of this second account of moral responsibility have several motivations, but among them is the fact that they believe it allows us to hold agents morally responsible for actions that are caused by implicit attitudes. An agent's implicit attitudes are authentically his own, and thus reflect his evaluations, "even if he disapproves of, rejects, and controls them, and would eliminate them if he could" (Scanlon 2002:171). Because in acting on our implicit attitudes, we express our evaluative agency, we express who we are—even if only in part—and can therefore justifiably be held responsible for our actions.

Because the evaluative view is motivated, in part, by the belief that the reactive attitude view lets people off the hook too easily, and especially in cases in which they fail to be aware of the attitudes that cause their behavior, we should not be surprised to find that the reactive attitude view has problems explaining how agents can be responsible in such cases. The basic problem is this: when actions are caused by agents' implicit attitudes, they do not seem to be expressions of their will—good, bad or indifferent—at all. It is one thing for an action to express an attitude, and another altogether for it to express the agent's will. An agent's will, in the relevant sense, might be identified with where she stands on questions of value. To have good will toward someone is to have attitudes that are expressed in a broad range of dispositions to behavior (dispositions to aid them, to express support for them, to be pleased when they succeed at something and when their life goes well, and so on); we can understand ill will in analogous ways (indifference might best be understood as the absence of such dispositions). Attributing good or ill will is therefore much like attributing beliefs: just as we can measure the degree to which agents accept some claim by measuring the degree to which they conform to the *dispositional stereotype* associated with that belief, where a dispositional stereotype is "the cluster of dispositions that we are *apt* to associate with the belief that *p*" (Schwitzgebel 2002: 251), so we discover an agent's will by measuring the degree to which they conform to the associated dispositional stereotype. Beliefs are not all or nothing; we can conform more or less closely to the associated dispositional stereotype. We are all, to use Schwitzgebel's phrase, in-between believers (Schwitzgebel 2001). But we are not in-between believers with regard to *every* belief we hold. Sometimes, we manifest sufficient conformity to a dispositional stereotype to make it correct to attribute the relevant belief to us. To exhibit "sufficient" conformity to the stereotype, it is not necessary that we exhibit *perfect* conformity to the stereotype (indeed, perhaps we *never* exhibit perfect conformity to the stereotype). Someone can be an atheist, say, without being an atheist in a foxhole (especially if they are unlikely ever to find themselves in a foxhole).

The moral should be clear. The fact that I have implicit attitude *a* does not entail that if *a* causes me to act, my action expresses my good or ill will. My will is to be identified with a broad swathe of my dispositions. Implicit attitudes do not generate a broad swathe of dispositions. Instead, they reveal themselves only in a narrow range of circumstances, and in ways that are characteristically inflexible and stereotyped. Implicit attitudes are associated with what psychologists call *System 1*;—the set of psychological mechanisms that operate below the threshold of conscious awareness. System 1 processes are automatic, fast, massively parallel, and effortless. They are also typically encapsulated, which is to say both that they have access to only a small subset of the agent's representations and that the agent has no access to their inner workings. They are also often, perhaps always, modular (Stanovich 1999), which is to say that they operate in isolation from one another and from domain-general processing. It is their encapsulation and modularity that makes the behavior they generate inflexible and stereotyped: a System 1 process cannot respond to a broad range of information and therefore lacks flexibility of response. System 1 processes are therefore narrow, both in the cues to which they respond and in the behavior they cause. These facts make it implausible to identify an agent's implicit attitudes with their will. On the other hand, these same facts entail that the presence of a conflicting implicit attitude need not prevent us from identifying an attitude with the agent's will. If an agent has an explicit attitude that is manifested in a broad enough range of dispositions, then we can say that that attitude is his will, even if he also has a conflicting implicit attitude. Explicit attitudes are associated with *System 2*, which is slow, effortful, typically conscious, and serial. More important, in this context, System 2 processes are domain-general, which is to say that they are sensitive to a very broad range of the agent's representations. It is this feature that ensures that they give rise to a correlatively broad range of flexible behavior, and makes it correct, typically at least, to identify such attitudes with the agent's will.

We cannot identify an agent's implicit attitudes with her will; there is therefore no justification for holding her responsible for actions caused by such attitudes, on the reactive attitude account of moral responsibility. Let us turn now to the rival evaluative account. As we saw, this account is designed, inter alia, to remedy what its proponents see as a flaw in the reactive attitudes account, the flaw that prevents us from holding agents responsible for actions caused by their implicit attitudes. On the evaluative account, agents are responsible when their actions are caused by their attitudes, when these attitudes belong to the class of attitudes that are sensitive to agents' reasons. A particular agent need not be satisfied with having such an attitude—indeed, she may not even be aware that she has it—nor need the attitude be *actually* sensitive to her judgments. Insofar as she has the attitude, she takes its intentional object to be reason giving for her (whether she is aware of this fact or not), and her possession of it is an expression of her evaluative agency. Or so at least advocates of the evaluative account hold. I shall argue that they are wrong on both counts.

Let me begin with the claim that having an implicit pro-attitude toward a state of affairs entails valuing that state of affairs or seeing it as reason giving. Philosophers like Scanlon and Smith maintain that there is an internal connection between an agent's desires and other pro-attitudes and what that agent takes to be a reason for action. Scanlon understands a desire "as involving a tendency to see some consideration as a reason" (2002:177); similarly, Smith argues that "the desire to do X involves the judgment that X is good in some way" (2005:270). The claimed internal connection between a desire and what the agent takes, if only momentarily, to be a reason makes *all* our desires constitutive of our evaluative outlook. However, there is good evidence that the supposed link actually fails to hold. Robinson and Berridge (2003; see also Balfour 2004) have argued that neurophysiology supports a dissociation between what agents "want"—that is, what has high incentive value for them—and what they "like." What agents take to be reasons for actions seems to be linked only to the latter, but the term "desire," as used by Scanlon and Smith, seems to cover both states. Implicit pro-attitudes seem better identified with agents' wants, insofar as they are manifested only in what might be called *narrow revealed preferences*. They do not entail liking the state of affairs or otherwise approving of it.

In fact, there is reason to go further still. Robinson and Berridge seem to have in mind two different propositional attitudes toward a state of affairs. Showing that they dissociate is an important contribution to moral psychology and is sufficient to show that inferring the second from the first (or vice versa) is illegitimate. It is therefore sufficient to block the claim that in having an implicit attitude, an agent reveals that she has a correlative value. But implicit attitudes may well be too thin to be identified even with wanting, or at any rate with a state that is as thick as the kind of state Robinson and Berridge have in mind. *Wanting* reveals itself in a broad range of circumstances and in a relatively broad range of responses, but implicit attitudes are thinner than that, in terms of their triggers and in terms of the responses they cause. They may not be thick enough even to be identified with a wanting-like state. They may be better regarded as subpropositional attitudes.

Let me turn now to the second claim, that agents are responsible for acting on their implicit attitudes because these attitudes are expressive of their activity as evaluative beings. These attitudes are allegedly in principle *judgment sensitive*: they are attitudes that, in ideally rational agents, are sensitive to reasons, such that these agents have them when, and only when, they judge there to be sufficient reason for them (Scanlon 1998:20). Insofar as we are rational agents, we are not simply "stuck" with our judgment-sensitive attitudes. Instead, they are the product and the expression of ourselves as agents; hence, they belong to us in a way sufficient to ground our moral responsibility. Proponents of the evaluative view acknowledge that some of our attitudes are not, in fact, sensitive to our judgments, but deny that we are excused from moral responsibility for acting on these attitudes. Because these attitudes *should* be "under the control of reason," their actual failure to be judgment sensitive reflects a failure of the agent to govern herself well or rationally (Scanlon 1998:272; Smith 2005:253) and therefore cannot function as an excuse.

The claim that agents are responsible for actions that are caused by their judgment-sensitive attitudes might be true. But we cannot reasonably conclude from this that agents are responsible for their implicit attitudes because these attitudes belong to the class of attitudes that are judgment sensitive (whether they are in fact themselves judgment sensitive or not). Implicit attitudes might be regarded as a psychological kind; that is, they can be categorized, for the purposes of explanation and prediction, as systematically different from explicit attitudes. This is precisely what psychologists do when they identify implicit attitudes with System 1 processes and explicit attitudes with system 2. Categorizing implicit attitudes as a kind gives us a narrower and better unified concept with which we can predict and explain human behavior. We might, if we like, regard the kind "implicit attitude" as a subset of the kind "belief," but we must be careful in doing so. Implicit attitudes are systematically different from other kinds of belief, and in categorizing them as beliefs we risk masking these differences.

One salient way in which implicit attitudes are systematically different from (other) beliefs is that they are *not* judgment sensitive. They are acquired in ways that bypass our capacities for cognitive control, and they are altered in similar ways. Because they are encapsulated, they are insensitive to our domain-general beliefs. Indeed, as Gendler (2008) suggests, insensitivity to belief is the *hallmark* of what she calls "aliefs," where alief is a broader category that includes implicit attitudes:

> Beliefs change in response to changes in evidence; aliefs change in response to changes in habit. If new evidence won't cause you to change your behavior in response to an apparent stimulus, then your reaction is due to alief rather than belief. (Gendler 2008:566)

Implicit attitudes are *not* judgment sensitive. It is misleading to regard them as a subcategory of the state "belief" because this risks masking the fact that judgment *in*sensitivity is the hallmark of such states. To hold agents responsible for behaviors caused by implicit attitudes on the grounds that these attitudes ought to be under the control of reason, such that their failure to be responsive to the agent's evidence reflects a failure to govern oneself appropriately, simply ignores the psychological realities. We cannot hope to exercise the kind of control over our implicit attitudes that these writers regard as rationally required. Instead, we can influence our implicit attitudes only indirectly: by the same kinds of methods whereby we acquired them in the first place (by attempting to form new associations). These methods are arduous, slow, and probably fail more often than they succeed (Devine 1989). Indeed, there is a debate within psychology about whether it is possible to alter our implicit attitudes *at all*: some psychologists believe that we can hope only to bypass these attitudes, not to rid ourselves of them (Wilson et al. 2000).

On neither major theory of moral responsibility, then, do we have good grounds for thinking that we are responsible for actions caused by our implicit attitudes. Our implicit attitudes do not constitute our will, and the behaviors caused by them do not express our good, bad, or indifferent will to others. They therefore do not justify reactive-attitude-based moral appraisal. Our implicit attitudes are not reflections of our evaluative or rational activity, either, and therefore do not belong to us in the kind of manner required by the evaluative account. There is therefore a strong prima facie case for holding that we are not responsible for actions caused by these attitudes. So far, the awareness thesis seems in good shape.

There is an obvious concern that needs to be addressed, however. I have claimed that we are not responsible for actions caused by our implicit attitudes because either these attitudes fail to constitute our will or because they are not judgment sensitive. However, the claim seems far more plausible when there is a conflict between agents' implicit attitudes and their explicit beliefs. Consider the reactive attitude account. Suppose that I have implicit attitude *a* and that *a* causes me to act. Suppose, further, that had I had the time to reflect, I would have performed precisely the same action. In that case, there is no conflict between my will and my action; should I not therefore be held responsible for it?

I don't think so. In the interests of space, I will set out my reasons briefly. I am responsible, on the reactive attitudes view, if my action is an *expression* of my will. For my action to express my will, it must actually be *caused by* my will (and caused in the right way); mere conformity between my will and my action isn't sufficient. Because my implicit attitudes do not constitute my will, the fact that I would approve of the action does not suffice to demonstrate that it is an expression of my will. Similar remarks apply, mutatis mutandis, to the evaluative view: mere conformity between my evaluative agency and my action will not suffice for expression.

It might be objected that this view would create perverse incentives for agents to act unreflectively because in so acting they avoid moral responsibility for their behavior. I don't think this is a serious worry. If agents act unreflectively, I have suggested, they avoid *direct* moral responsibility for their actions. But if they adopt a policy of acting unreflectively, their adoption of this policy (unless—implausibly—they adopt the policy unreflectively) will be something for which they are responsible. Recall that it is direct moral responsibility that is in question throughout this discussion: in claiming that the awareness thesis is true, I do not commit myself to claiming that agents are never responsible for their unreflective actions or omissions, only to claiming that they are not directly responsible for them. We might be responsible for our unreflective actions in virtue of adopting a policy of not reflecting. Equally, we might be responsible in virtue of having failed to avoid situations in which our implicit attitudes are likely to be expressed, or in virtue of having failed to attempt to discover and alter our implicit attitudes. My claim is that *if* we are responsible in one or more of these ways, our responsibility will bottom out in actions with regard to which we were aware of what we were doing. Because it is implausible that we are always directly responsible in one of these kinds of ways for unreflective actions and omissions, however, I am committed to claiming we are less often responsible than opponents of the awareness thesis think, and almost certainly less often than proponents of negligence think, too.

CONCLUDING REFLECTIONS

The awareness thesis is, I believe, safe. We have no reason to think that agents are (directly) morally responsible for actions caused by their implicit attitudes. Instead, we are responsible only for actions caused by our explicit beliefs, which constitute our wills and which are responsive to the evidence as we see it.

The truth of the awareness thesis represents both a challenge and a vindication of our moral responsibility practices. It is a challenge because it apparently entails that agents are not (directly) morally responsible for negligent actions. Insofar as this is true, the law must rethink the justification of negligence; if it cannot be justified (say, on the kinds of grounds that justify strict liability offences), it must abandon it.

To that extent, the awareness view requires us to rethink important aspects of our moral responsibility practices. But it allows us to retain its most central components, vindicating them against the skeptical challenge from the timing of mental states. I have argued that in order for us to be morally responsible for our actions, they must be caused (in the right way) by the mental states that constitute our will or our evaluative agency. Our actions must really be rooted in our will, where our will is constituted by a broad range of our dispositions, or they must really reflect our assessment of our reasons for action. But we can satisfy these conditions without being able to track action initiation in the fine-grained way that Libet and Wegner say we cannot achieve. It simply does not matter, for the purposes of moral responsibility, whether I consciously initiate my actions, or whether my consciousness that I have initiated an action is itself caused by mechanisms that initiate actions. What matters is the content of the mental states to which these mechanisms are responsive. Do the relevant mechanisms constitute my will (as I have defined it), or my evaluative agency? Nothing in the evidence presented by those scientists concerned with the timing of our mental states suggests that they do not. If my actions are responsive to my domain-general, explicit attitudes, then my action is the expression of my activity as an agent. Neither for our everyday responses to one another, nor for the purposes of the law, do we require anything further.

REFERENCES

Arpaly, N. (2002). *Unprincipled Virtue: An Inquiry into Moral Agency.* Oxford, UK, Oxford University Press.

Balfour, D. J. (2004). "The neurobiology of tobacco dependence: A preclinical perspective on the role of the dopamine projections to the nucleus accumbens." *Nicotine and Tobacco Research* 6: 899–912.

Bargh, J. A. (1997). The automaticity of everyday life. In: *The Automaticity of Everyday Life: Advances in Social Cognition, Volume 10.* R. S. Wyer, Jr. Mahwah, NJ, Erlbaum, pp. 1–61.

Baumeister, R., and E. J. Masicampo (2010). "Conscious thought is for facilitating social and cultural interactions: How simulations serve the animal-culture interface." *Psychological Review* 117: 945–971.

Baumeister, R., E. J. Masicampo, and K. D. Vohs (2011). "Do conscious thoughts cause behavior?" *Annual Review of Psychology* 62: 331–361.

Block, N. (1995). "On a confusion about a function of consciousness." *Behavioral and Brain Sciences* 18: 227–287.

Dasgupta, N. (2004). "Implicit ingroup favoritism, outgroup favoritism, and their behavioral manifestations." *Social Justice Research* 17: 143–168.

Devine, P. (1989). "Stereotypes and prejudice: Their automatic and controlled components." *Attitudes and Social Cognition* 56: 5–18.

Flanagan, O. (1996). Neuroscience, agency, and the meaning of life. In: *Self-Expressions.* Oxford, UK, Oxford University Press, pp. 53–64.

Gendler, T. (2008). "Alief and belief in action (and reaction)." *Mind & Language* 23: 552–585.

Libet, B. (1999). "Do we have free will?" *Journal of Consciousness Studies* 6: 47–57.

McConnell, A. R., and J. M. Leibold (2001). "Relations among the Implicit Association Test, discriminatory behavior, and explicit measures of racial attitudes." *Journal of Experimental Social Psychology* 37: 435–442.

Morsella, E. (2005). "The function of phenomenal states: Supramodular interaction theory." *Psychological Review*, 112, 1000–1021.

Nosek, B. A., Banaji, M. R., & Greenwald, A. G. (2002). "Harvesting implicit group attitudes and beliefs from a demonstration website." *Group Dynamics* 6: 101–115.

Payne, K. (2001). "Prejudice and perception: The role of automatic and controlled processes in misperceiving a weapon." *Journal of Personality and Social Psychology* 81: 181–192.

Robinson, T. E., and K. C. Berridge (2003). "Addiction." *Annual Review of Psychology* 54: 25–53.

Scanlon, T. M. (1998) *What We Owe to Each Other*. Cambridge, MA, Harvard University Press.

Scanlon, T. M. (2002). Reasons and passions. In: *Contours of Agency: Essays on Themes from Harry Frankfurt*. S. Buss and L. Overton. Cambridge, MA, MIT Press, pp. 165–183.

Schwitzgebel, E. (2001). "In-between believing." *Philosophical Quarterly* 51: 76–82.

Schwitzgebel, E. (2002). "A phenomenal, dispositional account of belief." *Noûs* 36: 249–275.

Sher, G. (2009). *Who Knew? Responsibility Without Awareness*. New York, Oxford University Press.

Smith, A. (2005). "Responsibility for attitudes: Activity and passivity in mental life." *Ethics* 115: 236–271.

Stanovich, K. (1999). *Who Is Rational? Studies of Individual Differences in Reasoning*. Mahwah, NJ, Lawrence Erlbaum Associates.

Strawson, P. (1962). "Freedom and resentment." *Proceedings of the British Academy* 48: 1–25.

Wegner, D. (2002). *The Illusion of Conscious Will*. Cambridge, MA, MIT Press.

Wilson, T. D., S. Lindsey, and T. Y. Schooler (2000). "A model of dual attitudes." *Psychological Review* 107: 101–126.

PART 3

Assessment

8

Translating Scientific Evidence into the Language of the "Folk"

Executive Function as Capacity-Responsibility

KATRINA L. SIFFERD[1]

There are legitimate worries about gaps between scientific evidence of brain function and legal criteria for determining criminal culpability. Behavioral evidence (such as arranging a getaway car) appears easier for judges and juries to use to generate verdicts because it triggers the application of commonsense psychological concepts that guide responsibility assessments. In contrast, scientific evidence of neurological processes will not generally lead a judge or jury to make direct inferences regarding criminal culpability (Vincent 2008). In these cases, an expert witness will be required to indicate to the fact finder what scientific evidence means with regard to mental capacity; and then another inference must be made from this possible lack of capacity to the legal criteria for guilt. In this chapter, I argue that formulating the relevant mental capacities as executive functions within the brain can provide a reliable link between neuroscience and assessments of criminal culpability.

COMMONSENSE PSYCHOLOGY, CRIMINAL RESPONSIBILITY, AND CAPACITY-RESPONSIBILITY

The criminal law depends upon "folk" or "commonsense" psychology (CSP), a theory used by ordinary people as a means to understand and predict other humans' behavior (Sifferd 2006). CSP allows persons to postulate that behavior is causally related to mental states such as beliefs and desires, and to predict or interpret such behavior based on attribution of mental states.[2] In the criminal law, CSP is used to determine whether a defendant possesses both minimal mental capacity and the particular mental states necessary to qualify him for criminal responsibility.

Neuroscience provides substantial new and valuable information for determining criminal responsibility. However, there are legitimate worries about gaps between scientific evidence of brain states and function (e.g., as evidenced by functional magnetic resonance imaging [fMRI] data) and legal criteria for criminal culpability. In particular, there are gaps between CSP and neuroscientific data. Some would argue that these gaps are so severe that neuroscientific evidence should be kept out of the courtroom, which would mean that courts would fail to use potentially valuable information to generate verdicts that can deny a person liberty and even life.

I believe some neuroscientific evidence can be relevant to determinations of legal capacity. A conceptual mechanism is needed, however, to indicate to the fact finder what neuroscientific evidence means with regard to the legal criteria for full mental capacity cast in CSP terms. In this chapter, I argue that we can reconceptualize the mental competencies necessary for criminal responsibility—particularly those that are recognized as missing or compromised by the doctrine of "diminished capacity"—in a way that provides a bridge between direct evidence of brain function and structure and assessment of criminal responsibility. I claim that formulating these competencies as "executive functions" within the brain can provide this bridge.

The chapter begins by reviewing the criminal law's use of CSP mental criteria to define whether or not a defendant has the capacity to be legally liable for a crime. My analysis focuses on the legal concept of diminished capacity, which in US law may warrant the determination of lesser responsibility. I next discuss arguments against use of neuroscientific evidence in criminal trials, and find there is no convincing evidence that neuroscience cannot be mapped onto CSP concepts such that it can be used to generate criminal verdicts. However, such mapping must be done in a principled way for neuroscientific evidence to be used effectively by a judge or jury. I argue that conceptualizing the necessary mental competencies as "executive functions" within the brain can provide the needed bridge between a commonsense understanding of capacity responsibility and neuroscientific evidence. Finally, I review the value of this argument in reference to the criminal liability of juvenile and mentally retarded defendants because they have recently been deemed to be ineligible for the death penalty at least in part owing to the doctrine of diminished capacity. I conclude the chapter by noting that much work remains to be done to expand our understanding of the relevance of neuroscience to the various criteria for criminal responsibility cast in CSP terms. This work requires an interdisciplinary approach linking philosophy, neuroscience, and law.

Criminal Responsibility and Commonsense Psychology

To be found guilty and sentenced, criminal defendants must be found to possess particular commonsense psychological mental states at the time the crime was committed. Criminal verdicts thus depend upon commonsense attributions of mental states; for example, the mental state "intent to kill," which is comparable to a desire to cause an unlawful death. CSP is used by the criminal law both to categorize offenders (e.g., as guilty, not guilty, insane) and to determine the appropriateness of penalties, some of them very severe. In the United States, in the "guilt" phase of a murder trial, a prosecutor will ask a judge or jury to find (beyond a reasonable doubt) that the defendant committed an unlawful killing "knowingly"' or "purposely." [3] In the sentencing phase, the jury might be asked to determine whether the defendant meets certain mitigating or aggravating criteria, including the mitigating factor of diminished mental capacity—the incapacity to possess the requisite mental states in the "normal" way of adults with full cognitive capacity, for instance, and the aggravating factor of knowing one was to receive money for the killing.[4]

The criminal law's use of CSP mental criteria falls into two broad categories. The first concerns a defendant's general mental capacity for decision making, where a lack of such capacity might warrant lesser responsibility. The second category consists of the specific mental state attributions the criminal law must assign to a defendant for that defendant to be found to have committed a particular criminal act, often termed "mens rea." H. L. A. Hart called the former general qualifying conditions for criminal responsibility "capacity-responsibility," and the latter "legal liability-responsibility." (Hart 1968). Capacity-responsibility can be thought of as a precondition for legal liability-responsibility. In addition to requiring specific mental states, legal liability-responsibility also requires "causal or other connexions between a person and harm, or the presence of some relationship, such as that of master and servant, between different persons"[5] (Hart 1968:221).

Hart argues that, because capacity-responsibility concerns a general ability to understand and conform one's behavior to rules, it is a requirement for the efficacy of law. That is, if persons, or classes of persons, cannot understand the rule of law and make decisions that abide legal rules, then the institution of law as structured fails as applied to that person or class because the law cannot influence their behavior.[6] According to Hart, the specific capacities necessary to be responsible under the law are "understanding, reasoning, and control of conduct: the ability to understand what conduct legal and moral rules require, to deliberate and reach decisions concerning these requirements; and to conform to decisions when made" (Hart 1968:227). These capacities may be "diminished" or "impaired," as well as wholly lacking, "and persons may be said to be 'suffering from diminished responsibility' much as a wounded man may be said to be suffering from a diminished capacity to control the movements of his limbs" (Hart 1968:228). Traditionally, under English common law, one not guilty by reason of insanity was found to lack the legal capacity for a culpable act. Similarly, juveniles were thought to be disqualified from adult-level responsibility because of lack of capacity.

Diminished Capacity and Criminal Law

The way US law takes a defendant's mental capacity into account is complex. The diminished capacity doctrine allows a criminal defendant to either negate a mental element of the crime charged, thereby exonerating the defendant of that charge, or reduce the degree of the crime for which the defendant may be convicted, even if the defendant's conduct satisfies all the elements of a higher offense (Morse 1984; 2003). The former, termed the "mens rea variant" by Morse, claims that the defendant lacks the capacity to form the specific intent required for the crime (Morse 1984). That is, we needn't ask whether this defendant actually possessed the specific mental state necessary to be found guilty of a specific crime, such as knowingly causing a human death, because he is incapable of forming this mental state in a satisfactory way. In this case, the defendant cannot be convicted of the crime at issue. For example, the defense might present evidence that the defendant was sleepwalking when he killed his wife and thus could not have formed the intent to kill.[7]

Evidence of diminished capacity can also be used to reduce the degree of the crime for which the defendant may be convicted, even where the elements of a higher crime, including the possession of specific mens rea, seem to be met. This "partial responsibility" application of diminished capacity is theoretically more complicated than the mens rea variant. The justification for partial responsibility of those who suffer from diminished capacity is not that the prosecution didn't bear its burden in proving the defendant is guilty of a crime. Instead, the defendant claims that he is not fully responsible for the crime proved against him because he doesn't fully meet the preconditions for responsibility, including the ability to consider, plan, and control one's behavior in the "normal" way (Morse 1984). Thus, according to the doctrine of diminished capacity, some persons, such as children and the severely mentally retarded, cannot be held fully responsible for their acts because they do not have the capacity to form intentions in the way of normal adults, even though they are found to possess the requiste mens rea.[8]

In the United States, the partial responsibility variant grounds the rarely successful "extreme emotional disturbance" doctrine promulgated in the Model Penal Code.[9] Partial responsibility is also the theoretical grounds for the much more popular move to apply lesser sentences to defendants who are determined to have diminished mental capacity.[10] In addition, one can see the development of the US juvenile justice system as a logical outgrowth of the doctrine of partial responsibility.

EXECUTIVE FUNCTION AS MENTAL CAPACITY

When a criminal defendant attempts to show he or she lacks mental capacity, certain evidence is considered relevant to this claim. Standard offers of proof include the results of a psychological evaluation, intelligence testing, and school, work, and medical history. As scientific psychology has advanced, however, there has been a move to introduce neuroscientific evidence (such as results of a positron emission tomography scan or fMRI) as proof of lack of mental capacity (Seiden 2004), despite widely reported concerns that judges or juries may not be able to understand how such evidence comments upon criminal responsibility (Lelling 1992–1993; Morse 2006b; Vincent 2008).

In this section, I argue that some neuroscientific evidence of diminished capacity is relevant because the specific mental competencies necessary for capacity-responsibility can be conceptualized to interact with direct evidence of brain function and structure in a principled way. Later, I argue that conceptualizing these competencies as executive functions within the brain can provide a necessary bridge between a commonsense understanding of capacity-responsibility and neuroscientific evidence.

Worries About the Use of Neuroscientific Evidence in Determining Criminal Responsibility

There are two significant arguments why we might want to keep neuroscience out of the criminal court's capacity assessments altogether. The first argument states that CSP does such a good job of attributing the mental states necessary for criminal responsibility using behavioral evidence that neuroscientific evidence is just unnecessary—it provides no added value. Why fix the system of determining criminal culpability if it isn't broken? If judges and juries, using CSP, are currently correct in their attributions of criminal responsibility to defendants, the system does not need to be supplemented with neuroscience.

This argument fails to adequately recognize the shortcomings of the commonsense system of attributing mental states and capacity to defendants. CSP operates through something close to cognitive heuristics, whereby mental states are attributed based on theoretical assumptions regarding human behavior and outward behavioral cues.[11] Thus, as trial attorneys know well, it can be manipulated. When I was working at the juvenile courthouse in Chicago, I heard stories of prosecutors not allowing juvenile offenders to shave, and giving them clothes a size too small, in an attempt to make their request to transfer the juvenile to adult court more convincing. At the adult courthouse, defense attorneys told me of a catch-22 with their mentally retarded defendants: if they testify, they are more likely than normal defendants to admit to something terrible (and untrue) on cross-examination, but if they don't testify, the jury often claims that they look "perfectly capable" sitting at the defense table.[12]

Add to this family members sitting in the courtroom wearing giant buttons with pictures of the smiling murder victims; a process of voir dire where experts select jurors based upon guesses on how a person's life experience will affect their sympathies; and movies presented by the defense at sentencing that recounted the terrible abuse suffered by the defendant at the hands of his mother (set to a sweeping, tearjerker soundtrack). Criminal verdicts are subject to a variety of arguably unreliable influences and behavioral data that push and pull commonsense mental state attributions in different directions. Thus, I will argue that although it is true that neuroscientific evidence can also be manipulated to interact with CSP concepts in an unprincipled and unhelpful way, such evidence may be *more* reliable than behavioral evidence if presented using a good theory of how the science speaks to legal criteria—a theory I hope to provide in what follows.

A second argument in support of the claim that neuroscience should be kept out of the criminal court's capacity assessments is that there is an insuperable gap between the CSP concepts we use to attribute responsibility and neuroscientific evidence of psychological states. Paul Churchland, for example, argues that because CSP concepts are radically false, they should be eliminated and replaced with a neuroscientific theory of mind, a claim that has been widely refuted[13] (Churchland 1981). Others have made weaker claims that it is unlikely we will be able to *reduce* CSP concepts to neuroscientific concepts, or vice versa. For example, Stephen Morse is a nonreductive physicalist, which means that while he feels mental states are physical-causal entities, they cannot be reduced to neuroscientific objects or events (Morse 2007). [14] According to some versions of nonreductive physicalism, direct evidence of brain function or disorders—such as could be seen on an fMRI—should not be considered by courts seeking to judge criminal responsibility because the way they map onto mental states is "fundamentally mysterious."[15] Thus, the criminal law should embrace behaviorism as the primary means for attributing CSP states to defendants; that is, outward behavior should be used to attribute the mental states relevant to criminal responsibility.[16] I have argued elsewhere that the arguments that there is a necessary gap between CSP concepts and scientific evidence of brain states are unsound (Sifferd 2006). Although it is beyond the scope of this chapter to address the issues in adequate detail, there is no reason to think CSP is radically false and that its concepts fail to refer to real entities in the brain (Sifferd 2006). Further, this version of the nonreductive position regarding the relationship between the mind and brain fails as applied to the criminal law in one of two ways. If there is really such a loose connection between the mind and brain so that it remains "fundamentally mysterious," it would seem that the theory fails to preserve mental causation, an idea the criminal law relies upon. [17] As indicated previously, the criminal law demands that a defendant's mental states caused his criminal act (that his desire to kill caused the bodily act that resulted in an unlawful death). Any theory—non-reductive or otherwise—that failed to preserve this connection would undermine the existing system of attributing responsibility. If, however, a theory of nonreductive physicalism does preserve mental causation, it strengthens the relationship between the mind and brain to the point that mental states and brain states can at least be causally indentified (although the phenomenal aspects of a mental state need not be). [18] And in this case the modest claim that neuroscience may be relevant to mental state attributions couched in CSP terms is not threatened by non-reductivism.

At any rate, Morse's version of non-reductive physicalism cannot claim an "insuperable gap" between mental states and brain states, as he notes that a plausible theory of mind must preserve mental causation and claims that, in some cases, neuroscience may be relevant to criminal verdicts (Morse 2006b:400–401). It thus seems that Morse's worries about the use of neuroscience are grounded in worries about how certain scientific evidence can be reliably translated into an understanding of the mental capacities and mental states—couched in CSP terms—necessary for criminal responsibility.[19] As Nicole Vincent notes, even once one makes an inference from the results of an fMRI scan to attribute a disorder or a disease to a defendant, it is still a further inference to determine that one's mental capacities necessary for criminal responsibility are affected. "Neurological conditions do not undermine responsibility simply by virtue of being disorders, but rather they do so in virtue of the effect which they have on our mental capacities...which are required for moral agency" (Vincent 2008:200).

This worry regarding the interaction between neuroscientific evidence and commonsense concepts is legitimate. Principled use of such evidence requires careful work mapping neuroscientific evidence onto mental capacity framed in commonsense terms. In addition, introduction of neuroscientific evidence of mental capacity requires a bit of extra work on the part of the fact finder because it introduces an additional step in the process of attributing criminal responsibility.

For example, there is thought to be some overlap between the doctrines of legal insanity and diminished capacity because both can be used to indicate that a defendant was incapable of forming the requisite mental states for criminal guilt. Most US states use some version of what is known as the "M'Naghten rule" to determine whether a defendant is legally insane (Rex v. M'Naghten, 8 Eng. Rep. 718 [1843]). The M'Naghten rule states that a person is legally insane if he is "labouring under such a defect of reason, from disease of the mind, as not to know the nature and quality of the act he was doing; or, if he did know it, that he did not know he was doing what was wrong." Behavioral approaches to determining whether the defendant "knew the nature and quality" of what he was doing look to the defendant's behavior; for example, did he act in a way that makes it seem he didn't know what he was doing was wrong? Was he speaking out loud to no particular person right before committing the crime, or claim that the government had told him to kill someone? Behavioral approaches do not need to ask first whether the behavior indicates that the defendant is lacking one or more of a list of important mental capacities, such as the ability to normally process emotions or to consider future consequences of acts, and then ask if this lack means the M'Naghten criteria for legal insanity are met. Instead, the behavioral evidence speaks directly to the legal criteria for insanity. (If the jury determines the defendant truly thought he was saving his wife from being abducted by aliens when he threw her out the window, it is likely to decide he didn't know the nature or quality of his act.)

In contrast, evidence that the defendant has a large tumor in his orbital prefrontal lobe will not lead a judge or jury to infer anything about whether the defendant "knew the nature of his act" or was "laboring under a defect of reason." An expert witness will be required to indicate what this evidence means with regard to mental competencies and decision making, and then another inference will have to be made from this possible lack of mental capacity to the criteria for legal insanity, cast in CSP terms.[20] That is, only if the tumor resulted in the defendant not being able to make decisions in a normal way is it relevant to criminal responsibility.

Further, judges and juries tend to be completely dependent upon experts to make this inference from evidence of brain states and processes to mental capacities. This is reason enough to be worried about the use of neuroscience in criminal trials because attorneys seem to be able to find experts that will make almost any claim about the relationship between neuroscientific evidence and the mental capacity. Attorneys on both sides famously make hay of the fact finder's ignorance, arguing that chemicals as benign as sugar affect mental processes vital to responsibility (the "Twinkie defense") or that giant brain tumors leading to a wholesale lack of important mental processes mean nothing for criminal responsibility. For such expert testimony to be useful, it must express a deep understanding of the relationship between neuroscientific evidence and legal categories—or at least, be accompanied by good argument regarding this relationship.

In sum, the important difference between traditional behavioral evidence of a defendant's mental state and neuroscientific evidence is that behavioral evidence is just plain easier to relate to the CSP concepts that guide and structure criminal responsibility. Behavioral evidence is likely to directly trigger attribution of mental states required by the legal criteria for guilt (or not), whereas scientific evidence of brain states or processes needs a bridge to link it to the mental competencies that comment upon legal criteria for guilt. Thus, a "translation" of the evidence into CSP concepts—or at least information about how the science interacts with the CSP concepts—is necessary through testimony of an expert who understands this link, or through other means (possibly closing arguments or jury instructions). When an accurate translation is not made for the judge or jury, they are left to make their own unprincipled guess as to how scientific evidence of a particular mental defect or disease may affect attribution of mental capacity or the mental states necessary for criminal guilt.[21]

Using Executive Function to Understand Mental Capacity

Scientific research suggests that executive processes reside in the frontal lobes of the brain, primarily in the prefrontal lobes, including the dorsolateral frontal lobes on the side of the brain, the ventrolateral frontal lobes below them, and the orbitofrontal lobes located on the brain's undersurface just above the eye sockets (Moscovitch and Winocur 2002). Much of what is known about the function of the frontal lobes is known through loss of function due to traumatic brain injuries. Frontal lobe damage is associated with an erosion of foresight and judgment, and people with such damage often fail to delay gratification or experience remorse for their acts (Stuss and Knight 2002). Thus, it is thought that these abilities—the ability to plan and execute acts in accordance with that plan—are dependent upon executive processing.

Although neuroimaging of executive functions is still in its infancy, current research suggests that executive processes are distinguishable. One study by Miyake et al. (2000) found that for three target executive functions—shifting between mental sets or tasks, updating and monitoring working memory contents, and inhibiting prepotent responses—a full three-factor model in which the correlations among the three latent variables were allowed to vary freely produced a significantly better fit to the data than any models that assumed unity among two or all three of the variables. Thus, the study found that three processes were separable, although they also shared some underlying commonality (Miyake et al. 2000).

Most agree that although executive processes work in tandem with and operate upon the representations stored in memory, the two can be dissociated: there are neurological patients who have intact storage systems but defective executive processing, and vice versa (Smith 1999). Many also believe that the frontal lobes do not contain our mental representations, and that they instead reside in the posterior cortical regions, in the temporal and parietal lobes (Hirstein 2009). The prefrontal lobes contain the executive processes that monitor and manipulate these representations (Hirstein 2009). In addition, they can correct incorrect or improbable perceptions (Hirstein 2009).

Thus, executive processes control mental activity by directing attention and rearrange items held in working memory (Smith 1999). In general, although there is a lack of consensus about an exact taxonomy of executive processes, there is basic agreement that they contain certain core functions, such as those articulated by Stuss and Knight (Smith 1999): (1) working memory and related attentional processes (the "online" holding and mental manipulation of information); (2) the inhibition of distractibility, perseveration, and immediate gratification (being able to continue to orient oneself toward current goals); (3) the active pursuit of choice and novelty; (4) the conditional mapping of emotional significance (moderation and use of emotional responses); and (5) the encoding of context, perspective, and mental relativism (including consideration of an inferred future, including potential future consequences of actions) (Stuss and Knight 2002).

The mental capacities Hart claimed qualified one for legal responsibility—those that "enable understanding, reasoning, and control of conduct; the ability to understand legal and moral rules; to deliberate and reach decisions concerning these rules; and to conform to decisions when made" (Hart 1968:227)—are strikingly similar to Stuss and Knight's characterization of "executive processes." Banich claims that executive processes "can be thought of as the set of abilities required to effortfully guide behavior toward a goal, especially in non-routine situations" (Banich 2009:89). According to Hart, legal responsibility is the capacity to understand how one ought to behave and the ability to conform one's behavior to these rules (Hart 1968:227). The same pre-frontal processes crucial to management of attention, planning, and initiation or inhibition of action also ground legal capacity and responsibility (Hirstein and Sifferd 2011).[22]

To understand the extent of the overlap between legal capacity and executive processes, it is useful to explore the following high-profile example of a person lacking legal capacity. In 2003, Burns and Swerdlow reported a case of a 40-year-old schoolteacher in an otherwise normal state of health, who developed an increasing interest in pornography, including child pornography. The patient also began soliciting prostitution at "massage parlors," which he had not previously done. An MRI revealed a large tumor in his right orbitofrontal lobe. The patient's symptoms disappeared after the tumor was removed (Burns 2003).

Interestingly, the researchers postulated that the tumor had not created the patient's interest in child pornography, but that it had exacerbated a preexisting interest in pornography, manifesting as sexual deviancy and pedophilia (Burns 2003). The tumor in the frontal lobes had affected several executive processes, including inhibition (denial of immediate gratification), the moderation of emotional responses, and consideration of the potential future consequences of actions. Burns and Swerdlow hypothesized that the tumor had affected the patient's ability to conform his behavior to societal norms and laws. This sounds rather similar to Hart's characterization of the capacity "to deliberate and reach decisions concerning these rules; and to conform to decisions when made" (Hart 1968:227).

Burns' and Swerdlow's patient was arrested for sexual assault of his stepdaughter. I would argue that, before his operation to remove the tumor, this patient was a person who was lacking capacity-responsibility in Hart's sense; and that although his behavior was bizarre, behavioral accounts would not necessarily lead a jury to a determination of a lack of mental capacity. The patient's "symptoms," for example, included a keen interest in pornography, visiting prostitutes, and soliciting unknown females for sexual favors (Burns 2003). Certainly, when contrasted with the patient's earlier behavior, this pattern seems odd, but I see no reason why this behavior, in the absence of any further evidence, would lead a judge or jury to conclude that the defendant lacked mental capacity. Indeed, given his ongoing responsibilities (including his job as a teacher and role as a parent), without any indication of brain disease or trauma his actions might be labeled particularly egregious and thus deserving of a stronger penalty. In this case, it seems identification of the man's tumor would be extremely helpful in achieving an accurate understanding and categorization of his behavior. Indeed, it would appear to be unjust to deny the fact finder a chance to consider the neuroscientific evidence of the presence of a tumor in the man's orbitofrontal lobe as evidence of diminished capacity.

Note, however, that presenting evidence of the tumor, and its location, would be pretty much useless to a fact finder without information about how the tumor may have affected the mental capabilities necessary to form criminal intent. But if we assume that the functioning of executive processes is directly relevant to mental capacity, the following instructions could allow a jury to use evidence of the tumor properly: that the tumor affects the operation of executive processes; that executive processes represent mental capabilities that are crucial to mental capacity; and that significantly reduced capacity can negate mens rea, or indicate that a lesser sentence is appropriate. Specifically, an expert witness with a good understanding of the connection between executive processes and the mental capabilities necessary for mental capacity could explain the way in which the tumor affected executive processes, including the impact upon (1) working memory and mental manipulation of information; (2) the defendant's ability to orient toward current goals; (3) the moderation and use of emotional responses in decision making; and (4) consideration of an inferred future, including consideration of potential future consequences of actions.

One might object that, at present, science could not provide such a detailed account of the way in which the tumor affected executive processing. Fair enough. But remember, particular executive functions are already thought to be distinguishable on an fMRI, and progress is to be expected regarding mapping executive functions onto the frontal lobes (Elliott 2003); thus, it seems increasingly likely that a future neuroscience could allow us to understand the way in which a tumor, or traumatic brain injury, affects executive function.

EXECUTIVE FUNCTION AS A MEANS OF BETTER UNDERSTANDING DIMINISHED CAPACITY IN JUVENILE AND MENTALLY RETARDED DEFENDANTS

I have attempted to provide a framework for reliably bridging the gap between neuroscientific evidence of brain function and the specific mental competencies necessary for capacity-responsibility by conceptualizing these competencies as executive functions within the brain. In this section, I will show how this framework can be usefully employed in the context of individualized assessments of mental capacity in juveniles and mentally retarded individuals. Both have recently been deemed to be ineligible for the death penalty by the US Supreme Court, in part because of diminished capacity-responsibility in Hart's sense.[23]

Juveniles

In the 2005 case of *Roper v. Simmons*, the Supreme Court barred capital punishment of juveniles who killed while they were under the age of 18. Amicus curiae briefs filed before the decision strongly encouraged the court to use evidence from neuroscience in their assessment of juvenile culpability. The

American Medical Association, the American Bar Association, the American Psychiatric Association, and the American Psychological Association all filed or subscribed to amicus briefs urging abolition of the juvenile death penalty based in part on the neuroscience findings (Morse 2006a).

Morse, however, has argued that the neuroscientific evidence offered by the amici added nothing to the behavioral evidence already available to the court (Morse 2006a). Everyone, Morse says, knows juveniles are less rational than adults. Scientific evidence just distracts the court away from the real problem of assessing juvenile capacity for rationality in commonsense terms. Plus, "[a]s a normative matter, the Court could decide that sixteen and seventeen year olds are responsible enough to be executed despite all of them being less responsible than older murderers. Assuming the validity of the findings of behavioral and biological difference, the size of that difference entails no necessary moral or constitutional conclusions" (Morse 2006a:409).

Morse is right that an understanding of the differences between adult and juvenile decision making doesn't necessarily compel a policy decision regarding whether the difference matters for criminal responsibility. However, given that the policy decision has already been made—in the United States, established case law and federal and state legislation indicate that juveniles deserve special consideration of their status as juveniles for the determination of criminal responsibility and decisions regarding punishment—it seems appropriate to use all the reliable tools available to define this category in a principled way. Neuroscientific evidence of executive function should be used as an instrument to better define the class of juveniles who are less culpable owing to diminished capacity. Age alone is a fairly crude measure of a juvenile's cognitive capacity, as is an assessment of one's capacity using behavioral cues. Even evidence that a juvenile makes unwise, risky decisions at *every opportunity* would not necessarily show that the juvenile was incapable of adult-level decision making because it could be the case that he just enjoyed taking risks (as some adults do) (Vincent 2008). As Vincent argues, "[t]o be a responsible moral agent one must have the right mental capacities, but since mental capacities are implemented in brain mechanisms (in brain "hardware"), to be a responsible moral agent one must have the right brain mechanisms, and that—i.e. whether the person whose responsibility is being assessed has those brain mechanisms—is precisely what neuroimaging would be used for in this high-tech approach to the individualized assessment of responsibility" (Vincent 2008:4).

Making matters worse, most American states use age plus severity of offense, in conjunction with other factors, when determining whether a juvenile defendant should be transferred to adult court (Butts and Mitchell 2000). Consideration of severity of offence is clearly not a good indicator of mental capacity, and it appears to be used because states consider juveniles who commit particularly violent crimes more deserving of retribution, despite their mental capacity. However, other factors used to transfer juveniles appear to be far-from-satisfactory heuristic methods of determining mental capacity. For example, under California's Welfare and Institutions Code 707(a), the juvenile judge must evaluate the degree of "criminal sophistication" exhibited by the child. Evidence of "criminal sophistication" could in theory indicate mental capacity, if the judge interpreted it to mean sophistication in decision making, including the appropriate use of emotional and consideration of potential future consequences. However, most often, judges use behavioral evidence (such as evidence of premeditation) to determine whether a juvenile is "sophisticated."

The juvenile system is meant to address the unique concerns of dealing with offenders who are different in their capacity to understand the nature of the crimes they commit. Offenders who lack decision-making capacity because of juvenile status are to be treated not only as less responsible for their acts but also as "decision makers in development." One of the goals of the juvenile system is to try to keep juveniles from becoming career offenders by keeping legitimate life options open (e.g., by not giving them a felony record) and by stopping them from being negatively influenced into a life of crime by putting them in prison where they will be surrounded by adult offenders. Hence, the juvenile justice system aims to identify those offenders whose cognitive system is still in development, and it is plausible that neuroscience might be our best tool for identifying these offenders.

Interestingly, the court in *Roper* seemed to reject Morse's idea that only behavioral evidence matters for determining a juvenile's mental capacity. In the *Roper* majority opinion, Justice Kennedy cited much scientific psychology, some of it seemingly based on neuroscience. Noting that juveniles are more subject to peer pressure and that they have less control over their environment, Kennedy then cited an article by Steinberg and Scott (2003). Here is a section of the article Kennedy cited:

> What is most interesting is that studies of brain development during adolescence, and of differences in patterns of brain activation between adolescents and adults, indicate that the most important developments during adolescence occur in regions that are implicated in processes of long-term planning. (Spear 2000)... [P]atterns of development in the prefrontal cortex, which is active during the performance of complicated tasks involving long-term planning and judgment and decision-making, suggest that these higher order cognitive capacities may be immature well into late adolescence (Giedd 1999)." (Steinburg and Scott 2003:1013)

Steinberg and Scott are referring to the slow development of gray matter in the prefrontal cortex, which occurs into late adolescence. Development of gray matter has been identified by Giedd as locationally specific: "These changes in cortical gray matter were regionally specific, with developmental curves for the frontal and parietal lobe peaking at about age 12 and for the temporal lobe at about age 16, whereas cortical gray matter continued to increase in the occipital lobe through age 20" (Giedd 1999:861). It is at least possible that an instrument could be developed that could determine whether an individual lacked mental capacity because portions of their brain relevant to executive function were still under construction.

In sum, if presented in a principled way, by way of analysis of executive function, neuroscientific evidence seems relevant to designation of certain juveniles as belonging to the class of offenders who are less culpable owing to diminished mental capacity.

Mental Retardation

In the case of mentally retarded offenders, neuroscientific evidence of executive function could be used to help inform the criteria for a category of defendants who should be deemed ineligible for the death penalty because of diminished mental capacity, and to help place individual offenders within, or outside of, this category.

In *Atkins v. Virginia*, the US Supreme Court held that the Eighth Amendment forbids states from executing mentally retarded individuals. Although the court discussed several characteristics of mentally retarded people that make application of the death penalty unacceptable, including diminished mental capacity, the court refrained from endorsing any particular definition of mental retardation, leaving it to the states to develop procedures for assessing mental capacity due to mental retardation[24] (Editors 2003).

The *Atkins* court opined that deficiencies in reasoning and communication diminished the personal culpability of mentally retarded offenders and provided two reasons consistent with the legislative consensus that mentally retarded people should be categorically excluded from execution (Atkins 2002:318). The first reason is that putting a mentally retarded person to death doesn't serve the purposes of retribution or deterrence because of their diminished capacity. The second reason, however, is particularly interesting because it identifies a particular case in which CSP attributions of responsibility can be misleading. The court noted that mentally retarded defendants may be less able to give meaningful assistance to their counsel and are typically poor witnesses, and their demeanor may create an unwarranted impression of lack of remorse for their crimes (320–321). As a result, mentally retarded defendants face a special risk for wrongful execution (321). In creating a categorical exclusion from the death penalty, the Court argued it was protecting against these limitations in the commonsense system of guilt attribution.

Persons with mental retardation do not represent a homogeneous, discrete biological or psychological category of persons (Mossman 2003). Similar to the case of juvenile culpability, it would seem that, given a Supreme Court opinion indicating that it is unconstitutional to execute mentally retarded people, it is appropriate that science should step in to help clarify appropriate criteria to determine a particular class of defendants as mentally retarded for the purposes of criminal responsibility. As a note in the *Harvard Law Review* puts it: "[A] general, shared understanding that it is 'wrong' to execute mentally retarded offenders offers no assurance that a specific individual or set of individuals charged with determining whether offenders are mentally retarded would implement that understanding accurately. Scientific research could assist in the creation of objective standards and offer legislators tools to make the determination accurate and non-arbitrary" (Editors 2003:2571).

The American Association on Intellectual and Developmental Disabilities (AAIDD) definition states that a person is considered mentally retarded if:

1. He or she has an IQ test score of approximately 70 or below (but considering Standard Error of Measurement [SEM], the ceiling may go up to and IQ of 75).
2. He or she has deficient adaptive behavior as expressed in conceptual, social, and practical adaptive skills.
3. The age of onset was before age 18.[25]

Most U.S. states adhere to this definition and use IQ score, behavioral evidence and age of onset to establish who counts as mentally retarded for the purposes of sentencing. However, as Donald Bersoff, President of the American Psychological Association, has noted, "IQ…is not the factor that renders the imposition of the death penalty against those with mental retardation unjust. Rather, IQ is a proxy, and an imperfect one at that, for a combination of factors, such as maturity, judgment, and the capability of assessing the consequences of one's conduct, that determine the relative culpability of a mentally retarded killer" (Bersoff 2002:568).

Thus IQ score seems to be an inadequate way to determine whether a person lacks the specific mental capabilities relevant to capacity-responsibility. A neuroscientific understanding of the executive function of those deemed mentally retarded could help "clean up" the category and allow the law to focus on those offenders who have deficits relevant to criminal responsibility. For example, there are inevitably "close" cases where a defendant IQ score falls just above or below the designated cut-off to be determined mentally retarded. Evidence of executive function could help the court determine whether defendants within a range of scores (say, from 65 to 80) have the mental capacity necessary for full criminal responsibility.

CONCLUSION

Douglass Mossman, an MD familiar with Morse's work who is worried about the inconsistent way in which mental retardation is being used as evidence of diminished capacity post-*Atkins*, notes that Professor Morse believes that psychiatric diagnoses of any kind should play little or no role in legal proceedings. However, he argues against this position, stating:

> [T]o the extent that diagnostic schemes are valid and an individual's diagnoses has been properly rendered—a diagnosis invokes the consensus of mental health professionals, not just the opinion of the expert who might be testifying...Identifying a set of symptoms and behaviors with a diagnosis (especially when scientific evidence links biology to patterns of thought and behavior) may serve the additional, important purpose of helping jurors and judges consider other explanations for a defendant's behavior besides such common-but-psychologically-naïve explanations as being the result of sheer evil, carelessness, faking, or not trying hard enough. (Mossman 2003)

Mossman here is speaking primarily of psychiatric diagnoses, not neuroscience. However, I think his argument can be expanded to include all scientifically grounded and reliable evidence of mental states or mental capacity. I take Mossman to mean that science is an important tool that should be used to supplement, and occasionally correct, the CSP way of attributing responsibility. But only good science should be used, and it should be used carefully. Insofar as neuroscience is relevant to capacity-responsibility, and it can reliably speak to the legal criteria for mental capacity—as I have argued—it should be considered just such a complementary tool.

In this chapter, I have argued that neuroscientific evidence can be relevant to determinations of mental capacity. Within the context of rapid advances in neuroscience, I believe that it is possible to reconceptualize the mental capabilities necessary for responsibility—particularly those that are recognized as missing or compromised by the doctrine of "diminished capacity"—in a way that provides a bridge between the direct evidence of brain function and structure and assessment of criminal responsibility. I have argued that formulating these capacities as executive functions within the brain can provide this bridge.

Admittedly, this paper represents a small step toward an understanding of how neuroscientific evidence can be relevant to criminal verdicts. Much more work remains to be done. Such work will require an interdisciplinary approach linking philosophy, neuroscience, and law.

NOTES

1. Associate Professor of Philosophy, Elmhurst College. The author would like to thank Nicole Vincent for her comments on an earlier draft of this paper, and gratefully acknowledges the research assistance of Amanda Bonanotte and Paul Shakeshaft (Elmhurst College graduates 2010).
2. The theory of CSP places mental states in a privileged role in the explanation of human action, where such states are seen as the source or cause of behavior. CSP is thus thought to work in the following way: my CSP uses anything I can learn of your behavior (using my perceptual apparatus and information or assumptions manifest in the relevant mechanisms in my brain) to generate interpretations and predictions of further behavior.
3. See the Model Penal Code (MPC) §2.02. The MPC acts as a guide for US state criminal codes. (Proposed Official Draft 1962).
4. See, for example, 720 ILCS 5/9–1, for both the legal criteria for first-degree murder and the aggravating and mitigating factors for capital punishment in the state of Illinois.
5. In the case of vicarious responsibility, an agent can cause criminal harm for which his employer is responsible.
6. Antony Duff claims Hart's capacity-responsibility "specifies a minimal condition of liability, which is satisfied alike by a willful murderer and by one who negligently causes death. That minimal condition could, however, be taken to require the possibility of effective choice: a person has the capacity to obey the law, we can say, only if she would obey the law if she chose to do so, and has a fair opportunity to obey only if she has a real choice of whether to obey or not." Duff 1993.
7. See, for example, R. v Parks 2 S.C.R. 871 (1992), where the Canadian Supreme Court upheld Parks' acquittal for murder and attempted murder based upon a defense of automatism.
8. An important question then becomes, how low (or high) do we set the bar for "normal adult" mental capacity? The law tends to be quite conservative in answering this question. See Hart 1968, where he claims that "[l]egal systems left to themselves may be very niggardly in their admission of the relevance of liability to legal punishment of the several capacities, possession of which are necessary to render a man morally responsible for his actions" (p. 228). However, others, such as Morse, argue that legal systems should be more conservative in considering legal capacity. See Morse 1984, where he states that the partial responsibility variant of the diminished capacity defense should be rejected. Instead, the law should focus on "whether defendants actually formed mens rea rather than on whether they had capacity to form it..." (p. 55).

9. Morse, and many feminist scholars, have recommended jettisoning the "extreme emotional disturbance" defense. See Morse 2003; Victoria Nourse, "Passion's Progress."
10. Morse has argued that the "partial responsibility" variant should be expressed in adoption of a new verdict, "guilty but partially responsible." *Ibid.* In this case, evidence of diminished capacity could be used to argue that a defendant who clearly possessed the mental states necessary to be found guilty of a higher crime (possessed the requisite mens rea) should be convicted of a lesser offense.
11. See Dennett 1991; Fodor 1987; and Sellars 1956.
12. The Supreme Court noted this worry about a fact finder's commonsense assessment of the mentally retarded in the *Atkins* opinion as a justification for exemption from the death penalty. See Atkins 2002.
13. I provide a detailed explanation as to why this claim fails in Sifferd 2006.
14. Morse claims that a good theory of mind will support folk psychological explanations and should be "thoroughly material but non-reductive and non-dualist. It hypothesizes that all mental and behavioral activity is the causal product of lawful physical events in the brain, that mental states are real, that they are caused by lower level biological processes in the brain, that they are realized in the brain (mind-brain) but not at the level of neurons, and that mental states can be causally efficacious." (Morse 2007: 2545–2575). Then Morse cites John Searle's "biological naturalism" as such a theory.
15. Morse states that "[m]ost fundamentally, action and consciousness are scientific and conceptual mysteries. We don't not know how the brain enables the mind, and we do not know how action is possible" (Morse 2007:2555).
16. Legal scholar and prominent Federal Judge Richard Posner has made such claims. See Posner (1985).
17. For an interesting discussion of criteria a theory of mind must meet to preserve mental causation, see Segal and Sober (1991).
18. See Kim (1995)
19. Nicole Vincent seems to agree with this assessment. See Vincent 2008.
20. Interestingly, an attempt by US courts to use a more "scientific" test for legal insanity, the Durham test, failed. The 1950s saw a surge in interest in the field of psychology, and the invention of new psychiatric drugs forced the public to reconceive mental problems as a medical condition. The Durham test focused on the medical evidence of mental disorder by requiring that the criminal act be the "product of a mental disease or defect." (Durham v. United States, 214 F.2d 862 [1954]). Under the Durham test, a jury was required to answer two questions: (1) did the defendant have a mental disease or defect? and if so, (2) Was the disease or defect the reason for the unlawful act? Both of the answers had to be "yes" to return a verdict of not guilty by reason of insanity.The test never received wide acceptance in the United States; 30 states and five federal circuits examined the test and ultimately rejected it.
21. Some have discussed how couching scientific evidence in CSP terms can produce "framing effects" that affect responsibility assessments. For example, Marga Reimer has argued that when structural and functional differences between psychopaths' brain and normal people's brain are characterized as "biological

disorders," we tend to view them as less culpable than the average person. However, when such differences are presented as "mere biological differences"—not as a deficit or disorder—we view psychopaths as more culpable than the average person. See Reimer 2008.

Similarly, Nichols and Knobe found that information that our universe was "deterministic"—that everything that happens, including human behavior, is "determined" in some sense by the laws of physics—affected assessment of responsibility differently depending on how the information was presented. When asked if one could be held responsible in a deterministic universe, most subjects said "no." However, when presented with a case in which a man killed his wife and kids to be with his secretary, the majority held him responsible despite the fact that his acts were "determined." Nichols and Knobe postulate that the factual scenarios illicit strong emotional responses that are crucial to attribution of responsibility. See Nichols and Knobe 2007.

22. For a detailed argument in support of the claim that executive function grounds legal responsibility, see Hirstein and Sifferd (2011).
23. See Roper 2005, where the court deemed juveniles ineligible for the death penalty, and Atkins 2002, (where the court deemed mentally retarded defendants ineligible for the death penalty).
24. In his dissent, Justice Scalia argued that no general definition of mental retardation should act as a disqualifier from capital punishment and that, instead, mental capacity should be assessed on a case-by-case basis at sentencing. Note that this position is perfectly compatible with the relevance of neuroscientific evidence at trial: it would be used to assess an individual's legal capacity with regard to the legal criteria for execution, instead of placing a defendant into a particular category of persons ineligible for the death penalty.
25. See http://www.aamr.org/content_104.cfm.

REFERENCES

Atkins v. Virginia 536 U.S. 304 (2002).

Banich, M. T. (2009). "Executive function: A search for an integrative account." *Current Directions in Psychological Science* 18(2): 89–94.

Bersoff, D. N. (2002). "Some contrarian concerns about law, psychology, and public policy." *Law and Human Behavior* 26(5): 565.

Burns, J., and R. H. Swerdlow (2003). "Right orbitofrontal tumor with pedophilia symptom and constructional apraxia sign" *Archives of Neurology* 60: 434–440.

Butts, J. A., and O. Mitchell (2000). Brick by brick: Dismantling the border between juvenile and adult justice. *Boundary Changes in Criminal Justice Organizations* (NIJ, US Department of Justice, Office of Justice Programs) 2: 167–213.

Churchland, P. (1981). "Eliminativist Materialism and the Propositional Attitudes." *Journal of Philosophy* 78: 67–90.

Dennett, D. (1991). "Real patterns." *Journal of Philosophy* 88: 27–51.

Duff, R. A. (1993). "Choice, character and criminal liability." *Law and Philosophy* 12(4): 345–383.

Editors (2003). "Implementing 'Atkins.'" *Harvard Law Review* 116(8): 2565–2587.

Elliott, R. (2003). "Executive functions and their disorders." *British Medical Bulletin* 65: 49–59.

Fodor, J. (1987). *Psychosemantics: The Problem of Meaning in the Philosophy of Mind*. Cambridge, MA, MIT Press.

Giedd, J. N., J. Blumenthal, N. O. Jeffries, et al. (1999). "Brain development during childhood and adolescence: A longitudinal MRI study. *Nature Neuroscience* 2(10): 861–863.

Hart, H. L. A. (1968). *Punishment and Responsibility: Essays in the Philosophy of Law*. Oxford, UK, Clarendon Press.

Hirstein, W. (2009). Confabulations about people and their limbs, present or absent. In: *The Oxford Handbook of Philosophy and Neuroscience*. J. Bickle. Oxford, Oxford University Press, pp. 474–512.

Hirstein, W. and K. Sifferd (2011). "The Legal Self: executive processes and legal theory." *Consciousness and Cognition* 20(1): 156–171.

Kim, J. (1995). The myth of nonreductive materialism. In: *Contemporary Materialism*. P. K. Moser and J. D. Trout. New York, Routledge, pp. 134–149.

Lelling, A. E. (1992–1993). "Eliminative materialism, neuroscience and the criminal law." *University of Pennsylvania Law Review* 141(1471): 1471–1564.

Miyake, A., N. P. Friedman, M. J. Emerson, A. H. Witzki, and . Howerter (2000). "The unity and diversity of executive functions and their contributions to complex "frontal love" tasks: A latent variable analysis." *Cognitive Psychology* 41: 49–100.

Morse, S. J. (1984). "Undiminished confusion in diminished capacity." *Journal of Criminal Law and Criminology* 75(1): 1–55.

Morse, S. J. (2003). "Inevitable mens rea." *Harvard Journal of Law & Public Policy* 27: 51–64.

Morse, S. J. (2006a). "Brain overclaim syndrome and criminal responsibility: A diagnostic note." *Ohio State Journal of Criminal Law* 3: 397–412.

Morse, S. J. (2006b). Moral and legal responsibility and the new neuroscience. In: *Neuroethics: Defining the Issues in Theory, Practice, and Policy*. J. Illes. Oxford, New York, Oxford University Press, p. 33.

Morse, S. J. (2007). "Criminal responsibility and the disappearing person." *Cardozo Law Review* 28: 2545–2575.

Moscovitch, M., and G. Winocur (2002). The frontal cortex and working with memory. *Principles of Frontal Lobe Function*. D. T. Stuss and R. R. Knight. New York, Oxford University Press, pp. 188–209.

Mossman, D. (2003). "Atkins v. Virginia: A psychiatric can of worms." *University of New Mexico Law Review* 33: 255–291

Model Penal Code (Proposed Official Draft, 1962).

Marga Reimer (2008). Psychopathy Without (the Language of) Disorder. *Neuroethics* 1 (3).

Nichols, S., and J. Knobe (2007). "Moral responsibility and determinism: The cognitive science of folk intuitions." *Nous* 41(4): 663–685.

Nourse, Victoria, Passion's Progress: Modern Law Reform and the Provocation Defense, 106 Yale Law Journal 1331 (1997).

Posner, R. (1985). "An economic theory of criminal law." *Columbia Law Review* 85: 1193–1134.

Roper v. Simmons 543 U.S. 551 (2005).

Segal, G., and E. Sober (1991). "The causal efficacy of content." *Philosophical Studies* 63(1): 1–30.

Seiden, J. A. (2004). "The criminal brain: Frontal lobe dysfunction evidence in capital proceedings." *Capital Defense Journal* 16(2): 395–420.

Sellars, W. (1956). Empiricism and the philosophy of mind. In: *The Foundations of Science and the Concepts of Psychoanalysis, Minnesota Studies in the Philosophy of Science*. H. Feigl and M. Scriven. Minneapolis, MN, University of Minnesota Press, pp. 127–196.

Sifferd, K. L. (2006). "In defense of the use of commonsense psychology in the criminal law." *Law and Philosophy* 25: 571–612.

Smith, E. E., and J. Jonides (1999). "Storage and executive processes in the frontal lobes." *Science* 283(5408): 1657.

Spear, P. (2000). The adolescent brain and age-related behavioral manifestations. *Neuroscience and Bio-behavioral Reviews* 24: 417–463.

Steinburg, L., and E. S. Scott (2003). "Less guilty by reason of adolescence: Developmental immaturity, diminished responsibility, and the juvenile death penalty." *American Psychologist* 58(12): 1009–1018.

Stuss, D., and R. Knight, Eds. (2002). *Principles of Frontal Lobe Function*. New York, Oxford University Press.

Vincent, N. A. (2008). "Responsibility, Dysfunction, and Capacity." *Neuroethics* 1(3): 199–204.

9

Neuroscience, Deviant Appetites, and the Criminal Law

COLIN GAVAGHAN[1]

Attempts directly to measure sexual interest or preference for legal purposes, while by no means novel, have been rendered particularly topical by some high-profile controversies. In December 2010, Czech immigration authorities were criticized by European human rights advocates (EU Agency for Fundamental Rights, 2010; ORAM, 2011) after it emerged that asylum seekers putatively fearing anti-homosexual persecution were being "hooked up to a machine that monitors blood-flow to the penis and then shown straight porn. Those applicants who become aroused are denied asylum" (BBC, 2010).

In April 2011, the same technology—phallometric testing—was the subject of a report by the British Columbia (BC) Representative for Children and Youth (2011) The report, which followed criticism from the BC Civil Liberties Association (2010) concluded that such testing should no longer be used on young males in BC's sex offender treatment program.

Some of the criticism levied at these initiatives focused on doubts about the reliability of the technology in question, whereas other concerns related to less tangible matters of dignity and respect for the subjects. Indeed, in terms of legal and ethical acceptability, as well as scientific accuracy, there is considerable cause for concern about the use of phallometric testing in a legal context, a concern that has been discussed in a series of court decisions across several jurisdictions.[2]

In this chapter, I consider the possible role of sexual appetite testing in a neurolegal future. Several papers (Ponseti et al. 2011; Sartorius et al. 2008; Schiltz et al. 2007; Walter et al. 2007) have demonstrated the potential for functional magnetic resonance imaging (fMRI) technology to identify and measure sexual interest with potentially greater accuracy than phallometric tests. Furthermore, because fMRI would not involve the potentially undignified techniques involved in phallometry, at least some of the dignity-based objections (ORAM, 2011: 18) would potentially be obviated.

Certainly, it requires little by way of imagination to foresee how a safety-based case could be advanced for seeking to identify, for example, potential pedophiles or other sexually predatory or abusive individuals in various settings. Most obviously, this could be used in parole decisions for convicted sex offenders, where predictions of likely recidivism are already an established part of the process. But such techniques might also be expected to have popular support as a precondition for entry into some sorts of employment with vulnerable charges (such as a preschool teacher).

On the other hand, such technologies cause a measure of unease. How could they be reconciled with the increasingly influential notion of a right to cognitive privacy? Is the probing of our innermost appetites and desires not inherently undignified, regardless of whether the means of doing so involves genital contact? And—perhaps most important of all—does the conflation of appetite with propensity to act not threaten to compress the space in which authentically moral decisions are made, that is, the space in which we opt not to act on our base appetites, but instead to be guided by our higher order faculties, such as ethical imperatives and even enlightened self-interest?

PHALLOMETRIC TESTING: HISTORY, USE, AND CONCERNS

Given the recent controversial uses of such techniques in the Czech Republic, it is interesting to note that the origins of phallometric testing can be traced there, or more precisely, to the former Czechoslovakia, where physician Kurt Freund pioneered such testing in 1957. Freund's innovation, the penile plethysmograph (PPG)—which derives from the Greek words *plethysmos* ("enlargement") and *graphos* ("to write") (Odeshoo 2004–2005:6)—involves "the measurement...of changes in penile circumference or volume occasioned by images of persons who vary in age and sex, or audiotaped stories concerning sexual interactions with persons who vary in age and sex.

An increase in either penile circumference or volume is assumed to indicate sexual arousal, thereby indicating sexual desire" (Camilleri and Quinsey 2007:185).

Although "early versions of the volumetric device were cumbersome, expensive, and tended to break down rather frequently" (Marshall and Fernandez 2000:808), the Czechoslovakian government required Freund "to use it to determine the sexual orientation of military recruits claiming to be homosexual for the purpose of avoiding military service" (a use more than a little reminiscent of the uses to which the Czech government put the technique last year). Freund's own views on homosexuality were considerably more progressive; he later advocated its decriminalization and became one of the first opponents of the characterization of homosexuality as a disorder (Odeshoo 2004–2005:7). Nonetheless, his technique was used in conjunction with aversion therapy in an attempt to "cure" homosexuals by "administering a shock to subjects whose plethysmographic tests indicated arousal to homoerotic stimuli" (Odeshoo 2004–2005:7).

Attitudes to homosexual preferences may have changed, at least among the medical profession, but the PPG has continued to be used in the detection and treatment of sexual disorders—although it has been said: "It is difficult to ascertain just how widespread use of the technique is at the current time" (Odeshoo 2004–2005:7). Most commonly, it has been used in assessing convicted sex offenders, although research on nonoffenders has also taken place.[3] Although other techniques purporting to offer objective information have been used, including "measuring penile temperature, penile surface blood volume, skin conductance responses, and pupillometry," their predictive value has been reported to be poorer than that of the PPG (Camilleri and Quinsey 2007:188).

Objections to the use of PPG have come in various forms. Some have concerned the accuracy of the technique, and particularly, its vulnerability to faked responses, with some studies seeming to show that subjects "can significantly inhibit their arousal by using mental activities to distract themselves, despite a clear indication that they were attending to the stimuli . . . " (Marshall and Fernandez 2000:810) Clearly, if the PPG does not yield accurate results, and if it is indeed "impossible to prevent or detect stimulation," (Marshall and Fernandez 2000:810), then the argument in favor of its continued use is seriously undermined.

Other concerns have related to what is claimed to be the inherently undignified nature of the technique. As one critic of the technique has argued, "with the possible exception of cavity searches, it is difficult to imagine a more drastic form of interference with a person's bodily integrity than that entailed in PPG" (Odeshoo 2004–2005:21).

The description of the technique presented to the European Court of Human Rights in *Toomey v. United Kingdom* (discussed later) provides some insight into this aspect of the technique:

> On 3 October 1997 the applicant underwent his first PPG test and it lasted 1 hour and 20 minutes. A female trained technician conducted the tests. The applicant was put in a small room without windows and with bolts both inside and outside the door. Two electrodes were attached to his left index and middle finger. A video recorder was adjusted to the level of his face and the operator left the room although monitoring of the applicant was possible via a camera and a microphone. The applicant then had to attach a sensor clip to his penis and to leave his underpants and trousers removed throughout the test.

In some instances (though this was not referred to in *Toomey*), the subject is next "instructed to become fully aroused, either via self-stimulation or by the presentation of so called 'warm-up stimuli,' in order to derive a baseline against which to compare later erectile measurements" (Odeshoo 2004–2005:9). The subject is then shown a variety of potentially arousing material. In *Toomey*, this comprised

> ...three categories of material. The first category...was a set of slides comprising nude images of young children, pubescent and adult males and females. The images were produced in the United States for the purpose of the PPG assessment and all images were single frontal nudes, either sitting, standing or prone but not posed in a deliberately erotic fashion. The second category was a set of video sequences depicting consensual sex, rape and non-sexual violence...The third category was a set of slides depicting young men and women in more erotic poses than in the first set of slides, elderly naked women in relatively non-erotic poses and women in bondage poses...The slides were left on for about 20 seconds and each was shown approximately six times...The television on which the slides and videos were shown was at eye level about 18 inches from his face and the applicant's head was kept steady by a headrest on the back of the chair.'

In view of the nature of the PPG, it is perhaps unsurprising that it has been severely criticized on the grounds of its "extreme invasiveness" (ORAM 2011:17) and "intrusive and degrading" nature (Odeshoo 2004–2005:9). Furthermore, a number of other concerns, ranging from the possibility of traumatizing participants by exposing them to violent sexual stimuli[4] to the nature of some of the images used,[5] have been advanced.

PENILE PLETHYSMOGRAPH AND THE LAW

Challenges to the use of PPG testing have been raised in a number of jurisdictions. Typically, these have taken two forms: challenges to the process of testing per se, and challenges to the admissibility of evidence derived from PPG evidence in trials or other legal proceedings. An example of the former type of challenge can be seen in *Toomey v. United Kingdom* (No. 37231/97, ECHR [1999]).

Here, the applicant was an individual convicted of assaulting (in a serious, but nonsexual manner) two young women. Having been released on license, T was subsequently recalled to prison after aspects of his behavior caused concern to the Secretary of State. At the request of his own counsel, T underwent a range of tests, including electroencephalogram and computed tomography scans, as well as PPG. T subsequently complained that the PPG test constituted cruel, degrading, and inhuman treatment and torture in terms of Article 3 of the European Convention on Human Rights.

While accepting that the test was experienced as humiliating by T, the Court rejected his contention that it violated his Article 3 rights. In part, this decision was based upon the therapeutic nature of the test—it was argued by the UK Government to be a necessary precursor to T's participation in a sex offender treatment program. It would appear, however, that the Court's decision also rested upon the view that the treatment did not reach a minimum threshold to constitute a breach of Article 3.[6]

In *R. v. J.-L.J.*,[7] the Supreme Court of Canada considered the question of admissibility of PPG evidence. This case concerned an attempt by an individual accused of a series of sexual assaults on young children. The accused sought to lead evidence that he lacked the deviant personality traits that would be required to perform the acts in question and sought to use PPG results and expert psychiatric testimony to support this contention.

In upholding the trial judge's ruling that the evidence was inadmissible, the Supreme Court expressed some doubts about the reliability of the PPG[8] and about the contention that the crime could only have been committed by someone with the deviant appetites in question.

This mixed attitude to PPG tests—whereby it is deemed admissible for purposes of parole, but not for use in trials—has been mirrored in the United States: "where the test is required of individuals who have not been convicted of sex offences, it is virtually always struck down, whereas when a person has been convicted of a sex crime, it is virtually always upheld" (Odeshoo 2004–2005:20) As Odeshoo goes on to note, however, "greater sophistication of the technique, as well as recent changes in evidence law, may one day open the door to expert testimony based on the procedure."

A MOVE TO NEUROTESTING?

It can be seen, then, that objections to PPG testing typically take two forms:

1. Concerns regarding the putatively undignified or degrading nature of the treatment
2. Concerns regarding the accuracy of the technique

Although such concerns have not always found favor with the courts, they are certainly reflected in the responses of organizations such as ORAM and the BC Civil Liberties Association to recent uses of PPG testing, and they persist in academic commentary on the technique's use.

In this section, I consider the possibility that other forms of neurobiological testing may, to an extent, address these concerns by providing a mechanism for measuring deviant arousal that is both more accurate and less degrading than that offered by the PPG. In the final section, I turn my attention to some of the ethical and jurisprudential concerns raised by reliance, for legal purposes, on information about sexual preferences and appetites.

In both sections, I pay particular attention to pedophilic inclinations. This reflects, in part, the fact that attempts to measure such inclinations have featured in several of the legal controversies that have already arisen. In addition, though, I would suggest that such inclinations may be the most plausible candidates for the introduction of mandatory inclination testing, both within and outwith the criminal context. Both public and political concern with child sex abuse is, perhaps, quantitatively different from other sorts of crime, a reaction mirrored in the unique distaste directed at its perpetrators.[9]

Furthermore, I would suggest that this is perhaps the area in which inclination is most frequently conflated with action, with the terms "pedophile" and "child sex offender" being all but synonymous in everyday parlance.[10,11] For these reasons, I suggest that pedophilia presents the most plausible candidate for the criminalization of deviant appetites.

NEUROSCIENCE AND PEDOPHILIA

What prospects exist for detecting pedophilic inclinations by neuroscientific means? At the time of writing, "research in neurobiological correlates of pedophilia is scarce." (Sartorius et al, 2008:271; see also Dressing et al. 2008:9) Furthermore, it has been said that "[d]espite multiple investigations, evidence of a causal relationship between abnormal brain functioning and pedophilia has remained elusive" (Schiffer et al. 2007:753).

Nonetheless, some research on individuals convicted of sexual abuse of children has been conducted, typically in the form of fMRI. Some of these have reported finding differences in brain structure between pedophiles and control groups,[12] whereas others—of more relevance for the purposes of this discussion—have detected differences in brain activity when exposed to relevant kinds of erotic stimuli. One study reported that, "relative to adults, controls show less amygdala activation and pedophiles more activation while viewing children" (Sartorius et al, 2008: 274). Another reported:

> Pedophilic patients showed significantly lower signal intensities…in the hypothalamus…to dorsal midbrain…the dorsolateral prefrontal cortex…the right lateral parietal…the right ventrolateral…and the right occipital cortex…as well as the left insula…for the sexual arousal condition than healthy subjects. (Walter et al. 2007:698)

This, the investigators claimed, "demonstrate[d], for the first time, abnormal neural activity in subcortical and cortical regions in pedophilia during sexual arousal" (Walter et al. 2007:699), a finding that "could…be considered a complementary tool to investigate the pedophilic patient's 'true' feelings of sexual arousal" (Walter et al. 2007:700). Most recently, Ponseti and colleagues claimed to have used fMRI technology to distinguish pedophiles from "healthy controls with a high degree of accuracy" (Ponseti et al. 2011).

A body of evidence is, therefore, emerging in support of at least the possibility that pedophilic inclinations may be detectable by techniques such as fMRI. Nonetheless, some caution should be exercised before accepting these claims at face value. The authors of one report acknowledged a potential methodological limitation of their study, specifically that "our results may be restricted to pedophiles who are exclusively attracted to children and are not able to discourage themselves from the sexual abuse of children" (Schiffer et al. 2007:760). In fact, this is a potential issue with all but one[13] of the studies discussed here, all of which have taken convicted child sex offenders as their study groups; perhaps what is really being measured is not so much the deviant appetite, but rather the inability to refrain from acting upon it.

If this is so, then a possible danger in overreliance on neurotesting would lie in the prospect of potential false negatives; that is, the tests would fail to identify individuals who harbored pedophilic inclinations that were not either exclusive or coupled with irresistible urges to act upon them. Whether such potential false negatives should be seen as a problem, of course, very much depends on what were considered legitimate uses for such techniques.

The potential concerns about the use of neuroimaging to detect pedophilic inclinations, however, do not lie exclusively—or perhaps even predominantly—with methodology. The more pressing question, it seems to me, concerns whether and when the state is justified in demanding information about sexual inclinations.

Some of these concerns may be specific to particular uses of the information. As noted earlier, those courts that have had occasion to consider PPG testing have tended to distinguish between, on the one hand, the technique's use for the purposes of assessing the suitability of convicted felons for parole and, on the other, its use in determining guilt or innocence. Although the latter use has generally been rejected, the use of phallometric testing in the parole context seems to be fairly widely accepted.

This may reflect a sense that the convicted party has, by virtue of his previous offence, to some extent forfeited some of the rights and liberties to which the rest of society is entitled. Alternatively, it may derive from the intuition that, while a "fair trial" is a universal entitlement, the same cannot be said of parole; provided the individual has been fairly tried and sentenced, he has no *right* to have that sentence reduced, and therefore cannot complain if certain conditions are attached. It is not, after all, uncommon for parole boards to be able to demand answers to questions that would never be permitted in a trial setting.

Nonetheless, even if we accept that the parole hearing is the least objectionable setting for the use of sexual inclination testing, it may be argued that a number of legitimate concerns about such use exist.

ACCURACY

In *R. v. J.-L.J.*, the Supreme Court of Canada expressed doubts as to the accuracy of the phallometric evidence. Even if neurotesting were to be much more accurate at measuring responses to particular stimuli, however, we may still be legitimately concerned that the technology will not provide accurate answers to the questions asked of it.

If we consider the prospect of neuroevidence being used in a trial setting, perhaps in the manner that the defense sought to admit it in *R. v. J.-L.J*, then the danger soon becomes apparent. Roughly speaking, the defense case rested on the following reasoning:

1. Only someone with a sexual interest in children could have committed the crime in question.

2. The defendant did not possess a sexual interest in children.
3. Therefore, the defendant could not have committed the crime.

In asking whether neurotesting could provide us with greater "accuracy," the temptation may be to concentrate on the second premise of the syllogism. Yet even if we allow that neurotesting—now or in the future—could gauge the defendant's response to children with greater accuracy than PPG technology, a potential problem remains with regard to the first premise, a problem that derives from an apparently erroneous assumption about the relationship between pedophilia and child abuse. As Fagan and associates explain: "Not all who sexually abuse minors are pedophilic. For example, some who sexually abuse children may opportunistically select minors simply because they are available" (Fagan et al. 2002:2459). If the first premise is false, then the use of "exclusionary" neuroevidence raises the prospect of wrongful acquittals; the fact that the defendant does not fit the profile of a pedophile does not mean that he cannot have committed a sex offence against a child.

Even more troubling, however, would be the prospect of a prosecution case seeking to rely on such evidence. The strategy would presumably be to demonstrate that, since (1) only a small percentage of the population harbored a particular Appetite X, and (2) the accused was one of that minority, then (3) it is more likely that the accused committed Offence Y. Of course, this strategy could be undermined by the observation I have already made about the type of argument attempted in Canada; indeed, it may be that the prosecution's syllogism would require the insertion of an additional, and highly contestable, premise, to the effect that *only* individuals with Appetite X ever commit Offense Y.

These examples show that, however accurately it is possible to gauge the presence or absence of pedophilic urges, serious limitations will exist with regard to what we can safely infer from this about the presence or absence of guilt. More accurate measurements will not invariably yield more accurate answers to the important questions of criminal law.

DIGNITY

Whether neurotests are perceived to pose less of a threat to the dignity of the subjects than PPG tests is open to argument. If what is seen as most undignified about phallometry relates to the exposure of genitals and direct measurement of genital responses, then the switch to fMRI would certainly obviate such concerns. Yet it is not at all clear that it is the physical nature of the PPG procedure that is the only, or indeed the most, undignified aspect.

Indeed, it is quite conceivable that, for many people, the truly humiliating aspect of the procedure would lie with the mere fact of having their most intimate sexual fantasies revealed before strangers. One might expect such feelings to be compounded when the reactions are of a nature that the subject himself finds repellant, as may not uncommonly be the case with pedophilic appetites.

Analogous concerns have been expressed about the use of neurotechnologies to aid the detection of deception: "The idea that our bodies can be reduced to a *means* by the state—that the human body itself can be a crime control technology—offends human rights at its very roots in human dignity" (Bowling et al. 2008:66).

Whether this would reach a level sufficient to constitute a violation of any particular legal guarantees of dignity is uncertain; as we have seen, the European Court in *Toomey* did not consider that use of the PPG constituted such a violation. For those concerned with the ethical acceptability of future generations of tests, however, the implications for the dignity interests and rights of the subjects cannot easily be dismissed.

PRIVACY

Much the same could be said with regard to concerns about privacy. As Odeshoo has persuasively argued, "If there is any aspect of our lives that most of us wish to keep private, it is our sexual fantasies and desires" (Odeshoo 2004–2005:1). Whether it follows that neurotechnological examination of these areas of life would constitute a legal violation of any privacy right is, at this time, uncertain. In those jurisdictions that have legal or constitutional guarantees of privacy—such as the United States and the states covered by the European Convention on Human Rights—courts may be invited to consider whether neurotesting for deviant appetites would constitute infringements of those guarantees.

In the meantime, neuroethicists have begun to consider the notion of a right to—or at least, an interest in—what has variously been described as cognitive privacy (Gazzaniga 2005:107), thought privacy (Illes and Racine 2006:149), or simply "the privacy of my inclinations and my thoughts" (Kennedy 2006:60). As the author of one article on the subject has argued:

> ... it is time for the privacy debate to acknowledge, and make explicit, that a person's mind and mental processes must be protected as private ... Certainly, a person's thoughts and thought processes belong to himself or herself, and not to society, the government, or any other meddlesome external force. (Boire, 2000)

As with concerns about dignity, this is of course merely a *pro tanto* objection, and I do not mean to beg the question as to whether such concerns could ever be outweighed by competing concerns about public safety. Nonetheless, some serious reflection on the implications for privacy would seem appropriate before neurotesting becomes a realistic option.

THE RISK OF LINE BLURRING

Whether or not a move to neurotesting would address the concerns posed by PPG testing may depend, to an extent, on the form that technology takes, but it seems unlikely that *all* concerns about dignity and accuracy—and perhaps *any* concerns about privacy—will be addressed to the satisfaction of all critics. In this part of the paper, I will turn to a different concern that would be posed by an increasing reliance on sexual appetite testing. It seems to me that a danger exists of conflating appetites with intentions, a danger that would blur certain important distinctions that have profound implications for law.

Again, the example of pedophilia and its relation to child sex abuse may be helpful in illustrating my concern. The Fourth Edition of the *Diagnostic and Statistical Manual of Mental Disorders* (DSM-IV) defines the "disorder" as being

> ...characterized by either intense sexually arousing fantasies, urges, or behaviors involving sexual activity with a prepubescent child (typically age 13 or younger). To be considered for this diagnosis, the individual must be at least 16 years old and at least 5 years older than the child.

Pedophilia is listed in DSM-IV among the "paraphilias," which are defined as follows:

> Paraphilias all have in common distressing and repetitive sexual fantasies, urges, or behaviors. These fantasies, urges, or behaviors must occur for a significant period of time and must interfere with either satisfactory sexual relations or everyday functioning if the diagnosis is to be made. There is also a sense of distress within these individuals. In other words, they typically recognize the symptoms as negatively impacting their life but feel as if they are unable to control them.

It should be clear from the definition, then, that neither pedophilia itself, nor the wider class of paraphilias to which it belongs, includes among its diagnostic criteria that the affected individual has ever, is likely to, or intends to act on his deviant inclinations. Rather, "Possessing such fantasies and being distressed by them is sufficient to meet diagnostic criteria" (Fagan et al. 2002:2460).

Whether such a subjective test of "mental disorder" is an appropriate one is something of a moot point. Many individuals who harbor homosexual inclinations are profoundly distressed by them, particularly if they seem to clash with other cultural or religious values. The question of the ethical acceptability of attempting to "cure" such individuals of their sexual desires that are incongruous with their other values has been the subject of discussion in academic literature for some time (Murphy 1991; Yarhouse 1998).

(Of course, we may seriously consider that the ethical objections to acting on a homosexual urge are of a wholly different order from the objections to acting on a pedophilic urge. My point here is simply that, if the clinical criterion relates only to the subjective distress experienced by the individual, it is difficult to see how one could be included within and the other excluded from the definition of paraphilia.)

For present purposes, however, the concern is rather different. First, the use of the results of such tests may risk conflating urges with wants or intentions, in a confusing and possibly dangerous manner. Because there is already a rich philosophical and psychological literature dealing with each of these concepts, I should define the sense in which I will use them here. By an "urge," I mean simply a positive or rewarding disposition accompanying a particular thought. I may feel urges when I see a succulent confection advertised on television—or indeed, when I see it advertised by a particularly lovely model or actress!

Urges need not be overwhelming; many urges can be resisted with a greater or lesser degree of effort. Urges may also be incompatible, either with other urges or with our higher order desires or values; the urge generated by the sight of the confectionary may be impossible to reconcile with my desire to lose weight, for example, or indeed, my inclination not to be manipulated by television adverts. Indeed, some of our urges may be repellant to us, at an ethical or aesthetic level.

When I speak of wants,[14] I refer to the dispositional state wherein an urge is associated with a wish to bring about a particular state of affairs.[15] My urge generated by the sight of the chocolate bar would become a want if it led to the mental state wherein I wished to have the chocolate bar, when I thought something like, "if I could have that chocolate bar in my hand right now, I would will it thus." Wants may be fanciful or realistic, compelling or easily resisted, persistent or fleeting. And as with urges, they may sit more or less well with our other tastes and values. The main difference between a want and an urge, though, lies in the disposition toward bringing about a particular state of affairs. (Although as Michael

Bratman has pointed out, it is possible that "I might know that any realistic means to what I desire would itself be something I do not desire" [Bratman 2009:231].)

The notion of intention is one of the most discussed and disputed in legal literature,[16] and it is not my purpose to revisit it here. Suffice to say that, for the purposes of my argument, I speak of intention (at least in the context of what is generally referred to as "direct intention") as conveying not only a wish that some state of affairs could come to pass, but a purposeful setting of the mind toward bringing it about. My urge for the chocolate bar becomes a want when I think something like, "if I could have that in my hand, I would make it thus." It becomes an intention when I formulate some sort of strategy for bringing about that state of affairs, when I think something like, "I will walk to the local shop now and buy one of those chocolate bars."

Intentions can be more or less definite—instead of, "I will walk to the local shop now," I could have formulated an intention along the lines of, "I must try one of those chocolate bars next time I am passing the local shop," or even, perhaps, "someday, I will try one of those chocolate bars." Equally, they may vary in terms of their resoluteness (we all surely formulate countless intentions that are never turned into actions), in their specificity (I may determine to travel to London in June, without yet knowing the details of that trip), and in the extent to which they fit with our broader system of values and beliefs.

It is the class of intentions, rather than urges or wants, that is typically relevant to questions of criminal accountability. A wish that a state of affairs could be so must be wedded to some sort of plan to bring it about before someone's mental state becomes of interest to the criminal law. Because intention is a (usually[17]) necessary but not sufficient prerequisite for such attribution—it must also be accompanied by some act or omission—the prospect of the introduction of some sort of "thought crime," based on the results of fMRI scans, seems fairly remote. That is, we would not (under anything resembling our current legal system) look to punish someone merely for intending or wanting to carry out a particular action, far less from the mere fact of feeling urges toward it (it being only the latter state that can apparently be detected by the sorts of fMRI technology discussed in this chapter).[18]

Yet the presence of deviant urges may be used to justify deprivations of liberty all the same. As discussed earlier, PPG evidence has been admitted, in several jurisdictions, in parole hearings, and it is surely not improbable that neuroevidence could be used in a similar manner. Furthermore, the possibility exists that it will form the basis for post-sentence detention in those jurisdictions—including several Australian and American states—that provide for this.[19]

Why should this be a cause of concern? Is it not justifiable for a society, when deciding how to deal with someone already convicted of a serious offence, to seek to satisfy itself that he no longer harbors the urges that led him to offend in the first place? For one thing, it should be noted that the urges in question may bear no relation to the original offence. In some jurisdictions, it is possible that such evidence could be admissible in decisions about parole and post-sentence detention, even where the original offence did not concern offenses of a sexual nature or offenses against children.

When discussing the possible use of neurotesting to inform a verdict in a criminal trial, I emphasized the point that it is not *only* pedophiles who sexually abuse children. It is equally important to remember that it is not *all* pedophiles who abuse children. Because the diagnostic criteria can be met by the mere presence of "fantasies" or "urges," the possibility at least must be considered that there exists a cohort of "pedophiles" who have never abused, nor even formulated an intention (or perhaps even a want) to abuse, children.

As almost all of the research to date has been conducted on convicted sex offenders, reliable data on the size, or even the existence, of such a cohort of "harmless pedophiles" is elusive.[20] However, in the absence of some reason to assume that pedophilic urges are typically or invariably accompanied by impaired self-control, it seems reasonable at least to share Joel Feinberg's view:

> There is no a priori reason why the desires, impulses, and motives that lead a person to do bizarre things need necessarily be more powerful or compulsive than the desires that lead ordinary men to do perfectly ordinary things. It is by no means self-evident, for example, that the sex drives of a pedophiliac, an exhibitionist, or a homosexual must always be stronger than the sexual desires normal men and women may feel for one another. (Feinberg 1970:282)

To return to my previous terminology, someone experiencing a pedophilic *urge* may never develop a pedophilic *want*, if he does not identify[21] with that urge in such a way that he would wish it to be fulfilled. And that person may not develop an *intention* to abuse a child unless he sets his mind toward bringing about the satisfaction of that urge.

These, I argue, are important distinctions not only for reasons of legal culpability but also for our ideas of fair attribution of moral praise or blame. Before trying to articulate the exact nature of my concern, however, I should perhaps distinguish what I consider to be a kind of ethical red herring. It may be thought that conflating urges and intentions risks blurring an important line between mental states that we can, and those that we cannot, control. On this argument, it may be legitimate to hold someone responsible for an intention he actively formulates, or even for a want that he actively cultivates or with which he identifies, but not for an urge that he passively experiences.

As Walter Glannon argues, however, although we do not choose to have mental states such as beliefs and desires ab initio, it is not irrational to hold someone responsible to the extent that he retains "the combined cognitive (rational) and affective (emotional) capacity to critically reflect on [his] mental states and eliminate, or else modify or reinforce them and come to identify with them as [his] own" (Glannon 2002:18). If this is correct, then it would not be irrational to follow Jonathan Glover in thinking that, although it may not be fair to blame an individual for possessing a "bad desire," it may be fair to hold him responsible when he does not desire to change that bad desire. (Glover 1970:80).

If urges may, to some extent, be chosen (at least insofar as we choose to embrace them, or choose not to challenge or try to change them), and intentions may sometimes be unchosen (when they are irresistible responses to compelling urges with which we do not identify), then this provides a weak platform for any argument that only intentions, and never urges, can attract justifiable blame. Yet the idea that someone could be detained merely because of the presence of an urge seems intuitively problematic.[22]

Perhaps we might seek to account for this intuition by pointing out that urges, alone, present no danger of the sort that would justify depriving someone of liberty. Without evidence that an individual finds an urge to be irresistible, there is no reason to suppose that he will necessarily, or even probably, act upon it. Yet this, too, fails adequately to distinguish intentions from urges. Someone may formulate an intention that is, for practical purposes, wholly unrealistic (some of those arrested during New Zealand's now infamous Urewera police raids in 2007 are alleged to have plotted to assassinate George W. Bush by catapulting a cow on top of him!), whereas an individual's circumstances may mean that an urge would be easy to act upon, should his resolve ever slip. It does not invariably follow, then, that the possessor of an intention will always be more dangerous than the possessor of an urge.

If neither responsibility for the mental state, nor the threat presented by that state, serves to distinguish urges from intentions, then how can I account for the intuition that it is more problematic to detain people for urges than for intentions? Perhaps the intuition derives from a concern that detaining people on the basis of urges rather than intentions risks disrespecting them as moral agents. The argument would look something like this: it is precisely in the space between urges and intentions (or perhaps between urges and wants) that we are able to operate as moral agents. It is precisely when I elect not to act on my urges—or to formulate intentions to act upon them, or maybe even to wish that they could be made real—that I conduct myself morally.

This sort of neo-Kantian approach would lead us to say that it is not morally praiseworthy of me to refrain from eating crate-raised veal if the very thought of such meat makes me nauseated, but only if I refrain for legitimate moral reasons (such as an aversion to cruelty). I believe that a similar concern may exist here: that a policy of removing or excluding those who harbor "deviant" urges from any situation in which they could potentially act on those urges denies them the status of moral agents.

In the case of someone who has previously demonstrated an inability or unwillingness to control those urges, such a denial may have considerable justification. Even in the parole or post-sentence detention scenario, though, we may worry that individuals are being assessed on the wrong criteria—that in requiring evidence of changed appetites, rather than improved moral capacities such as self-restraint or an appreciation of the wrongness of hurting others, something significant is being lost. At the risk of making light of a serious issue, we would not typically require that, in order to be "reformed," a burglar had lost interest in DVD players or expensive jewelry; rather, we would want to satisfy ourselves that he had come to appreciate the extent that his activities distressed his victims and had undertaken not to inflict similar distress again. Rehabilitation, perhaps, should aim at creating better, not only safer, people.

CONCLUSION

I have shown that technologies for measuring patterns of "deviant" arousal are nothing new. Although the use of techniques such as the PPG remains controversial, it is clear that at least some governments of Western democracies consider them to be justified. Although human rights groups have lobbied—with some success—to have at least the most controversial uses halted, it is not entirely clear that any legal or constitutional rights are violated by such uses.

My suggestion is that new generations of such techniques, making use of technologies such as fMRI, may offer a more reliable, and arguably less invasive and degrading, means of making such determinations. Where it is considered justifiable to determine whether an individual harbors some deviant sexual inclination, a *pro tanto* case will emerge for using the most accurate technologies available.

It is not my purpose in this chapter to argue that such uses could never be justified. The balance to be struck between, for example, privacy and public safety must be determined with reference to particular sets of facts.[23] It may be that, all things considered, the very grave potential harms that may be[24] avoided by neurotesting as a parole condition, or as a basis for post-sentence detention, are deemed to outweigh the ethical and legal concerns I have identified here. Nor am I suggesting that society is enriched by having frustrated pedophiles, sadists, and the like in our midst. All else being equal, it may well be preferable if no person had to carry the burden of a permanently frustrated urge, and preferable if society did not have to fear that—for some of them—that burden may one day become too weighty. At least as absurd would be the notion that we should create situations of temptation wherein "moral pedophiles" can test their resolve.[25]

I should also, perhaps, be clear that I am not convinced that—as Stephen Garvey recently argued (Garvey 2009)—child-abusing pedophiles should be "given a break" because they usually resist their unlucky urges. Unless their urges can be shown to be literally irresistible, I suspect that they should be culpable for failing to resist them (though, with Vera Bergelson, I am certainly open to the possibility that their generally successful struggle against temptation may properly be reflected in sentencing. (Bergelson 2009)).

My purpose, rather, is to highlight a range of concerns that could easily be overlooked amid an (understandable) clamor to protect children from "pedophiles." First, we should, I suggest, be *very* concerned about any move toward precrime or thought crime, that is, penalizing (in whatever way) those who have done no more than possess deviant urges or appetites. The possession or retention of urges may, in some circumstances, be appropriate targets of moral blame, but it would be a remarkable extension of the reach of criminal law were it ever to regard urges alone as sufficient for the infliction of punishment. Indeed, any such extension is probably highly unrealistic under anything like the criminal justice systems in at least the common law world.[26]

Second, I think that we should be *somewhat* concerned about use of appetite evidence in determinations of innocence or guilt—at least until we have a more nuanced appreciation of the neuroscience and have given due consideration to exactly what questions we are seeking accurately to answer. Pedophilia, it would appear, is neither a necessary nor a sufficient condition for the commission of child sexual abuse, but while "pedophile" and "child abuser" are commonly used synonymously, a real danger exists that juries could be inappropriately swayed by neuroevidence pertaining to the presence (or, indeed, the absence) of pedophilic urges.

Finally, I suggest that we should be *somewhat* concerned about (over) reliance on neuroevidence of urges in decisions regarding parole or post-sentence detention. Society should, I think, be open to the possibility that someone has developed other moral or psychological capacities, such as empathy or self-control, such that—while he retains the same "deviant" sexual preferences—he presents very little danger. Rehabilitation, I have suggested, may (at least in part) be about making better, and not just safer, people.

We have no obligation to praise the "moral pedophile" merely for adhering to minimal moral and legal duties. Neither are we obliged to forgive him if, on a rare occasion, his resolve proves too weak, and he acts on his urges. But if we recognize that such people exist, that the presence of deviant urges and appetites is not inconsistent with the presence of fully developed moral capacities, then we should be wary of treating urges alone as a sufficient reason to deprive someone of their liberty. We are not—at least, most of us—doomed slavishly to follow our urges, and it risks diminishing our status as legitimate moral agents when it is implied that we are.

NOTES

1. This chapter has benefited greatly from thoughtful discussions with and helpful suggestions from Anson Fehross and Mike King. Thanks are also due to the participants at the Neurolaw workshop organised by Nicole Vincent at McQuarrie University, and Grant Gillett's neuroethics class at Otago University, who responded kindly and helpfully to earlier, rougher versions.
2. See, for example, US v. Powers, 59 F.3d 1460 (4th Cir. 1995); Doe ex. rel. Rudy-Glanzer v. Glanzer, 232 F.3d 1258 (9th Cir. 2000); R. v. J.-L.J. (2000) 2 S. C. R. 600 (Can.); and Toomey v. United Kingdom, App. No. 37231/97 (1999).
3. "It has been used to determine, for example, sexual preferences for forced or consenting sex among university students who have a proclivity to rape…Much of this type of research with non-offenders is aimed at clarifying the role of various factors in instilling a propensity to be forceful in a sexual context." Marshall and Fernandez, 2000:807.
4. As the report into the use of PPG in British Columbia pointed out, many of the sex offenders subjected to the test had themselves as children previously been victims of sexual abuse. The applicant in *Toomey v. UK* claimed to be traumatized by being shown images of a rape, having himself been the victim of sexual assault in childhood.
5. Odeshoo argues that the inherent "sexualisation" of children involved in the production and use of the photographs is a legitimate cause for ethical concern.
6. A similar result was reached *Walrath v. United States* 830 F. Supp. 444 (N.D. Ill. 1993), where the court determined that PPG examination was not "exceptionally more intrusive" than other physical or psychological tests.
7. [2000] 2 S.C.R. 600.
8. "[P]enile plethysmography has received a mixed reception in Quebec courts," at paragraph 35

9. For some examples of such reactions, see McSherry, B., and P. Keyzer (2009). *Sex Offenders and Preventive Detention: Politics, Policy and Practice*. Leichhardt, The Federation Press, particularly chapter 1.
10. A particularly confused-seeming example of this conflation can be found on the website of the New Zealand-based Sensible Sentencing Trust, which at some points seems to distinguish "pedophiles" from "sex offenders," but at others treats the former category as de facto child abusers; see http://www.safe-nz.org.nz/sxdb/sxdb.htm.
11. A similar conflation, it must be said, is found not uncommonly in academic writing. See, for one of myriad examples: "paedophiles, rapists and murders [sic] are supposed to be treated without reference to their crimes or moral character." Draper, H., and T. Sorrell. "Patients' responsibilities in medical ethics." *Bioethics*, 16(4), 335–352.
12. "We found that the pedophilic perpetrators had a significantly smaller amygdalar volume, with the difference predominantly on the right side" (Schilz et al. 2007:742). "Compared with healthy controls, patients with pedophilia showed a significantly lower amount of gray matter volume in the bilateral orbitofrontal cortex, the bilateral insula, the bilateral ventral stratium (putamen) and some limbic gyri (cingulated and parahippocampal)" (Schiffer et al. 2007:757).
13. The exception, the study by Ponseti et al. (2007), used self-confessed pedophiles attending outpatient sexual medicine departments.
14. I refer to "wants" instead of the more commonly used "desires" because the latter term may be somewhat ambiguous as to whether it describes a motivational state or a mere appetite.
15. Again, I readily accept that this use may diverge from others in the literature, but I hope that at least the concepts are intelligible, if not the terms I use to describe them.
16. See, for a few among many examples, Duff, R. A. (1989). "Intentions legal and philosophical." *Oxford Journal of Legal Studies* 9(1): 76–95; Lacey, N. (1993). "A clear concept of intention: Elusive or illusory?" *Modern Law Review* 56(5): 621–643; Moore, M. S. (2011). Intention as a marker of moral culpability and legal punishment. In: *Philosophical Foundations of Criminal Law*. R. A. Duff and S. P. Green. Oxford, UK, Oxford University Press.
17. We can ignore, for present purposes, crimes of strict and absolute liability.
18. Whether such a prospect is wholly fanciful, though, is less certain; certainly, a growing body of commentary is pointing to a "temporal shift" in criminal justice policy. Lucia Zedner, for example, has predicted that "we are on the cusp of a shift from a post- to a precrime society," which will shift "the temporal perspective to anticipate and forestall that which has not yet occurred and may never do so." Zedner, L. (2006). "Pre-crime and post-criminology?" *Theoretical Criminology* 11(2):261–281, at p. 262. Quite where this trend is likely to end is impossible to predict, but it is probably fair to say that it has a distance more to travel before it leads to the criminalization of mere thoughts alone.
19. For an excellent overview of these schemes and the issues of policy and principle that they raise, see McSherry, B., and P. Keyzer (2009). *Sex Offenders and*

Preventive Detention: Politics, Policy and Practice. Leichhardt, The Federation Press.

20. "Research inventories on psychiatric disorders or sexual behaviors in the general population have not inquired about pedophilic fantasies or behaviors" (Fagan et al. 2002:2460).
21. By "identify," I mean something like what John Christman describes thus: "Identification takes place when an agent reflects critically on a desire and, at the higher level, approves of having the desire." Christman, J. (1988). "Constructing the inner citadel: Recent work on the concept of autonomy" *Ethics* 99(1): 109–124, at p.112.
22. At least, the three audiences before which I have presented earlier versions of this paper seem to share that intuition.
23. I have discussed this sort of balancing act in more detail elsewhere; q.v. "Dangerous Patients and Duties to Warn: a European Human Rights Perspective," *European Journal of Health Law* (2007); 14: 113–130; "A *Tarasoff* for Europe? A European Human Rights perspective on the duty to protect" *International Journal of Law and Psychiatry* (2007); 30: 255–267.
24. Though, as I have argued, we should be wary about assuming too much in this regard.
25. Situations analogous, perhaps, to the practice Mahatma Gandhi is said to have adopted in his later life of "sleeping naked next to nubile, naked women to test his restraint." Adams, J. "Thrill of the chaste: The truth about Gandhi's sex life," *The Independent*, 7 April 2010.

25. Though it is interesting to consider, in this context, the approach of two eminent criminal law commentators to the decision of the English Court of Appeal in *R v. Kingston* (1994) QB 81. In arguing that New Zealand courts should decline to allow a defense for weakened inhibitions due to involuntary intoxication, A. P. Simester and W. J. Brookbanks point to the fact that "the propensity to commit the crime *must already have been present*" (*Principles of Criminal Law*, 3rd Edition, Thompson Brookers, 2007, at p. 354, original emphasis). Given that they are discussing a situation wherein it is accepted that the defendant would not have committed the offence but for the involuntary intoxication, then it seems that by "propensity," Simester and Brookbanks are talking about something like a pre-existing urge—or, at most, a want—rather than a prior intention. Unfortunately, they do not develop this line of reasoning further, but it does appear as though they are invoking some sort of concept of "culpable urge," whereby the presence of the appetite goes some way toward justifying punishment for the involuntarily disinhibited conduct.

REFERENCES

BBC Online (2010). "Czech gay asylum 'phallometric test' criticised by EU." 8 December.

BC Civil Liberties Association (2010). "Exploitive child testing must be stopped." Retrieved July 28, 2010, from http://bccla.org/news/2010/07/exploitive-child-testing-must-be-stopped/

Bergelson, V. (2009). "The case of weak will and wayward desire." *Criminal Law and Philosophy* 3: 19–28.

Boire, R. G. (2000). "On cognitive liberty." *Journal of Cognitive Liberties* 2(1): 7–22.

Bowling, B., A. Marks, and C. C. Murphy (2008). Crime control technologies: Towards an analytical framework and research agenda. In: *Regulating Technologies: Legal Futures, Regulatory Frames and Technological Fixes*. R. Brownsword and K. Yeung. Oxford, Hart Publishing (pp. 51–78).

Bratman, M. E. (2009). "Intention rationality." *Philosophical Explorations* 12(3): 227–241.

Camilleri, J. A., and V. L. Quinsey (2007). Pedophilia: Assessment and treatment. In: *Sexual Deviance: Theory, Assessment and Treatment*. R. D. Laws. New York, Guilford Press.

Dressing, H., A. Sartorius, and A. Meyer-Lindenberg (2008). Implications of fMRI and genetics for the law and the routine practice of forensic psychiatry. *Neurocase* 14(1): 7–14.

EU Agency for Fundamental Human Rights (2010). The practice of "phallometric testing" for gay asylum seekers. Retrieved September 12, 2010, from http://www.fra.europa.eu/fraWebsite/lgbt-rights/infocus10_0912_en.htm.

Fagan, P. J., T. N. Wise, C. W. Schmidt, Jr., and F. S. Berlin (2002). "Pedophilia," *Journal of the American Medical Association* 288(19): 2458–2465

Feinberg, J. (1970). *Doing and Deserving: Essays in the Theory of Responsibility*. Princeton, NJ, Princeton University Press.

Garvey, S. P. (2009). "Dealing with wayward desire." *Criminal Law and Philosophy* 3: 1–17.

Gazzaniga, M. (2005). *The Ethical Brain*. New York, Dana Press.

Glannon, W. (2002). *The Mental Basis of Responsibility*. Aldershot, UK, Ashgate.

Glover, J. (1970). *Responsibility*. London, Routledge & Kegan Paul.

Illes, J., and E. Racine (2006). Imaging or imagining? In: *Bioethics and the Brain*. W. Glannon. New York, Oxford University Press.

Kennedy, D. (2006). Neuroscience and Neuroethics. In: *Bioethics and the Brain*. W. Glannon. New York, Oxford University Press.

Marshall, W. L., and Y. M. Fernandez (2000). "Phallometric testing with sexual offenders: Limits to its value." *Clinical Psychology Review* 20(7): 807–822.

Murphy, T. F. (1991). "The ethics of conversion therapy." *Bioethics* 5(2): 123–138.

Odeshoo, J. R. (2004–2005). "Of penology and perversity: The use of penile plethysmography on convicted child sex offenders" *Temple Political & Civil Rights Law Review* 14(1): 1–44.

Organisation for Refuge, Asylum and Migration (ORAM) (2011). "Testing sexual orientation: A scientific and legal analysis of plethysmography in asylum and refugee status proceedings."

Ponseti, J., O. Granert, O. Jansen, S. Wolff, K. Beier, J. Neutze, G. Deuschl, and H. Mehdorn (2011). "Assessment of pedophilia using hemodynamic brain response to sexual stimuli." *Archives of General Psychiatry*, e-publication ahead of print edition, Oct 3. Now published in 2012 Feb;69(2): 187–94.

Representative for Children and Youth (British Columbia) (2011). Report: Phallometric testing and B.C.'s youth justice system. Available at: http://www.rcybc.ca/Images/PDFs/Reports/PPG%20Report%20FINAL%20Updated%20April%2014.pdf.

Sartorius, A., M. Ruf, C. Kief, T. Demirakca, J. Bailer, G. Ende, F. A. Henn, A. Meyer-Lindenberg, and H. Dressing (2008). "Abnormal amygdala activation profile in pedophilia." *European Archives of Psychiatry and Clinical Neuroscience* 258: 271–277.

Schiffer, B., T. Peschel, T. Paul, E. Gizewski, M. Forsting, N. Leygraf, M. Schedlowski, and T. H. C. Krueger (2007). "Structural brain abnormalities in the frontostratial system and cerebellum in pedophilia." *Journal of Psychiatric Research* 41: 753–762.

Schiltz, K, J. Witzel, G. Northoff, and K. Zierhut (2007). "Brain pathology in pedophilic offenders: evidence of volume reduction in the right amygdala and related diencephalic structures." *Archives of General Psychiatry* 64:737–746

Walter, M., J. Witzel, C.Wiebkig, U. Gubka, M. Rotte, K. Schilz, F. Bermpohl, C.Templemann, B. Bogerts, H. J.Heinze, and G. Northoff (2007). "Pedophilia is linked to reduced activation in hypothalamus and lateral prefrontal cortex during visual erotic stimulation" *Biological Psychiatry* 62: 698–701

Yarhouse, M. A. (1998). "When clients seek treatment for same-sex attraction: ethical issues in the 'right to choose' debate." *Psychotherapy: Theory, Research, Practice, Training* 35(2): 248–259.

PART 4

Disease and Disorder

10

Is Psychopathy a Mental Disease?

THOMAS NADELHOFFER AND
WALTER P. SINNOTT-ARMSTRONG

The notion of psychopathy has a long and contentious history (Andrade 2008). The term "psychopathic" was first coined by German psychiatrist J. L. Koch (1891) to refer to individuals who had previously been diagnosed with "moral insanity" (Prichard 1835) or "madness without delirium" (Pinel 1801/1962). In the beginning, "psychopath" was a fairly vague blanket term used to label an otherwise heterogeneous group of individuals who shared a proneness to violence and other forms of immoral behavior. Consequently, some researchers have suggested that the label "psychopath" was really used only as a moral judgment of character (Karpman 1948; Kernberg 1975). In contrast, others have seen psychopathy as a medical diagnosis and have worked long and hard to sharpen its boundaries and distinguish it from conditions, such as antisocial personality disorder (ASPD), that are often confused with psychopathy (Cleckley 1941/1976; Hare 1991). Given this controversy among clinicians, it is not surprising that lawyers and moral philosophers have also debated whether psychopaths are immoral or mentally ill—"bad or mad" (Deigh 1995; Fine and Kennett 2004; Glannon 1997; Greenspan 2003; Haji 2003; Levy 2007a, 2007b; Maibom 2005; Malatesti and McMillan 2010).

This distinction makes a difference. If psychopaths are just bad and not mad, then it is not clear why psychiatrists should treat them. The job of psychiatrists is to make people well, not good. Clinical psychologists might try to treat disruptive behaviors that are not caused by mental illness, but psychiatrists are supposed to be treating mental illnesses or diseases. Even if psychiatrists did treat some people who are bad but not mad, which treatment is appropriate might depend on whether psychopaths are seen as bad or mad. These views of psychopathy also make a difference to the law. If psychopaths are merely immoral, then they should (or at least may) be held fully responsible for their criminal conduct, just like normal criminals. On the other hand, if psychopathy is a mental disease, then individuals who commit crimes as a result of their psychopathy arguably should be eligible for an insanity defense, much like individuals who commit crimes as a result of schizophrenia or perhaps kleptomania (Levy 2007a, 2007b; Morse 2008; Litton Forthcoming).

Traditional formulations of the insanity defense make the issue clear. The two most common versions in modern legal systems are variations on the M'Naghten rule:

> To establish a defense on the grounds of insanity, it must be clearly proved that, at the time of committing the act, the party accused was labouring under such a defect of reason, from *disease of the mind*, as not to know the nature and quality of the act he was doing; or, if he did know it, that he did not know that what he was doing was wrong. (*Regina v. M'Naghten*, 10 Cl. & Fin. 200, 9 Eng. Rep. 718 [1843]; our emphasis)

and the Model Penal Code (MPC) of the American Law Institute (ALI):

> a person is not responsible for criminal conduct if at the time of such conduct as a result of *mental disease or defect* he lacks substantial capacity either to appreciate the criminality (wrongfulness) of his conduct or to conform his conduct to the requirements of the law. (ALI 1962, §4.01[1]; our emphasis)

Both formulations mention features that are usually assumed to characterize psychopaths, who supposedly do not "know" or "appreciate" the "wrongfulness" of their acts. This assumption is empirically controversial (see Schaich Borg and Sinnott-Armstrong Forthcoming). Moreover, the crucial terms "know," "capacity," "appreciate," and "wrongfulness" are subject to multiple interpretations (see Sinnott-Armstrong and Levy 2011), only some of which apply to psychopaths. Nonetheless, the crucial point here is that neither formulation has any chance of applying to psychopaths unless psychopathy is a "disease of the mind" (M'Naghten) or a "mental disease" (ALI/MPC).

Admittedly, the drafters (and ratifiers) of the MPC added, "the terms 'mental disease or defect' do not include an abnormality manifested only by repeated criminal conduct or otherwise anti-social conduct" (ALI 1962, §4.01[2]). Many commentators claim that this clause was intended to exclude psychopathy. However, anyone who uses this clause to exclude psychopathy faces at least two problems. First, this clause is controversial. It was added in only 17 of the jurisdictions that adopted the MPC insanity defense. One reason why other states omitted this clause was, presumably, that it seems ad hoc and unjustified to exclude psychopaths if psychopaths really do lack the very capacity whose lack excuses nonpsychopaths. Second, even if this clause applies to ASPD, which is largely behavioral, psychopathy is an affective disorder. Recent science has discovered many manifestations of psychopathy other than criminal or antisocial behavior, including callousness as well as neural indicators of psychopathy (discussed later). Hence, psychopathy is no longer (if it ever was) "manifested only by repeated criminal conduct or otherwise anti-social conduct." Thus, the MPC clause that was supposed to exclude psychopathy does not really exclude psychopathy if taken literally. We need to look beyond that clause to determine whether or not psychopathy is a mental disease.

The natural place to look is at general definitions of mental disease or illness. Our goal in this paper is to argue that psychopathy counts as a mental disease on any plausible account of mental disease. We will begin (in section 1) by quickly outlining some relevant discoveries about the nature of psychopathy. Next (in section 2), we will survey the most prominent definitions of mental disease. We will not endorse any particular definition. Instead, we will argue that each of these definitions of mental disease either includes psychopathy or is implausible on independent grounds. Finally (in section 3), we will draw our conclusion and return to the issue of why it matters.

THE NATURE OF PSYCHOPATHY

By far the most widely used and reliable tool for diagnosing psychopathy is the Psychopathy Checklist–Revised (familiarly known as the PCL-R), developed by Robert Hare over the last few decades (Hare 1991).[1] The PCL-R relies on a semistructured interview and documented case history (when available) to assign individuals scores of 0, 1, or 2 on each of 20 items. Total scores range from 0 to 40 and reflect the degree to which the individual matches a prototypical psychopath. Eighteen of the 20 items can be divided into four facets:

Interpersonal facet 1: glibness/superficial charm, grandiose self-image, pathological lying, cunning/manipulative

Affective facet 2: lack of guilt or remorse, callousness/lack of empathy, shallow affect, refusal to accept responsibility

Lifestyle facet 3: need for stimulation/proneness to boredom, parasitic lifestyle, failure to have realistic long-term goals, impulsivity, irresponsibility

Antisocial facet 4: poor behavioral control, early-onset behavioral problems, juvenile delinquency, revocation of conditional release, criminal versatility

Facets 1 and 2 are often combined into factor 1, and facets 3 and 4 are often combined into factor 2. Two of the 20 items (sexual promiscuity and many "marital" relationships) do not load statistically on any of the facets or factors (Blair et al. 2005), but they still contribute to an overall score on the PCL-R. An individual is officially diagnosed with psychopathy if and only if that person's overall score is 30 or higher (although a lower cutoff is used sometimes).

This technical diagnosis of psychopathy needs to be distinguished from ASPD. According to the current *Diagnostic and Statistical Manual* (DSM-IV-TR) of the American Psychiatric Association (APA 2000), individuals are properly diagnosed with ASPD if and only if they satisfy at least three of the following seven criteria:

1. They fail to conform to social norms with respect to lawful behavior.
2. They are frequently deceitful and manipulative in order to gain personal profit or pleasure.
3. They exhibit a pattern of impulsivity that may be manifested by a failure to plan ahead.
4. They tend to be irritable and aggressive and may repeatedly get into physical fights or commit acts of physical assault.
5. They display a reckless disregard toward the safety of themselves or others.
6. They tend to be consistently and extremely irresponsible.
7. They tend to show little remorse for the consequences of their acts.

Because only three of these seven criteria are required, two individuals can both be properly diagnosed with ASPD even if they share no symptoms at all in common; for example, one of them might meet criteria 1 to 3, whereas the other meets criteria 4 to 7. As a result, ASPD covers a much more diverse group than psychopathy. The ASPD group is also larger: experts estimate that 80% or more of medium-security inmates meet the ASPD diagnosis, whereas 20% or fewer from the same population should be diagnosed with psychopathy (Serin 1996). In addition, whereas psychopathy is a useful predictor of violent recidivism (see later), ASPD is of very limited use when it comes to violence risk assessment (Hart and Hare 1996).

These differences between ASPD and psychopathy are crucial here. The diversity within the diagnosis of ASPD and its failure to predict future behavior are strong reasons to doubt that ASPD is a unified mental disease. The great numbers of criminals with ASPD also provides a strong argument against making them all eligible for the insanity defense. However, these considerations do not have the same force against counting psychopathy as a mental disease or against making psychopaths eligible for the insanity defense.

Of course, even though psychopathy is narrower and less diverse than ASPD, psychopathy still might not be unified enough to count as a mental disease. No matter how narrow, an arbitrary list of symptoms is not a disease. However, recent studies have shown that the items on the PCL-R are not just an arbitrary list of symptoms. Their coherence is suggested by their statistical interrelations.

Moreover, this cluster of items is useful for predicting behavior. High scores on the PCL-R have been repeatedly shown to confer an increased risk for violence,[2] and the PCL-R (or some derivative) has been included as a predictor variable in several prominent actuarial models of violence risk assessment.[3] In a recent review of the sprawling literature, Leistico, Salekin, DeCoster, and Rogers (2008) present the results of a meta-analysis that integrates the effect sizes from 95 non-overlapping psychopathy studies. Their primary finding was that "psychopathy was similarly predictive across different ages (adolescents vs. adults), study methodologies (prospective vs. retrospective), and different types of outcomes (institutional infractions vs. recidivism)" (Leistico et al. 2008).

Perhaps the best evidence that the items on the PCL-R are more than an arbitrary list comes from psychology and neuroscience. Cognitive psychologists have found that individuals with psychopathy

(a) often act in their own worst interest (Hare 1991; Blair, Colledge, and Mitchell 2001),
(b) exhibit cognitive-perceptual deficits in the recognition of certain emotions in others' faces and voices (Blair, Colledge, Murray, and Mitchell 2001; Blair and Coles 2000),
(c) have deficits in attention, have exaggerated views of their own capabilities, and are intransigent to aversive conditioning[4] (Hare 1978),
(d) exhibit shortcomings when it comes to the so-called gambling task (Bechara, Damasio, Damasio, and Anderson 1994),
(e) exhibit deficits in response reversal—the inhibition of previously rewarded responses that are now punished (Newman and Kosson 1986), and
(f) fail to pass the moral-conventional task (Blair 1995; but see Aharoni, Sinnott-Armstrong, and Kiehl 2011)—which is a basic moral cognition task that both young children and individuals with autism are able to successfully pass.

More recently, there is gathering evidence that psychopaths display the following functional neural deficits:

(g) reduced amygdala and vmPFC activity during aversive conditioning tasks (Veit et al. 2002),
(h) impairment in passive avoidance learning tasks[5] and differential reward-punishment tasks[6] associated with amygdala activity (Blair et al. 2004; Blair, Leonard, and Blair 2006),
(i) reduced amygdala activation during emotional memory (Kiehl et al. 2001),
(j) reduced activation in the anterior and posterior cingulate gyri, left inferior frontal gyrus, amygdala, and ventral striatum when encoding, rehearsing, and recognizing negatively valenced words (Kiehl et al. 2004), and
(k) reduced activity in the ventromedial prefrontal cortex, anterior temporal cortex, and amygdala when rating severity of moral violation in images (Harenski, Harenski, Shane, and Kiehl 2010; also see Glenn, Raine, and Schug 2009).

In addition to deficits in neural function (see also Anderson and Kiehl Forthcoming) and structure (see Boccardi Forthcoming), researchers are also discovering a lot about the neurochemistry of psychopathy:

(l) Blair, Mitchell, and Blair (2005) show that the neurotransmitter noradrenaline plays an important role in the deficits associated with psychopathy.
(m) Rogers, Lancaster, Wakeley, and Bhagwager (2004) show that administering noradrenaline antagonists reduces the impact of aversive cues when making decisions.
(n) Strange and Dolan (2004) show that amygdala activity in response to emotional stimuli is also reduced by the administration of a noradrenaline antagonist.
(o) Cima, Smeets, and Jelicic (2008) show differences in psychopathic and nonpsychopathic inmates with respect to cortisol function.

Complementary research is being done on the heritability of psychopathy. Two recent studies suggest a genetic contribution to the disorder, especially when it comes to the callous-unemotional components of psychopathy (Blonigen, Hicks, Krueger, Patrick, and Iacono 2005; Viding, Jones, Frick, Moffitt, and Plomin 2007). Moreover, a large adolescent twin study by Larsson and colleagues (2007) found that the four facets of psychopathy load onto a single genetic factor.

Of course, many questions about the psychological, neural, and genetic underpinnings of psychopathy, as well as its behavioral consequences, remain unanswered. Nonetheless, the best recent research points toward strong connections of psychopathy to future criminal behavior, to certain kinds of psychological and neural function and structure, and perhaps also to genetics. These discoveries suggest that psychopathy as defined by the PCL-R is far from an arbitrary list of unconnected symptoms and, indeed, might even have a unified physical basis. If the PCL-R items were arbitrary, it would be hard to explain why psychopaths share so many behavioral, psychological, and neural features (other than the items on the PCL-R). Thus, the fact that researchers have found so many added commonalities among psychopaths suggests that the PCL-R symptoms are unified at some level, even though many details remain unknown.

THEORIES OF MENTAL DISEASE

With this understanding of psychopathy, we can ask our main question: Is psychopathy a mental disease? The answer might vary with different accounts of what a mental disease is. Accordingly, we will survey a variety of prominent accounts of mental disease. We do not endorse any of these particular accounts. Instead, for each account, we will argue that either psychopathy is a mental disease on that account or that the account is indefensible for reasons independent of psychopathy.

Before we begin our survey, a short terminological aside is needed. Although some theorists talk about "disease," others refer to "illness," "disorder," or "malady." People with mental diseases are also sometimes called insane, crazy, mad, and even sick. These terms are not all synonymous. Distinctions can be drawn and are useful in other contexts. Nonetheless, these distinctions do not make any difference to the present issue. What matters here is whether psychopathy is relevantly similar to schizophrenia, clinical depression, obsessive-compulsive disorder, and other paradigms of mental disease. Even if some disorders, maladies, and illnesses are not diseases, if psychopathy is a mental disease, then it is also a mental illness, disorder, and malady. The issue of whether psychopathy is a mental disease is what matters to law and to psychiatry. Hence, we will talk primarily about whether psychopathy is a mental disease, although we will sometimes use the other terms when commenting on theorists who use those other terms.

Eliminativism

During the mid-20th century, psychiatrists debated not just how to classify specific mental diseases but also whether mental diseases exist at all. Thomas Szasz famously claimed that mental illness is a myth (1961; 1999). According to

Szasz and other eliminativists (including Pickard 2009; Sarbin 1967), so-called mental illnesses are not really illnesses because they don't satisfy the proper scientific criteria for illnesses or diseases—namely, "a derangement in the structure or function of *cells, tissues, and organs*" (Szasz 1999:38). According to this view, the common analogy between mental and bodily diseases is inapt because psychiatrists cannot specify the underlying neurological deficits that cause mental diseases (Szasz 1961:113–114). Szasz admits, of course, that there are brain diseases as well as brain damage, but what are typically classified as mental illnesses are not, in his view, really illnesses at all.

In contrast with genuine physical diseases, so-called mental illnesses are really just natural reactions to the "problems of living" associated with navigating an increasingly complex society, according to Szasz (1961:114). The only difference between "normal" ways of living and so-called mental illnesses is that mental illnesses involve deviations from norms that are "psycho-social, ethical, and legal" (Ibid.). To diagnose so-called mental illnesses, then, psychiatrists must make veiled moral and sociopolitical judgments, often for the purposes of social coercion and control. Because these judgments are inherently subjective and value laden rather than objective and value neutral, psychiatrists are not acting like real doctors or scientists.

Because labeling someone as mentally ill can be stigmatizing and counterproductive (Sarbin 1967:451), eliminativists claim that explaining people's behavior in terms of mental illnesses is "the proper heir to the belief in demonology and witchcraft" (Szasz 1961:117). They also claim that the notion of mental illness leads to theoretical confusions, inappropriate methodology, and dead ends. To avoid such practical, legal, moral, and theoretical problems, eliminativists argue that psychiatrists should abandon talking about mental illnesses much as they have abandoned talking about demonic possession, witches, and the like. Szasz concludes, "the notion of mental illness has outlived whatever usefulness it might have had" (Szasz 1961:118). Mental illness is not only a myth but also a harmful myth.

Of course, since eliminativists think that all so-called mental diseases are mythical, they would also think that psychopathy is mythical. If nothing properly counts as a mental disease, then neither does psychopathy.

The fact that eliminativism implies this conclusion does not, however, show that psychopathy (or anything else) is not a mental illness, unless eliminativism is correct, justified, or at least defensible. However, eliminativism is implausible for several reasons. First, even if scientists do not know the physical basis for schizophrenia, for example, that need not show that schizophrenia has no basis in a derangement in the structure or function of cells, tissues, and organs. All it might show is our current ignorance of which cells, tissues, and organs are deranged. Second, scientists have discovered a lot about the physical basis for some mental illnesses. Addiction is perhaps the best example (Robbins, Everitt, and Nutt 2010). There is a lot that we do not yet know, of course, but what we do know suggests that addiction does have a neural basis, even if we do not know exactly what it is.

Of course, our current concept of mental illness in general or of a particular mental illness, such as bipolar disorder, might turn out to result from a group of brain disorders that cause similar symptoms but still do not seem unified from the perspective of biology or physics. Some eliminativists (e.g., Pickard 2009) infer that schizophrenia, for example, is not unified enough or at the right level to count as a single mental illness. However, that disunity would not show that schizophrenia is a myth. All it would show is that our current understanding of schizophrenia is imperfect. Even if we need to reform our concepts of mental illnesses in order to bring them in line with biology and physics, the reformed concepts might be close enough to our current concepts that the reforms improve our understanding without showing that we were not talking about anything real before we got all of the details right.

The same goes for psychopathy in particular. As we discussed (in section 1), neuroscientists have discovered something and are learning more and more about the physical basis for psychopathy. That should not be surprising, unless you think that minds are nonphysical substances. When we learn more, neuroscience might teach us that some of the items on the PCL-R are not really essential to psychopathy or its neural basis in the way that the current PCL-R suggests. We might need to trim some items or add others in order to strengthen the relation between the items and their neural basis. Moreover, future neuroscience might teach us that psychopathy is really a bunch of separate mental illnesses (as Hervé 2007 proposes), each with its own distinctive neural basis. Psychopathy is then like anemia, which can result from many distinct causes. Nonetheless, that future neuroscience still would not show that psychopathy does not exist in the way that eliminativists claim. It just turns out that there are more forms of psychopathy than we imagined, and their boundaries are different than we believed, but the acts of psychopaths can still result from a mental disease, even if the mental disease that causes their acts is not quite what we thought.

Of course, we do not know for sure exactly where research into the neural basis of psychopathy is headed. Nonetheless, the research cited previously seems to be headed toward finding some neural basis for something like what the PCL-R classifies as psychopathy. Even if that optimism sounds premature, at least eliminativists have not yet ruled it out. Their argument against the status of psychopathy as a mental disease depends on their claim that psychopathy does not correspond to any "derangement in the structure or function of *cells, tissues, and organs*" (Szasz 1999:38). But how can they know that no neural basis will ever be discovered? Until they justify that radical claim, they cannot show that psychopathy is not a mental disease.

Social Constructionism

As we saw, eliminativists argue that real diseases can be defined in a value-neutral way, but mental illnesses cannot be defined in a value-neutral way, so mental illnesses are not real diseases. One popular response rejects the first premise and denies that even paradigm physical diseases can be defined in a value-neutral way. When values and hence diseases are then taken to be social constructions, the result is called *social constructionism*. Peter Sedgwick, for example, says, "a careful examination of the concept of illness in man himself will reveal the same value impregnation, the same dependency of apparently descriptive, natural-scientific notions upon our norms of what is desirable. To complain of illness, or to ascribe illness to another person, is not to make a descriptive statement about physiology or anatomy" (Sedgwick 1973:32). If so, then the value-laden quality of mental illnesses need not make them any less real than physical diseases. There is a sense in which neither mental nor physical diseases are real according to this view because they are socially constructed, but that is not supposed to be a problem for diagnoses of mental illness or psychopathy.

Unlike eliminativists, who believe that labeling people as mentally ill is confused and pernicious, social constructionists think that as long as we acknowledge the value-laden nature of the diagnosis and treatment of both physical and mental illnesses, diagnoses of mental illnesses can have important reformative roles to play both within the field of psychiatry and within broader policy debates about mental health and the law. Sedgwick, for instance, claims that:

> Mental illness, like mental health, is a fundamentally critical concept or can be made into one provided that those who use it are prepared to place demands and pressures on the existing organization of society. In trying to remove and reduce the concept of mental illness, the revisionist [i.e., eliminativist] theorists have made it that much harder for a powerful campaign of reform in the mental-health services to get off the ground. The revisionists have thought themselves, and their public, into a state of complete inertia: they can expose the hypocrisies and annotate the tragedies of official psychiatry, but the concepts which they have developed enable them to engage in no public action which is grander than that of wringing their hands. (1973:39)

Thus, despite their shared belief that mental illness is value laden, eliminativists and social constructionists part ways when it comes to the role that mental illnesses ought to play both within psychiatry and within society more generally. Social constructionists don't want to expunge "mental illness" from the psychiatric lexicon—they simply want psychiatrists to acknowledge the inherently evaluative and sociopolitical nature of their undertaking.

As a result, social constructionism need not exclude psychopathy from the realm of mental diseases. Even if psychopathy were defined purely by culturally variable antisocial behaviors (which is questionable), that variability might make it a paradigmatic example of the kind of disorder that social constructionists have in mind. Just as one's culture can determine whether one is shamanistic rather than schizophrenic, similarly one's culture can determine whether someone is a psychopath rather than a fearless warrior or a cunning liar. According to constructionism, diagnosing someone with psychopathy will be an inherently value-laden enterprise, but psychopathy can still be just as real as any other mental or physical disease. Indeed, if the behaviors that characterize psychopathy are disvalued in almost any stable society, then a diagnosis of psychopathy might be less relative to culture than diagnoses of other mental illnesses that depend on more variable values.

In any case, even if social constructionist accounts of mental disease did exclude psychopathy, that would not show that psychopathy is not a mental disease unless those social constructivist views were correct. However, constructionists face a host of problems. First, pure social constructionism seems to imply that any mental condition can properly count as a mental disease as long as a given society deems it so. By this view, no society could ever be incorrect about what is a mental illness. The problem is that societies do make mistakes. Slaves who ran away were not really suffering from the mental illness drapetomania, regardless of whether this behavior violated an existing social norm in the American South during the days of slavery (Wakefield 1992b:374). A second problem is that, according to social constructionism, mental diseases can be cured merely by relocating to a new culture. If schizophrenia is a mental disease in one society but not in another, then moving from the former to the latter will make one cease to have schizophrenia, even if one's symptoms did not change at all (Graham 2010, chapter 5). Conversely, moving from the latter to the former will give one the mental illness of schizophrenia, even without any change in behavior or thought. That implication of social constructionism is hard to stomach. And if constructionism cannot be defended against these (and other) objections, then it cannot show that psychopathy is not a mental illness.

The Biomedical View

Eliminativists and social constructionists share the belief that the diagnosis and treatment of mental illness is thoroughly value laden. Some opponents reject this view and claim that we both can and should strive to discuss mental illnesses or at least mental diseases without any reference to or assumptions about values or norms. The most common attempt to avoid evaluation refers to some purely scientific and descriptive kind of dysfunction (e.g., Boorse 1975:90; Scadding 1990:245).

Assuming that the relevant science is biology, this view claims that all diseases—physical or mental—can and should be understood in terms of biological dysfunctions. When an organism is functioning properly according to its biological design, it is healthy. When an organism is not functioning properly according to its biological design, it has a disease (or at least it meets one condition of disease). The relevant notions of biological function and dysfunction are supposed to be evaluatively neutral.

This general biological model of disease is then extended to cover mental diseases. Scadding puts the rationale in this way:

> Since psychology and behavioral science are aspects of the study of living organisms, they are subsumed by biology, in its widest sense; but I will take it that "biology" is being used here to mean the study of living organisms directed towards explanation in physico-chemical terms... Some disorders of behavior can already be explained in this way; and it is to be expected that with advances in knowledge, more and more of psychology and behavioral science will become biological. (Scadding 1990:245)

Instead of referring to physicochemical processes, another kind of biomedical view can cite mental functions:

> The health of an organism consists in the performance by each part of its natural function. And as Plato also saw, one of the most interesting features of the analysis is that it applies without alteration to mental health as long as there are standard mental functions. In another way, however, the classical heritage is misleading, for it seems clear that biological function statements are descriptive rather than normative claims. (Boorse 1975:58)

Just as medical doctors identify and then treat physical dysfunction, psychiatrists should strive to identify and treat mental dysfunction understood in purely scientific and value-neutral terms.

Proponents of the biomedical view often appeal to biological factors such as survival and reproductive fitness in order to determine what counts as a dysfunction and, thus, as a mental disease. Boorse, for instance, defines the normal function of a part of an organism as its species-typical contribution to the survival and reproduction of organisms in the relevant reference class (something like an age group or a sex of a species). A type of condition of either the body or the mind then counts as a disease only if it typically tends to reduce an individual's longevity or likelihood of reproducing. Because determining whether a condition typically reduces reproduction or longevity is supposedly a straightforwardly empirical matter, proponents of the biomedical view insist that psychiatry can be placed on a value-neutral footing.

It is not clear whether or how these theorists can succeed in defining mental or neural functions without reference to any value or norm. After all, many kinds of mental states (such as desires to race cars or climb ice cliffs) reduce average expected longevity without counting as mental diseases if they do not go too far, and how far is too far seems to depend on values or norms. Psychiatrists cannot avoid assuming norms if, as Wakefield puts it, "disorder is in certain respects a practical concept that is supposed to pick out only conditions that are undesirable and grounds for social concern, and there is no purely scientific non-evaluative account that captures such notions" (1992a:237).

Even if this biomedical view could somehow achieve value neutrality, it is also not clear how it applies to psychopathy. Some experts argue that psychopathy is not dysfunctional from an evolutionary viewpoint because several of its traits actually increase reproduction. Men who score high on psychopathic traits have been found to engage more frequently in short-term mating behaviors, to have more flexible sociosexual attitudes, and to have a higher number of sexual partners (Jonason et al. 2009). In short, they are promiscuous. Psychopathy differs in this respect from maladaptive mental disorders, such as schizophrenia, that clearly result in decreased reproductive fitness in modern environments (see, e.g., Cosmides and Tooby 2000; Haukka, Suvisarri, and Lonnqvist 2003). In contrast with paradigm mental diseases, psychopathy might seem to enhance fitness, at least in certain narrow environmental niches. This advantage makes psychopathy seem less like an illness and more like an adaptive "life strategy."[7]

One problem for this view is that other psychopathic traits also lead to risky behaviors that decrease longevity (or land the person in prison, where reproduction is curtailed). Psychopaths tend to die earlier than their nonpsychopathic counterparts. This disadvantage is not surprising in light of correlations between psychopathy and violence, crime, drug and alcohol abuse, and stimulation seeking. The fact that psychopathy decreases longevity could be enough to make it count as a dysfunction and a mental disease, according to the biomedical view.

It is crucial here to distinguish psychopathic traits from full-blown psychopathy. Some psychopathic traits might be adaptive, even if psychopathy is not. It is also important to contrast different environments. Psychopathy might be adaptive in environments where psychopaths will go undetected and unpunished, even if the same condition is not adaptive in environments where psychopaths are more likely to be caught and imprisoned. Our question for now is whether full-blown psychopathy is an adaptive lifestyle or a dysfunction in modern circumstances.

This issue seems to depend on whether the increased reproduction due to promiscuity and manipulativeness outweighs the decreased longevity due to violence and stimulation seeking. If it did, then we would expect to find a lot more psychopaths than we do. The best estimates put the prevalence of psychopathy at less than 1%. Psychopathy might increase reproduction for some people in some circumstances. However, studies (cited earlier) suggest that psychopathy is passed on genetically. It is hard to see how generations of psychopaths in a subpopulation would continually find themselves in unusual circumstances where promiscuity maximizes reproductive chances despite reduced longevity.

One possible story is that the low percentage of psychopaths constitutes an adaptive equilibrium. Maybe psychopaths do well enough on average as long as not too many other people are psychopaths. If their numbers get too high, then there is selection *against* the condition; but if their numbers get too low, then there is selection *in favor of* it. However, even if this story is accurate, it does not rule out biological dysfunction and disease. Compare a physical disease, gigantism, which is abnormally large growth due to excessive growth hormone, often because of abnormalities in the pituitary gland. Imagine that this condition increases reproduction in youth because size attracts mates, but only when there are not too many giants, so a low percentage of giants is an adaptive equilibrium. Nonetheless, if gigantism leads to early death and results from pituitary abnormalities, then it still involves biological dysfunction and disease. Analogously, even if a low percentage of psychopaths is an adaptive equilibrium, psychopathy could still involve a biological dysfunction and could still be a mental disease.

The same point applies to other accounts of function. Some proponents of a biomedical view might define normal functioning not in terms of what is currently adaptive but instead in terms of whether a system is able to do that for which it was selected by evolution (see Neander 1991). According to this etiological theory of proper functioning, psychopathy could still count as a mental disease by the biomedical view even if it increases overall fitness in some circumstances or on average currently. The reason is that particular parts of the brain function abnormally in psychopaths. The amygdala presumably evolved in its current form in order to perform certain functions, including producing fear in reaction to risk of punishment that aims to prevent risky and harmful behaviors. Psychopaths do not fear punishment in the usual way, probably because of hypoactivity in their amygdala (Blair, Mitchell, and Blair 2005). This amygdala hypoactivity might have advantages in enabling psychopaths to be promiscuous. Nonetheless, it has disadvantages in decreasing longevity. However these considerations weigh against each other, their amygdala is not performing the general function for which the amygdala evolved or, at least, not at the level for which the amygdala was selected. That failure of the amygdala to fulfill its function is enough to make psychopathy a dysfunction and a mental disease even if it did increase reproduction overall. It should not be surprising that a disease can have some advantages in special circumstances, but that does not keep it from being a disease when it arises from an abnormally low level of activity in a bodily organ that evolved to perform a certain function that requires a higher level of activity in that organ.[8]

These issues are far from settled. Much work remains to be done on the epigenetic etiology of psychopathy. Only once this work is done will we be in the position to determine whether psychopathy is pathological or adaptive. In the meantime, whether psychopathy counts as a mental illness according to the biomedical view will remain an open question, although there is no reason yet to deny that psychopathy is a mental disease according to the biomedical view.

As before, even if biomedical accounts did exclude psychopathy, that would not show that psychopathy is not a mental disease unless biomedical accounts were correct. However, biomedical accounts run into well-known troubles. In particular, if normal functioning is defined by adaptive history, and if sex organs and drives were selected by evolution for reproduction, then homosexuality seems to count as a dysfunction according to biomedical views. However, homosexuality is not a mental illness or disease, as almost all psychiatrists today recognize. One main reason is that nobody gets harmed in healthy loving homosexual relationships. Thus, the biomedical view seems to need to be supplemented by some reference to harm.

The Harmful Dysfunction View

Although the biomedical view aspires to value neutrality, one could give up that aspiration and incorporate values but still define diseases partly by reference to biological dysfunctions. The rationale for including both aspects is stated by Jerome Wakefield:

> Disorder lies on the boundary between the given natural world and the constructed social world; a disorder exists when the failure of a person's

> internal mechanisms to perform their function as designed by nature impinges harmfully on the person's well-being as defined by social values and meanings. The order that is disturbed when one has a disorder is thus simultaneously biological and social; neither alone is sufficient to justify the label disorder. (1992b:373)

More formally, Wakefield defines disorders as a conjunction:

> A condition is a disorder if and only if (a) the condition causes some harm or deprivation of benefit to the person as judged by the standards of the person's culture (the value criterion), and (b) the condition results from the inability of some internal mechanism to perform its natural function, wherein natural function is an effect that is part of the evolutionary explanation of the existence and structure of the mechanism (the explanatory criterion). (1992b:384)

Wakefield applies this hybrid model to both physical and mental diseases and calls it "the harmful dysfunction view."

This view is supposed to have several advantages over previous models. Wakefield uses the explanatory criterion—dysfunction—both to distinguish mental disorders from garden-variety "disvalued conditions" and also to explain why psychiatrists were mistaken when they classified conditions such as drapetomania as mental illnesses. On the other hand, Wakefield's value criterion enables us to acknowledge the important role played by value judgments and social norms in the diagnosis and treatment of mental illness (1992b:381). Thus, Wakefield's harmful dysfunction view tries to avoid the pitfalls while incorporating what he takes to be the respective truths captured by both social constructionism and the biomedical view.

As such, the harmful dysfunction view appears to be compatible with treating psychopathy as a mental disorder. We already discussed the ways in which psychopathy is associated with biological dysfunction. In addition, psychopathy clearly meets Wakefield's value criterion because it "causes some harm or deprivation of benefit to the person as judged by the standards of the person's culture." Psychopaths are unable to lead normal, productive, healthy lives. One obvious reason is that they usually live much of their lives in prison, so they are deprived of freedom, which society values. Even when not incarcerated, they are often driven to lead nomadic lifestyles devoid of normal relationships with friends, family, and romantic partners. Love and friendship are also valued by society, and psychopaths are deprived of those benefits. Admittedly, psychopaths often see themselves not as deprived but rather as not restricted by useless attachments, but the lack of friendship and love still counts as "harm or deprivation of benefit to the person as judged by the standards of the person's culture" (Wakefield 1992b:384). Furthermore, owing to psychopaths' impaired ability to deliberate, form long-term goals, and adopt effective strategies for achieving these goals, their lives are often frustrating and deprived of achievements that are valued by society. Finally, and perhaps most important, psychopaths tend to die earlier than their nonpsychopathic counterparts, probably because of the correlations between psychopathy and violence, drug and alcohol abuse, and various kinds of risky behavior. Thus, psychopathy perfectly exemplifies Wakefield's value criterion. Assuming that psychopathy also involves a dysfunction in neural mechanisms as well as cognitive and affective processes (as discussed earlier), psychopathy counts as a mental disorder on the harmful dysfunction view.[9]

The Objective Harm View

Wakefield's model agrees with social constructionism insofar as his value criterion refers to "the standards of the person's culture" (1992b:384). This social basis for evaluation creates difficulties when societies have unjustified standards. Consider the physical condition of being albino. As Bernard Gert and Charles Culver point out, albinism is a condition associated with underlying biological dysfunction, so Wakefield's view implies that "if a particular society negatively evaluates albinism, then it is a disorder in that society, but not a disorder in a society that does not negatively evaluate it" (2004:420). Thus, Wakefield's view arguably commits him to an unacceptable form of relativism whereby individuals can catch or be cured of diseases, including mental diseases, simply by moving from one society to another. In contrast, Gert and Culver want to ensure that mental disease is defined in such a way that mere conflicts between individuals and their specific societies don't count as mental diseases.

This goal was shared by the drafters of the DSM-IV-TR. Given that the DSM-IV-TR is currently the most widely used clinical tool for diagnosing mental illnesses—at least in the United States—it is especially relevant for our purposes here, so it is worth discussing this approach in detail. While acknowledging from the outset that "no definition adequately specifies precise boundaries for the concept of 'mental disorder,'" the drafters of the DSM-IV-TR explicated the nature of mental disorders in the following way:

> Each of the mental disorders is conceptualized as a clinically significant behavioral or psychological syndrome or pattern that occurs in a person and that is associated with present distress (a painful symptom) or disability (impairment in one or more important areas of functioning) or

> with a significantly increased risk of suffering death, pain, disability, or an important loss of freedom. In addition, this symptom or pattern must not be merely an expectable and culturally sanctioned response to a particular event, for example, the death of a loved one. Whatever its original cause, it must currently be considered a manifestation of a behavioral, psychological, or biological dysfunction in the person. Neither deviant behavior (e.g., political, religious, or sexual) nor conflicts that are primarily between the individual and society are mental disorders unless the deviance or conflict is a symptom of a dysfunction in the person, as described above. (APA 2000:xxx-xxxi)

Unlike Wakefield's view, this definition does not refer to "the standards of the person's culture" (Wakefield 1992:384). Instead, harms are merely listed—death, pain, disability, and loss of freedom (compare Gert and Culver 2004). This formulation suggests that whether these items are bad does not depend on culture, apart from "an expectable and culturally sanctioned response to a particular event." The disvalue of other things still might vary from culture to culture, but the items on this list are supposed to be universally or objectively bad.

Furthermore, the mental disorder must be a manifestation of a "dysfunction *in* the person" (APA 2000:xxxi; our emphasis). This word "in" is explicated by Gert and Culver's corresponding requirement that the harm not be due only to a "distinct sustaining cause" or to "the environment" (2004:419). The point is to avoid pathologizing individuals who merely display deviant behaviors, such as deviant sexual acts, that are deemed socially unacceptable. If such behaviors cause harm (i.e., death, pain, disability, and loss of freedom) only because of societal disapproval (a distinct sustaining cause), then the dysfunction is not *in* the person, and the individual should not be diagnosed with a mental disorder, according to DSM-IV-TR as well as Gert and Culver.

Nonetheless, this approach still might count a dysfunction as lying within the person (and, hence, the basis for a mental disorder) if the conflict with society is universal. Gert and Culver adopt this interpretation when they say, "traits that would result in conflict with any society count as a dysfunction in the person" (2004:421). By this view, the concept of mental disorder is purportedly "grounded in universal and universally agreed-upon features of human nature. This means that although values remain at the core of the concept . . . the specific values are objective and universal" (Gert et al. 2006:133). DSM-IV-TR is less explicit on this issue but can also be read in this way.

Does this approach include psychopathy as a mental disorder or disease? It might seem not to because psychopathy is not usually associated with distress in the psychopath. However, other mental illnesses, including some delusions and manias, also do not involve distress. During manic episodes, people can feel perfectly happy, although they are not able to control their mood or behavior. Someone who believes that he is Napoleon need not feel bad about this belief, although he is not able to adjust his belief to evidence. That is why the DSM-IV-TR requires only distress *or* disability *or* increased risk of harm—a triple disjunction. And psychopathy clearly does involve disability due to dysfunction in neural mechanisms as well as cognitive and affective processes.[10] Even more clearly, psychopathy creates a significantly increased risk of harm. As we said in the previous section, psychopaths are unable to lead normal, productive, healthy lives. They usually spend much of their lives in prison, where they are deprived of freedom. When not imprisoned, they are often driven to lead nomadic lifestyles devoid of love and friendship as well as the accompanying joys. Their lives are often frustrating and deprived of accomplishments. And, of course, psychopaths tend to die earlier than nonpsychopaths. These harms might not bother psychopaths, if they do not want friends or freedom, but that does not matter to a theory that defines mental disorder in terms of objective harms. Some critics might deny that any harms can be objective, but here we are merely describing implications of this approach, not endorsing it. If psychopathy significantly increases risks of suffering losses of freedom, friendship, and life, and if these things are really good, so that losing them is really bad or harmful, then psychopathy meets this condition in the DSM-IV-TR definition.

But is there a dysfunction *in* the person? Yes, because the typical behaviors of psychopaths are behaviors of which any and every society disapproves and punishes (Gert and Culver 2004:421). The conflict between psychopaths and their societies is not contingent in the same way as the conflict between societies and harmless private sexual practices. No society could survive with too many psychopaths. In contrast, societies with any number of homosexuals could survive because homosexuals can have children. The fact that all societies need to condemn psychopathy shows that the problems for psychopaths do not lie in any peculiarities of particular societies that condemn them but, instead, lie within psychopaths themselves. This absence of a distinct sustaining cause of harm makes psychopathy count as a mental disease on the accounts of the DSM-IV-TR definition (as well as Gert and Culver's nearby account).

Critics might ask why DSM-IV-TR does not list psychopathy as a separate mental disorder. The answer is that DSM-IV-TR mistakenly lumps psychopathy together with ASPD. This conflation is misleading because of the differences between ASPD and psychopathy that we discussed previously. Nonetheless, it still might make sense to think of psychopathy as a more extreme kind of personality disorder. To test this possibility, consider how the DSM-IV-TR views personality traits and disorders more generally:

> Personality traits are enduring patterns of perceiving, relating to, and thinking about the environment and oneself that are exhibited in a wide range of social and personal contexts. Only when personality traits are inflexible and maladaptive and cause significant functional impairment or subjective distress do they constitute Personality Disorders. The essential feature of a Personality Disorder is an enduring pattern of inner experience and behavior that deviates markedly from the expectations of the individual's culture and is manifested in at least two of the following areas: cognition, affectivity, interpersonal functioning, or impulse control (Criterion A). This enduring pattern is inflexible and pervasive across a broad range of personal and social situations (Criterion B) and leads to clinically significant distress or impairment in social, occupational, or other important areas of functioning (Criterion C). The pattern is stable and of long duration, and its onset can be traced back at least to adolescence or early adulthood (Criterion D). The pattern is not better accounted for as a manifestation or consequence of another mental disorder (Criterion E) and it not due to the direct physiological effects of a substance…or a general medical condition (Criterion F). (APA 2000:686)

From what we have said already, it should be clear that psychopathy meets these conditions of being a personality disorder.

Criterion A: Psychopathy is an enduring pattern of inner experience and behavior insofar as the PCL-R refers to inner experiences (lack of empathy and remorse, shallow affect, grandiose self-image, and so on) in addition to behaviors (lying, criminal versatility, sexual promiscuity, and so on), and these traits both endure, deviate markedly from culture's expectations, and are manifested in affectivity, interpersonal functioning, and impulse control.

Criterion B: Psychopathic traits are inflexible and pervasive.

Criterion C: Psychopathy "leads to clinically significant…impairment in social, occupational, or other important areas of functioning," as we discussed in the previous section. Psychopathy does not usually lead to distress, but distress is not necessary for personality disorder or mental illness or disease (for good reason: just think of mania).

Criterion D: The onset of psychopathy can usually be traced back at least to adolescence or early adulthood, as reflected in the PCL-R items of early onset behavioral problems and juvenile delinquency.

Criteria E and F: Psychopathy is not a manifestation or consequence of another mental disorder, a substance, or a general medical condition. Thus, psychopathy would count as a personality disorder under the criteria of DSM-IV-TR.

This result does not show that psychopathy is not a mental disease, of course. Indeed, it might show that psychopathy is a mental disease if personality disorders are mental diseases.

CONCLUSIONS

We have surveyed the most prominent accounts of mental disease. Some of these accounts imply that psychopathy is not a mental disease, but we argued that those accounts are implausible. Other accounts imply that psychopathy is a mental disease. We did not endorse any one of these accounts in particular. Nonetheless, these considerations together suggest that psychopathy is a mental disease on every plausible account of mental disease. Our arguments for this conclusion are not conclusive, and our survey of definitions of mental disease is not complete. Much more research and thought are needed. Still, we hope to have given at least some evidence for this conclusion.

So what? Why does it matter? As we said at the start, if psychopathy were not a mental disease, then psychopaths could not be eligible for the insanity defense on that basis under the M'Naghten or MPC rules. Thus, our conclusion shows that psychopaths can be eligible for the insanity defense if they meet the other conditions of those rules. We have not discussed those other conditions, so we cannot conclude that psychopaths are eligible for the insanity defense under current rules. Moreover, the law can always be changed (such as by removing the mental disease requirement), or its words can be left alone but redefined (for the legal system can simply declare that the term "mental disease" does not include psychopathy), so we cannot conclude that psychopaths *will* be eligible for the insanity defense under future rules. Nor can we argue that psychopaths *should* be eligible for an insanity defense because that normative conclusion would hinge on a number of factors that we have not discussed. All of these issues deserve further discussion, but all we have argued here is that one obstacle does not prevent psychopaths from being found not guilty by reason of insanity.

Our conclusion is also relevant to whether psychiatrists should treat psychopaths. As we said, the job of psychiatrists is to make people well, not good, so they have no business treating psychopaths if psychopathy is not a mental disease, illness, or disorder. This conclusion applies to psychiatric treatment of psychopaths both outside and inside prisons, juvenile detention facilities, and other institutions. Nonetheless, as with punishment, even if psychopathy is a mental disease, as we argued, it does not follow that psychiatrists should treat them. Psychiatrists might not be able to apply any effective treatment, and it is not clear why psychiatrists should treat people whom they cannot help. Nonetheless, our conclusion that psychopathy is a mental disease removes one obstacle that might seem to make it dubious for psychiatrists to treat (or try to develop treatments for) psychopaths.

Overall, then, our conclusion—that psychopathy is a mental disease—by itself will not revolutionize either the law of responsibility or the treatment of psychopaths. We hope, however, that it does at least clarify the status of psychopathy, and that it might help in some small way to deal with the problems created and faced by psychopaths.[11]

NOTES

1. For more on the development and psychometric properties of the PCL-R and alternative tools, see Forth, Bo, and Kongerslev Forthcoming.
2. See, e.g., Harris, Rice, and Cormier 1991; Heilbrun et al. 1998; Rice, Harris, and Quinsey 1990.
3. For instance, VRAG (Quinsey, Harris, Rice, and Cormier 1998) and HCr-20 (Webster, Douglas, Eaves, and Hart 1997) both used PCL-SV scores. PCL-SV was also used as a risk factor by the MacRisk researchers in developing the ICT approach. However, even though Monahan et al. (2001) found that the PCL-SV was the strongest predictor of violence, it was *not* included as one of the risk factors of COVR because the goal of the latter was to enable researchers to make quick decisions concerning future dangerousness in a forensic setting.
4. Through aversive conditioning, subjects learn to associate an unpleasant response—e.g., a mild shock—with an unwanted behavior that is supposed to discourage them from engaging in the behavior in the future.
5. Passive avoidance involves the inhibition of a previously exhibited response. In passive avoidance, a subject may freeze as soon as the stimulus is presented. In active avoidance, on the other hand, the subject flees when the stimulus is presented.
6. In differential reward-punishment tasks, sometimes subjects are exposed to both positive and negative reinforcement in response to the behavior under investigation.
7. See Glenn, Kurzban, and Raine (2011) for a more detailed discussion of evolution and psychopathy.
8. The point is even clearer if psychopathy involves abnormal activity throughout a system, the paralimbic system, which includes the amygdala (see Anderson and Kiehl, Forthcoming). If a genetic abnormality causes a large group of brain parts to function abnormally, and if some of these abnormalities increase fecundity and others decrease longevity, then the overall condition can count as a mental illness according to the biomedical view.
9. Wakefield's view might seem to misclassify homosexuality as a mental illness or disease, at least in societies where homosexuals are harmed or deprived of

benefits. However, Wakefield could respond to this objection by referring to "a distinct sustaining cause," which we will discuss in the next section.

10. Because psychopathy deprives psychopaths, it is a disability instead of just an inability: "According to a welfarist account of disability (Savulescu 2006; Savulescu and Kahane 2009a), a disability is a relatively stable physical or psychological condition *X* of person *P* counts as a disability in circumstances *C* if and only if *X* tends to reduce the amount of well-being that this person will enjoy in *C* (Savulescu and Kahane 2009b). This is a welfarist account of disability that relates disability to well-being. Disability, a normative term, is not the same as *inability*." (Savulescu 2009:63).
11. The authors are grateful to Kent Kiehl, Karen Neander, Alex Rosenberg, Julian Savulescu, and Nicole Vincent for helpful comments on earlier drafts.

REFERENCES

Aharoni, E., W. Sinnott-Armstrong, and K. Kiehl (2011). "Can psychopathic offenders discern moral wrongs? A new look at the moral/conventional distinction." *Journal of Abnormal Psychology* 47(10):1246–1247.

AmericanPsychiatric Association (2000). *Diagnostic and Statistical Manual of Mental Disorders*, Fourth Edition, Text Revision. Washington, DC, American Psychiatric Association.

Anderson, N. E., and K. A. Kiehl (Forthcoming). "Functional Neuroimaging and Psychopathy." In *Oxford Handbook of Psychopathy and Law.* K. Kiehl and W. Sinnott-Armstrong. New York, Oxford University Press.

Andrade, J. T. (2008). "The inclusion of antisocial behavior in the construct of psychopathy: A review of the research." *Aggression and Violent Behavior* 13: 328–335.

Bechara, A., A. Damasio, H. Damasio, and S. Anderson (1994). "Insensitivity to future consequences following damage to human prefrontal cortex." *Cognition* 50: 7–15.

Blair, R. (1995). "A cognitive developmental approach to morality: Investigating the psychopath." *Cognition* 57: 1–29.

Blair, R., and M. Coles (2000). "Expression recognition and behavioural problems in early adolescence." *Cognitive Development* 15: 421–434.

Blair, R., E. Colledge, L. Murray, and D. Mitchell (2001). "A selective impairment in the processing of sad and fearful expressions in children with psychopathic tendencies." *Journal of Abnormal Child Psychology* 29: 491–498.

Blair, R., E. Colledge, and D. Mitchell (2001). "Somatic markers and response reversal: Is there orbitofrontal cortex dysfunction in boys with psychopathic tendencies?" *Journal of Abnormal Child Psychology* 29: 499–511.

Blair, K. S., A. Leonard, and R. J. R. Blair (2006). "Impaired decision making on the basis of both reward and punishment information in individuals with psychopathy." *Personality and Individual Differences* 41: 155–165.

Blair, R. J. R., D. G. V. Mitchell, and K. S. Blair (2005). *The Psychopath: Emotion and the Brain.* Oxford, UK, Blackwell.

Blair, R. J. R., D. G. V. Mitchell, A. Leonard, S. Budhani, K. S. Peschardt, and C. Newman (2004). "Passive avoidance learning in individuals with psychopathy:

Modulation by reward but not punishment." *Personality and Individual Differences* 37: 1179–1192.

Blonigen, D. M., B. M. Hicks, R. F. Krueger, C. J. Patrick, and W. G. Iacono (2005). "Psychopathic personality traits: Heritability and genetic overlap with internalizing and externalizing psychopathology." *Psychological Medicine* 35(5): 637–648.

Boccardi, M. (Forthcoming). "Structural Brain Abnormalities and Psychopathy." In *Oxford Handbook of Psychopathy and Law*. K. Kiehl and W. Sinnott-Armstrong. New York, Oxford University Press.

Boorse, C. (1975). "On the distinction between disease and illness." *Philosophy and Public Affairs* 5(1), 49–68.

Cima, M., T. Smeets, and M. Jelicic (2008). "Self-reported trauma, cortisol levels, and aggression in psychopathic and non-psychopathic prison inmates." *Biological Psychology* 78(1): 75–86.

Cleckley, H. (1976). *The Mask of Sanity*, 4th Edition. St. Louis, Mosby. First published in 1941.

Cosmides, L., and J. Tooby (2000). "Consider the source: The evolution of adaptations for decoupling and metarepresentation." In: *Metarepresentations: A Multidisciplinary Perspective*. D. Sperber. New York, Oxford University Press, pp. 53–115.

Deigh, J. (1995). "Empathy and universalizability." *Ethics* 105(4): 743–763.

Fine, C., and J. Kennett (2004). "Mental impairment, moral understanding and criminal responsibility: Psychopathy and the purposes of punishment." *International Journal of Law and Psychiatry* 27: 425–443.

Forth, A., S. Bo, and M. Kongerslev (Forthcoming). "Assessment of psychopathy: The Hare Psychopathy Checklist Measures." In: *Oxford Handbook of Psychopathy and Law*. K. Kiehl and W. Sinnott-Armstrong. New York, Oxford University Press.

Gert, B., and C. Culver (2004). "Defining mental disorder." In: *The Philosophy of Psychiatry: A Companion*. J. Raden. Oxford, UK, Oxford University Press, pp. 415–425.

Gert, B., C. M. Culver, and K. D. Clouser (2006). *Bioethics: A Systematic Approach*, 2nd Edition. Oxford, UK, Oxford University Press.

Glannon, W. (1997). "Psychopathy and responsibility." *Journal of Applied Philosophy* 14(3): 263–275.

Glenn, A. L., R. Kurzban, and A. Raine (2011). "Evolutionary theory and psychopathy." *Aggression and Violent Behavior* 16(5): 371–380.

Glenn, A. L., A. Raine, and R. A.Schug (2009). "The neural correlates of moral decision-making in psychopathy." *Molecular Psychiatry* 14: 5–6.

Graham, G. (2010). *The Disordered Mind*. London, Routledge.

Greenspan, P. (2003). "Responsible psychopaths." *Philosophical Psychology* 16(3): 417–429.

Haji, I. (2003). "The emotional depravity of psychopaths and culpability." *Legal Theory* 9: 63–82.

Hare, R. (1978). "Electrodermal and cardiovascular correlates of psychopathy." In: *Psychopathic Behavior: Approaches to Research*. R. Hare and D. Schalling. New York, John Wiley & Sons, pp. 107–144.

Hare, R. (1991). *The Hare Psychopathy Checklist—Revised*. Toronto, Multi-Health Systems.

Harenski, C. L., K. A. Harenski, M. S. Shane, and K. Kiehl (2010). "Aberrant neural processing of moral violations in criminal psychopaths." *Journal of Abnormal Psychology* 119(4): 863–874.

Harris, G., M. Rice, and C. Cormier (1991). "Psychopathy and violent recidivism." *Law and Human Behavior* 15: 625–637.

Hart, S. D., and R. D. Hare (1996). "Psychopathy and antisocial personality disorder." *Current Opinion in Psychiatry* 9(2): 129–132.

Haukka, J., J. Suvisaari and J. Lonnqvist (2003). "Fertility of patients with schizophrenia, their siblings, and the general population: A cohort study from 1950–1959 in Finland." *American Journal of Psychiatry* 160: 460–463.

Heilbrun, K., S. D. Hart, R. D. Hare, D. Gustafson, C. Nunez, and A. J. White (1998). "Inpatient and postdischarge aggression in mentally disordered offenders: The role of psychopathy." *Journal of Interpersonal Violence* 13: 514–527.

Hervé, H. (2007). "Psychopathic subtypes: Historical and contemporary perspectives." In: *The Psychopath: Theory, Research and Practice*. H. Hervé and J. Yuille . Mahwah, NJ, Lawrence Erlbaum Associates.

Jonason, P. K., N. P. Li, G. D. Webster, and D. P. Schmitt (2009). "The dark triad: Facilitating a short-term mating strategy in men." *European Journal of Personality* 23: 5–18.

Karpman, B. (1948). "The myth of the psychopathic personality." *American Journal of Psychiatry* 104: 523–534.

Kernberg, O. (1975). *Borderline Conditions and Pathological Narcissism*. Northvale, NJ, Jason Aronson Inc.

Kiehl, K. A., A. M. Smith, R. D. Hare, A. Mendrek, B. B.Forster, J. Brink, J. Brink, and P. F. Liddle (2001). "Limbic abnormalities inaffective processing by criminal psychopaths as revealed by functional magnetic resonance imaging." *Biological Psychiatry* 50(9): 677–684.

Kiehl, K. A., A. M. Smith, A. Mendrek, B. B. Forster, R. D. Hare, and P. F. Liddle (2004). "Temporal lobe abnormalities in semantic processing by criminal psychopaths as revealed by functional magnetic resonance imaging." *Psychiatry Research* 130(3): 27–42.

KochJ. L. 1891. *Psychopathic Inferiority*. Ravensburg, Germany, Maier.

Larsson, H., C. Tuvblad, F. V. Rijsdijk, H. Andershed, M. Grann, and P. Lichtenstein (2007). "A common genetic factor explains the association between psychopathic personality and antisocial behavior." *Psychological Medicine* 37: 15–26.

Leistico, A. M., R. T. Salekin, J. DeCoster, and R. Rogers (2008). "A large-scale meta-analysis relating the Hare measures of psychopathy to antisocial conduct." *Law & Human Behavior* 32: 28–45.

Levy, N. (2007a). "The responsibility of the psychopath revisited." *Philosophy, Psychiatry, and Psychology* 1 (2), 129–138.

Levy, N. (2007b). "Norms, conventions, and psychopaths." *Philosophy, Psychiatry, and Psychology* 14(2): 163–170.

Litton, P. (Forthcoming). "Criminal responsibility and psychopathy." In: *Oxford Handbook of Psychopathy and Law*. K. Kiehl and W. Sinnott-Armstrong . New York, Oxford University Press.

Maibom, H. (2005). "Moral unreason: The case of psychopathy." *Mind & Language* 20(2): 237–257.

Malatesti, L., and J. McMillan (2010). *Responsibility and Psychopathy: Interfacing Law, Psychiatry and Philosophy.* New York, Oxford University Press.

Monahan, J., H. Steadman, E. Silver, P. S. Appelbaum, A. Clark-Robbins, E. P. Mulvey, L. Roth, T. Grisso, and S. Banks (2001). *Rethinking Risk Assessment: The MacArthur Study of Mental Disorder and Violence.* Oxford, UK, Oxford University Press.

Morse, S. (2008). "Psychopathy and criminal responsibility." *Neuroethics* 1: 205–212.

Neander, K. (1991). "Functions as selected effects: The conceptual analyst's defense." *Philosophy of Science* 58(2): 168–184.

Newman, J., and D. Kosson (1986). "Passive avoidance learning in psychopathic and nonpsychopathic offenders." *Journal of Abnormal Psychology* 96: 257–263.

Pickard, H. (2009). "Mental illness is indeed a myth." In: *Psychiatry as Cognitive Neuroscience: Philosophical Perspectives.* M. Broome and L. Bortolotti. New York, Oxford University Press.

Pinel, P. (1962). *A Treatise on Insanity.* D. Davis, trans. New York, Hafner. Original work published in 1801.

Prichard, J. C. (1835). *A Treatise on Insanity and other Disorders Affecting the Mind.* London, Sherwood, Gilber, and Piper.

Quinsey, V. L., G. E. Harris, M. E. Rice, and C. Cormier (1998). *Violent Offenders: Appraising and Managing Risk.* Washington, DC, American Psychological Association.

Rice, M. E., G. T. Harris, and V. L. Quinsey (1990). "A follow-up of rapists assessed in a maximum security psychiatric facility." *Journal of International Violence* 5 435–448.

Robbins, T. W., B. J. Everitt, and D. J. Nutt (2010). *The Neurobiology of Addiction: New Vistas.* Oxford, UK, Oxford University Press.

Rogers, R. D., M. Lancaster, J. Wakeley, and Z. Bhagwager (2004). "Effects of beta-adrenoceptor blockade on components of human decision making." *Psychopharmacology* 172(2): 157–164.

Sarbin, T. R. (1967). "On the futility of the proposition that some people be labeled 'mentally ill.'" *Journal of Consulting Psychology* 31(5): 447–453.

Savulescu, J. (2006). "Justice, fairness and enhancement." Ann N Y Acad Sci 1093: 321–338.

Savulescu, J. (2009). "Autonomy, well-being, disease, and disability." *Philosophy, Psychiatry, & Psychology* 16(1): 59–65.

Savulescu, J., and G. Kahane (2009a). "The welfarist account of disability." In: *Disability and disadvantage.* A. Cureton, and K. Brownlee. Oxford, UK, Oxford University Press, pp. 14–53.

Savulescu, J., and G. Kahane (2009b). "The moral obligation to create children with the best chance of the best life." *Bioethics* 23(5): 274–290.

Scadding, J. G. (1990). "The semantic problems of psychiatry." *Psychological Medicine* 20, 243–248.

Schaich Borg, J., and W. Sinnott-Armstrong (Forthcoming). "Do psychopaths make moral judgments?" In: *Oxford Handbook of Psychopathy and Law.* K. Kiehl and W. Sinnott-Armstrong. New York, Oxford University Press.

Sedgwick, P. (1973). "Illness: Mental and otherwise." *The Hastings Center Studies* 1(3): 19–40.

Serin, R. C. (1996). "Violent recidivism in criminal psychopaths." *Law and Human Behavior* 20: 207–217.

Sinnott-Armstrong, W., and K. Levy, K (2011). "Insanity defenses." In: *The Oxford Handbook of Philosophy of Criminal Law*. J. Deigh and D. Dolinko. New York, Oxford University Press.

Strange, B. A., and R. J. Dolan (2004). "Beta-adrenergic modulation of emotional memory-evoked human amygdala and hippocampal responses." *Proceedings of the National Academy of the Sciences of the United States of America* 101(31): 11454–11458.

Szasz, T. S. (1961). "The myth of mental illness." *American Psychologist* 15: 113–118.

Szasz, T. S. (1999). "Is mental illness a disease?" *The Freeman* 49: 38–39.

Veit, R., H. Flor, M.Erb, C. Hermann, M. Lotze, W. Grodd, and N. Birbaumer (2002). "Brain circuits involved in emotional learning in antisocial behavior and social phobia in humans." *Neuroscience Letters* 328(3): 233–236.

Viding, E., A. P. Jones, P. Frick, T. E. Moffitt, and R. Plomin (2008). "Genetic and phenotypic investigation to early risk factors for conduct problems in children with and without psychopathic tendencies." *Developmental Science* 11: 17–22.

Wakefield, J. C. (1992a). "Disorder as harmful dysfunction: A conceptual critique of DSM-IIIR's definition of mental disorder." *Psychological Review* 99(2): 232–247.

Wakefield, J. C. (1992b). "The concept of mental disorder: On the boundary between biological facts and social values." *American Psychologist* 47(3): 373–388.

Webster, C. D., K. S. Douglas, D. Eaves, and S. D. Hart (1997). *HCR-20: Assessing the Risk for Violence (Version 2)*. Vancouver, Mental Health, Law, and Policy Institute, Simon Fraser University.

11

Addiction, Choice, and Disease

How Voluntary Is Voluntary Action in Addiction?[1]

JEANETTE KENNETT

> *[I]t is unlikely that anyone chooses to be an addict, but what research shows is that everyone, including those who are called addicts, stops using drugs when the costs of continuing become too great.*
>
> *Just as the diurnal cool breezes at evening and the once in a lifetime hurricane are explained by the same physical principles, so... Everyday choices and addiction will be explained by the same motivational principles.*
>
> (Gene Heyman 2009:vii)

INTRODUCTION

Are drug addicts helpless in the face of their addiction, compelled by cravings too strong to resist? Or is drug taking voluntary activity that can be ceased at will? In his recent book, Gene Heyman (2009) challenges the dominant medical model of addiction, which, in its current form, characterizes addiction as "a chronic relapsing brain disease" (p. 17). Although Heyman does not focus on the responsibility of addicts, the disease model of addiction is often taken to at least partially excuse addicts from moral and criminal responsibility because it characterizes addiction not as mere weakness of will but as a disease of the mind that impairs the capacities required for moral responsibility.

Heyman argues, by contrast, that the behavior that characterizes addiction is voluntary and that the many ill effects of substance abuse result from a series of choices made by the addicted person. He aims to show that these choices conform to standard motivational principles; the choices of the addict, like all other choices, aim at reward and are responsive to incentives. Heyman argues that attending to these principles provides a more useful and more optimistic framework than the disease model for understanding what goes wrong in addiction and how to treat it. In this chapter, I examine Heyman's argument against the disease model and the assumptions that underpin it. Despite the many virtues of the account, I suggest that it does not succeed in ruling out the disease model, even on the assumption that addictive choices are voluntary. I then question that assumption and the account of motivation on which it rests. I conclude that there are significant involuntary aspects to addiction that could mitigate the responsibility of addicts for their choices.

THE DISEASE MODEL OF ADDICTION

According to Kalivas and Volkow (2005), "[a]mong the most insidious characteristics of drug addiction is the recurring desire to take drugs even after many years of abstinence. Equally sinister is the compromised ability of addicts to suppress drug seeking in response to that desire even when confronted with seriously adverse consequences, such as incarceration. The enduring vulnerability to relapse is a primary feature of the disorder" (p. 1403).

The brain disease model of addiction (as, e.g., outlined in chapter 12 of this volume) attributes this vulnerability to relapse to enduring changes in brain function, produced by chronic drug use, that seriously impair addicted persons' ability to control their drug use (Baler and Volkow 2006; Leshner 1997; Volkow and Li 2005). This view is supported by neuroimaging studies which indicate that the brains of addicted and nonaddicted persons differ in their responses to drugs and drug-related stimuli (Garavan et al. 2007). According to Antoine Bechara, neuroimaging reveals structural abnormalities in those parts of the brain thought to be important for the reflective system. For example, the ability to resist the intrusion of unwanted thoughts or memories is an important mechanism of impulse control, but neuroimaging studies appear to reveal deficits in this mechanism in addicts (2005:1460). Bechara claims that "substance related cues trigger bottom up mechanisms in substance abusers, influencing top-down mechanisms such as impulse and attentional control" (2005:1461). They can "hijack the top down goal-driven cognitive resources needed for the normal operation of the reflective system and exercising the willpower to resist drugs" (2005:1461). Addiction neuroscience also appears to confirm greater intensity of response to drugs in addicts, and abnormally low responsiveness to natural rewards. Research by Volkow indicates that addicts have fewer dopamine D2 receptors, which are found in parts of the brain involved in motivation and reward behavior. They are thus thought to be more vulnerable than others to the effects of addictive drugs, which send dopamine surging (reported in McGowan 2004). In a review of the brain neurocircuitry literature to date, Koob and Volkow conclude that "multiple brain regions and circuits are disrupted in drug addiction and are likely to contribute to the complex phenotype observed in addicted individuals" (2010:233). Furthermore, as Hall and Carter point out, genetic factors have been estimated to account for between 40% and 60% of addiction liability.

The disease model notes that addiction involves continued drug use in the face of "seriously adverse consequences" and suggests that there is a compulsive element to addiction that is at least in part explained by neurobiological vulnerabilities and progressive impairments to brain function. It also suggests that addiction, like other diseases, has a predictable and well-defined course. Over repeated cycles of intoxication, withdrawal, and anticipation and planning, there is a pattern of escalating use, loss of control over the amounts used, and progression to chronic preoccupation and drug-seeking behavior. While most people who try or use potentially addictive substances do not become addicts, the minority who do follow this course.

ADDICTION: DISEASE OR CHOICE?

Heyman is unconvinced by the disease model of addiction. First, he argues that the brain disease advocates are plain wrong in their characterization of addiction as chronic and relapsing. He produces evidence that for most people, it is not. Second, he argues that neurobiological evidence of neuroadaptation or of genetic predisposition doesn't show that addiction is a disease. He rightly points out that all behavior has neural correlates and so the mere fact of neuroadaptation is insufficient to ground the brain disease hypothesis. He thinks that an additional assumption is required to support the disease hypothesis: namely that the behavior that has such neural or genetic correlates be *involuntary*. The key question, he says, "is whether genes or neuroadaptations turn voluntary drug use into involuntary drug use" (2009:87). He claims that the issue of whether addiction is a disease "depends on the understanding of voluntary behavior." (2009:100).

Heyman argues that if addiction is a psychiatric disorder, it is a very unusual one, given its astonishingly high rate of spontaneous remission. He points out that more than 80% of individuals who have ever satisfied the criteria for substance dependence recover without treatment. Moreover, he argues that recovery from addiction is quite unlike recovery from other serious psychiatric disorders, for example, schizophrenia, in that it appears to be directly responsive to ordinary social incentives—namely the acquiring of life responsibilities, such as marriage, parenthood, and career, which are incompatible with continued heavy drug use. In support of this, he cites evidence that most users cease problematic use by age 30. In his view, the disease hypothesis arose and garnered support not because of any decisive medical or neurobiological evidence but because of a key flawed assumption. "Medical evidence did not turn alcoholism into a disease, but rather the assumption that voluntary behavior is not self-destructive turned alcoholism into a disease" (2009:99). Since he thinks it is quite apparent that voluntary behavior can be self-destructive, Heyman argues that we should apply what we know about the mechanisms of choice to the behavior of addicts.

Heyman explicitly denies that addicts choose addiction. Rather, addiction is the cumulative result of choices to use drugs on multiple separate occasions. Faced with a choice situation in which the immediate cost of refusing drugs is high and the benefits of abstinence lie well in the future, the drug user chooses the nearer good—the certain pleasure or relief that drug use offers. Considered from a local perspective where immediate pleasures and pains take center stage, the choice to use drugs right now might well make sense. Considered from a global perspective—the perspective from which we reflect upon how our lives are going overall and make choices in line with our long-term values, commitments, and welfare—continued heavy drug use is a very bad idea. The choices made by users threaten their health, their relationships, their education and career prospects, and their finances. The problem, according to Heyman, is that drug users are prone to impulsivity and tend to view their options from a local perspective.

The conflict between local and global perspectives on choice is familiar to us all. It is difficult to maintain a global view in the face of more immediately salient local demands and incentives. We have an empirically well-established tendency to apply a discount rate to future pleasures and pains (Ainslie 1975), a tendency noted as far back as Socrates, who argued that no one ever knowingly chooses the lesser good. The thought is that we are subject to a kind of evaluation illusion: the nearer good looks larger. The pleasures of staying at the bar with friends tonight are usually much more vivid to us than tomorrow's hangover, and many of the bad consequences of drug use are considerably more remote and uncertain than a hangover. Drug users have been found to apply a steeper discount rate to future pleasures and pains than controls, making it even harder for long-term considerations to compete for their attention (Ainslie and Monterosso 2003; Johnson and Bickel 2003).

Heyman says that addictive substances are those that are intoxicating, instantly rewarding "specious in that their costs are delayed or uncertain" and "behaviourally toxic" (2009:47). According to Heyman, "[a] substance is behaviorally toxic when it poisons the field making everything else relatively worse" (2009:145). So, for the drug user, the ordinary pleasures of eating, socializing, exercising, planning a holiday, or achieving at work no longer seem rewarding. Giving up drugs is thus especially difficult because all other options have been drained of value. It requires the individual to choose unrewarding options for a considerable period of time, maybe weeks or months, before they experience any benefit from abstinence. Given that the global perspective is cognitively more demanding to maintain than the local perspective (and the neuroscientific evidence suggests that drug users are less well equipped to take such a perspective), it is little wonder that relapse is common, at least in the early stages of quitting.

Full rationality requires us to bring local and global perspectives into equilibrium. Most of us are less than fully rational much of the time. Heyman thinks it is clear that behavior can be both voluntary and irrational, and he is surely correct. The dessert lover may be fooling herself that she won't put on weight and run the risk of heart disease, but she does not eat chocolate layer cake involuntarily. Likewise, the addicted person does not consume drugs involuntarily. Heyman points to the planning and versatility, and the responsiveness to circumstances and incentives, of those involved in drug seeking and consuming as evidence for this. Sustained drug abuse is *globally* irrational and self-destructive, but it is *locally* chosen behavior, so if we are to combat it effectively, Heyman says, we must target choice and the choice situations that addicts face.

The account predicts that changing the incentive structure will change the behavior, and this appears to be borne out by evidence cited by Heyman. In addition to evidence of users giving up as they acquire new interests and responsibility, and in the face of waning drug rewards, Heyman also argues that current successful treatments for addiction amount to changing the incentive structure by providing additional options and making drug taking less rewarding and other activities more rewarding. Examples of successful treatments given by Heyman include voucher schemes to reward abstinence, support groups that provide new social outlets, and medications that *block* or *replace* the rewarding effects of drugs.

Heyman's conclusion is that the choice model explains addiction better than the disease model. It provides a sounder basis for understanding *why* successful treatment programs are successful and for designing new programs, including medications that block or replace drug rewards.

DISEASE AND VOLUNTARY BEHAVIOR

> *If a key feature of a disease state is that the symptoms are involuntary* then we need to know how to distinguish between voluntary and involuntary behaviour. (Heyman 2009:90, my emphasis)

It is notable that we don't find significant disagreement between Heyman and addiction neuroscience about the facts. They agree that drug taking is hard to quit and is so both because drug pleasures undermine other natural pleasures and because drug users find it difficult to adopt and maintain a reflective stance. The disagreement is one of interpretation. Addiction neuroscientists focus on evidence of neurological dysfunction and impairment to support the view that addiction is rightly conceptualized as a brain disease. Heyman focuses on the voluntary nature of drug use and high rates of recovery to argue that it is not. His reasoning appears to go like this. If the behavior of addicts is voluntary (and it is), it cannot be the product of a disease or disorder.

- Premise 1: Voluntary behavior cannot be the product or symptom of a disease.
- Premise 2: The drug-seeking and drug-taking behavior of addicts is voluntary.

- -

- Therefore, addiction is not a disease.

Heyman's focus is on evidence for premise 2. Premise 1 is assumed. But neither is secure from challenge. Let's examine the evidence for these premises and the assumptions underlying them more closely.

Heyman cites the fact that most addicts reduce or cease using by their late 20s without treatment as evidence against the disease model in part because he thinks it shows that the "symptoms of this disorder" (i.e., drug use) are under direct voluntary control, whereas the symptoms of other illnesses (i.e., fevers) are not. He also seems to be using the high rates of recovery in addiction to argue that it is not an illness—even in the minority who do not stop using by their mid-30s. Neither prong of this argument works.

First, evidence of spontaneous recovery cannot impugn the disease model. The fact that a person ceases using drugs without treatment does not, of itself, show that they *could have* ceased using at some earlier point in time *or* that they were never ill. Many diseases and disorders are self-limiting; for example, the common cold has an almost 100% rate of natural recovery. High rates of spontaneous recovery may be less common in psychiatric disorders, but recovery without treatment does occur. For example, a study of 130 subjects suffering a depressive episode (Posternak et al. 2006) revealed a high rate of recovery among individuals receiving no treatment, with a median time to recovery of 13 weeks for the 84 subjects receiving no treatment compared with 23 weeks for the 46 subjects receiving drug therapy. It is also likely that there are many undiagnosed cases of depression which resolve without treatment. Should we conclude that depressed people who recover spontaneously were never ill? Could the figures on spontaneous recovery be used to argue that even those who don't recover without treatment are not ill? Does this mean that depression is not a bona fide disorder? Surely not. Perhaps addiction is not properly classified as a disease or disorder, but evidence of spontaneous recovery does not establish this.

Second, there may be important differences between addicts who cease using without treatment and those who don't. The common cold, influenza, and pneumonia have common symptoms in their early stages, but they are *different* illnesses. Something similar may be true of addiction. It may not be a single kind, and it may be that not all those who satisfy the current criteria for substance dependence are ill. Perhaps we need to distinguish between those cases that are more like a cold and those that are more like pneumonia in a person with a compromised immune system, that is, dangerous to life and requiring aggressive treatment. Addicts who cease using without treatment may turn out to have relevantly different biological or neurological responses to drugs than those who don't. Work by Bechara and colleagues (2002) suggests that this may be the case. A subgroup of substance abusers show normal behavioral and physiological responses to the Iowa Gambling Task, and this subgroup is also better functioning than other groups.

Heyman briskly dismisses the possibility that even the hardcore users who do not quit without treatment, or at all, fit the disease model by pointing to figures which suggest that most such individuals have comorbid mental illnesses. On his account, it is the presence of these other mental disorders that explains continued use. Other mental illnesses impair the capacity for work or intimate relationships, and so these important incentives to reduce use will not be available to sufferers. Heyman is undoubtedly correct that comorbidity is a significant factor in determining prospects for recovery, and much drug use in comorbid patients can be accounted for by his theory. I've argued elsewhere (Kennett and Matthews 2006; Kennett Forthcoming) that drug use can sometimes be a rational or understandable response to dire social circumstances and diminished life options and is also often to be understood in terms of self-medication of the distressing symptoms of other mental illnesses. However, the presence of other mental disorders in hardcore users does not establish that addiction is not a distinct disorder and that it does not follow a disease-like course in such cases. Where there is comorbidity, the disorders may have a common cause, or addiction may be the primary disorder. There is plenty of evidence of drug-induced psychosis among methamphetamine and cannabis users, heavy alcohol use is implicated in the onset of anxiety and depression, cannabis is thought to trigger schizophrenia in vulnerable individuals, and the dangers (of assault, rape, imprisonment, illness, and significant loss and grief) associated with the social circumstances of illicit drug use and the effects of intoxication and withdrawal may lead to the development of post-traumatic stress disorder (Holmwood 2003; Schadé 2003; Schatzberg et al. 2008; Stewart and Conrod 2008). While users beyond their early 30s often present with other mental health problems, these problems commonly emerge *after* heavy use is established and may be wholly or significantly caused by such use.[2] This is hardly surprising given the psychoactive nature of drugs of abuse and the recognized neurobiological changes brought about by abuse. A vicious cycle may then ensue. And there are also long-term users who have no obvious or significant comorbid mental illnesses but whose addiction is a source of significant distress and dysfunction. These hardcore addicts are the focus of clinical attention and, I suggest, are the target of the disease model. The fact that most problematic users do not fit the disease model as it is currently articulated does not show that it is misplaced when applied to a clinical population who do not cease using by their mid-30s.

Third, behavior that is generally accepted as the product of psychological disorder can be voluntary, as practices such as cutting and starving reveal. Self-harm, for example, is symptomatic of post-traumatic stress disorder and borderline personality disorder, but the young woman who engages in cutting rituals does not act involuntarily in relieving her distress in this way. Usually, she takes great care to keep her activities secret, which suggest a high degree of control. The anorexia sufferer engages in high levels of conscious control and planning of actions surrounding food. The depressed person plans, and may successfully complete, suicide. These are all voluntary behaviors that are relevantly responsive to local contingencies, just as drug use is. They are also familiar signs and symptoms of serious psychiatric disorders that can be fatal if left untreated. Would Heyman argue that post-traumatic stress disorder, borderline personality disorder, anorexia, and suicidal depression are not genuine disorders? The choices made by drug users could be both voluntary in Heyman's sense and substantially shaped by psychological disorder or impairments to brain function.

Fourth, the above examples of certain kinds of voluntary behavior as manifestations or key symptoms of psychiatric disorder or illness suggest that we should be cautious in accepting the view of human psychology that underpins Heyman's analysis of addiction in terms of a simple and universal motivational principle governing choice. I turn now to that picture and test it against the hard cases: those hardcore addicts who do not cease using in response to "ordinary social incentives."

HARD CASES AND THE REWARD ACCOUNT OF MOTIVATION

Heyman endorses the view that voluntary behavior always aims at some *reward*. On the view he endorses, if we want to explain behavior, we must understand it in terms (broadly speaking) of satisfactions gained or pains avoided. The assumption is that we always act to secure pleasure or satisfaction of some kind, or to avoid pain. All voluntary behavior is explicable in terms of expected utility—even that of the obsessive hand washer whose apparently pointless behavior provides temporary relief from anxiety. On this picture irrationality just consists, in making certain sorts of mistakes in calculating expected utility. So the addict *either* has a mistaken or distorted view of the benefits to be obtained from using drugs—typically because he adopts a local rather than a global view of the benefits (Heyman) or applies an overly steep discount curve to future rewards (hyperbolic discounting, Ainslie)—*or* drug taking *really is* the thing he finds most rewarding—as evidenced by his continuing to choose drugs. This last is the revealed preference view of classical economics (Samuelson 1938). Whenever we act, we act on the highest ranking of our desires that we believe to be available to us. If we don't choose something, it is because it is not preferred or not available.

This view of motivation in terms of (expected) reward looks on the surface to be simple common sense. It informs our everyday attributions of motives to each other and our interpretations of each other's behavior. When we seek to explain behavior, we ask what the point of it was. What was to be said for it from the agent's point of view? What did she want, hope for, or fear? But it is not obvious that the actions of addicts are always susceptible to this kind of rationalizing explanation, as the following first personal account illustrates.

> In brief: My father was an alcoholic, which broke our family when I was 10 or so. He died of his addictions at 52, which I believe is longer than his own father lasted. I lost a brother in 1983 to an incomprehensible murder fueled by PCP...I lost a brother in 1991 to suicide by heroin overdose, after watching him turn from a hopeful little kid to an utterly despairing addict, a liar and a thief. My third and last brother spent five years in the state pen for

> armed robbery. He was a junkie, crackhead and so on, and then a recovering junkie, crackhead, and so on for many years. He expired two years ago in his sleep, his body ravaged by hepatitis, diabetes, and heart disease. In our family, that's success…
>
> My every act of love, every home place, every hint of peace or happiness, is the premonition of another recapitulation of the endlessly-repeated loss that has no point or purpose, and has no end but in death…
>
> In my own case, once I start drinking, I don't know when or whether I'll stop. I've had periods from months to many years without alcohol, and I've found my way through various circuitous routes back to drinking. Then I become an evil idiot. I have no self-control or self-respect. I lie. I hide bottles around the house and drink the clock around. I puke my guts out. I'm alternately maudlin or consumed by rage.
>
> Putting it mildly, I'm not alone in these sorts of experiences, and a lot of people have been through even worse. What it's like being a meth addict or actually dying of alcohol poisoning, I don't literally know, yet. But if you think people are doing things like that in order to feel good, I say you're crazy. (Crispin Sartwell "Detritus" web link)

Heyman insists that: "everyone…stops using drugs when the costs of continuing become too great." This, apparently, is as true of those addicts who drink themselves to chronic illness, despair, and death (sometimes while begging for help), as it is of the majority who cease using of their own accord in their late 20s when the incentives to use change. Of course, it might be difficult for us to understand what benefit drug taking offers in some cases, but ex hypothesi, there is a perceived benefit or the person would not keep choosing to use drugs.

In my view, this claim is either trivial or false when applied to many chronic addicts. Heyman's explanation of addictive choices is that they aim at securing the greatest immediate benefit, but in common with many ordinary choices, they are globally irrational in that they fail to maximize, and may actively undermine, overall utility. On this analysis, Sartwell's case should puzzle us because there seems to be no obvious reward on offer from drinking *now*, even from *his own point of view*. Sartwell is only too aware of both the short- and long-term costs of drinking; he knows that drinking yields neither immediate pleasure nor relief from pain. There is not much to be said for the local prospect of puking your guts out and being maudlin or consumed by rage. Moreover, it is not the case that Sartwell claims as a benefit of drinking some feature, such as being intoxicated, that most of us would fail to find attractive. We don't know and Sartwell doesn't know why he drinks. It seems that there is nothing to be said for it even from his own point of view. To the extent that his family history explains his actions, it does so by postulating a cause (genetic determination perhaps or a sense of inevitability "this is who I am"), not a reason. Heyman's supposedly universal motivational principles don't account for this kind and degree of self-destructiveness. On this reading, the reward hypothesis is false of at least some voluntary actions.

Are we, then, pushed to a revealed preferences view to save the reward hypothesis? Sartwell just does often prefer drinking to not drinking. Therefore, the benefits—for Sartwell—outweigh the costs. This is what he currently finds most rewarding despite the fact that it gives him no pleasure and is indeed both emotionally and physically distressing. If we stipulate that voluntary behavior aims at some reward (or relief), then this conclusion follows from the fact that Sartwell's behavior is voluntary (if indeed it is). Of course, we can make this stipulation if we want. But as Camerer has recently pointed out, Pareto, to whom the revealed preferences view of utility is owed:

> ...advocated divorcing economics from psychology by simply assuming that unobserved utility is necessarily revealed by choice... The equation of utility with choice was not a scientific discovery on par with a powerful theorem or solid empirical regularity. Pareto simply asserted that, as a matter of convenience, it was okay to give up on understanding the "essence of things... Pareto's turn—the definition of utility as a quantity revealed by expressed preference equation—was an agreement on a convention for how to do economics, like the rules of tennis, or assuming away friction in physics. (Camerer 2006:89–90)

Such a convention, however helpful it may be in economics, sheds no light on Sartwell's behavior. In these hard cases, it stretches the notion of utility or reward beyond what commonsense and ordinary use can bear and so drains it of any meaning or usefulness. To say that Sartwell's actions aimed at increased utility or reward apparently means no more than to say that he was motivated to do what he did. And we already knew that. Such cases as his may be in the minority, but a disease model of addiction may get us closer to understanding them than a reward or choice model. The reward model assumes that the agent's drug use is *reasons responsive*,[3] it responds to incentives and disincentives in ways that render the agent's actions comprehensible in light of the considerations available to them. (Where the agent goes rationally astray is in assigning too great a value to immediate satisfactions.) The disease model makes no such assumption.

But could a version of the reward model be saved by recourse to motivations that are inaccessible to consciousness, of the sort postulated by psychoanalytic theory? This would be a desperate move. To be sure, such motivations may exist and be relevant to the explanation of behavior that can be so deeply puzzling even to those whose behavior it is. But they do not fit with the picture of motivation from behavioral economics and the psychology of choice that frames the challenge to the disease model. Heyman's account of addiction relies on the agent acting on the basis of expected benefit where the benefits are systematically diminishing owing to the toxic nature of the reward. The diminishing nature of drug rewards, combined with a bias toward the nearer benefit, results in the addicted agent acting in ways which undermine their overall welfare, but it is still expected benefit that motivates choice and rationalizes action. Actions are done for reasons, and it is that which distinguishes action from mere behavior. No such rationalizations are available for tics and reflexes. Tics and reflexes do not aim at anything. But this rationalizing framework, which relies on what there is to be said for an action from the agent's point of view, doesn't apply to subconscious or repressed motivation. In discovering or postulating such motivations, we provide explanations for action, not reasons for action. If, for example, we discover that a piece of self-destructive behavior was motivated by a subconscious fear or expectation of abandonment, we have identified a cause, not a goal. An action performed *out of* a subconscious fear of abandonment is not necessarily aiming at anything that could plausibly be called a benefit. It may simply be expressive of that fear.[4] Yet it may be done intentionally. The reduction of all psychological states that may be relevant to motivation to preferences, combined with the simplifying assumption that all preferences are concerned only with utility, ignores the distinct role that such states as moods, fears and imaginings, expectations, and so forth may play in the explanation of action. Actions motivated consciously or subconsciously by these psychological states may not be rationally responsive to changed incentive structures. They will not conform to the supposedly universal principles of motivation espoused by Heyman.[5]

VOLUNTARINESS, COMPULSION, AND CONTROL

It is important to note that by "voluntary behavior," Heyman just means behavior that counts as intentional—it is responsive to local contingencies, and the *execution* of the action is under the control of the agent.[6] And it is even more important to notice just how limited a claim this is. As Hannah Pickard points out (2012), mere behavior that is not the product of any intention nor under the control of the agent doesn't count as action. Tics and reflexes are not actions. Dropping and breaking a glass is something you do but it is not (usually) an action. Where established, automatism also undermines the claim that some piece of behavior was intentional and guided and can therefore count as an action of the agent. If Heyman just means to point out that drug use, even in addicts, is not like tics, accidents, and automatism, then we can agree with him. Addicts by and large intend to use drugs, and the particular actions that constitute such use—tightening the tourniquet, pressing the syringe—are under their executive control. Similarly, the obsessive compulsive disorder sufferer may intend to check the locks or clean the kitchen bench (again), and the actions of reaching for the disinfectant and wiping the bench are under her executive control. Unfortunately, the claim that some piece of behavior was voluntary in this sense—the merely intentional sense—is too easily conflated with the claim that that behavior was under the control of the agent in the sense that matters for moral, and ought to matter for legal, responsibility.

Do proponents of the disease model claim that the actions of addicts are *just like* tics or reflexes? If they do, they are mistaken, and Heyman would be right to criticize this kind of claim. But they are more likely invoking a notion of compulsion that, contra Pickard, does not require that the behavior fails to be action at all. To say that an action is compelled signifies that there are very significant automatic or involuntary elements involved that compromise the individual's capacity to choose and act differently. In my view, the correct analysis of the ordinary notion of compulsive motivation is of motivation that is largely impervious both to the agent's *values* and to common techniques of self-control. This is consistent with the claim that the behavior is intentional.

Techniques of Self-Control

Self-control may be exercised at the point of temptation or prior to the temptation. Techniques of *diachronic* self-control may aim at preventing temptation from arising or at arranging one's future circumstances so that when temptation inevitably arises, it will not be acted upon. In the case of addictive urges, examples of the first may be the decision to attend therapy to rid oneself of the disvalued desires or to take medication to block the onset of drug cravings, or to stay away from situations, such as Friday night work drinks, that will trigger cravings. An example of the second may be to put a lock on the liquor cabinet or to move to an area where the drug of choice will not be available.

Synchronic self-control, on the other hand, is exercised in situations of temptation. I've argued elsewhere (Kennett 2001) that attentional techniques are the central way in which we exercise such control, and this view is in line with the finding of Mischel's famous marshmallow experiments (1975; 1972). In brief, successful synchronic self-control often involves directing one's attention away from the stimulus or reconceiving the stimulus so that it becomes less attractive. In the marshmallow experiment, children were told that they could eat one marshmallow now, or if they waited until the experimenter reentered the room, they would be given two marshmallows. Children who successfully waited tended to distract themselves with other activities, such as singing songs and dancing, or else they thought of the marshmallows in cool rather than consummatory terms—as fluffy like clouds, rather than as sweet and delicious. Similarly, addicts might exercise control by narrowing or redirecting the focus of their attention. They might count their pulse or focus on their breathing until the craving passes. Or they might reconceive of the substance as poison, or remind themselves of all the reasons they have for not indulging. They might vividly picture the distress of loved ones and so forth. Sometimes, cravings might be subdued in the face of such reminders and imaginings. Or, they might simply resolve not to take drugs whatever the urge. They might exert willpower, and this too is a familiar enough experience, as when we resist the temptation to stay in bed and, so it seems, force ourselves to get up and go for a swim.

Unfortunately, however ,some desires and motivations may defy both diachronic and synchronic methods of self-control. They may be practically irresistible in the actual circumstances the person faces or can plausibly create for themselves.[7] If this is the case and such motivations are persistent, then in my view they are compulsive, and the person's responsibility for their subsequent actions is diminished.

OBSTACLES TO SELF-CONTROL IN ADDICTION

What is it about motivations for drug use that renders them impervious to ordinary techniques of self-control? In this section, I briefly examine why self-control fails when it does and point to some of the evidence that addiction creates a set of interrelated conditions that undermine or defeat the exercise of self-control. Taken together, these conditions suggest that drug use in at least the hard cases of addiction may often count as compelled. I also suggest along the way that this account of how and why self-control fails further calls into question the account of human motivation relied on by Heyman and others in their analysis of addiction.

Attention and Automatic Action Tendencies

As I've suggested, successful self-control involves controlling the focus of one's attention. If our attention is monopolized by thoughts of a particular course of action or outcome, then other options will not be deliberatively or phenomenologically salient. Intrusive thoughts and the monopolization of attention thus undermine self-control in the face of temptation. There is now a wealth of evidence that drug-related stimuli are differentially attended to by users and that this recruitment of attention happens automatically and involuntarily

(Robinson and Berridge 2003; Wiers and Stacey 2006). Breslin and colleagues (2002) point out that attention is a gateway to memory and that most drug and alcohol use is governed by memory-based action plans that operate automatically. The mere perception of the stimuli evokes the plan. Desire, or the anticipation of some reward, does not seem to play a key role here. Shaun Gallagher describes the relation between perception (of some familiar object) and action this way:

> The visual perception of such objects automatically evokes the most suitable motor programme required to interact with them and *the activation of motor preparation areas in the brain form part of what it means to perceive the object.* (Gallagher 2006:8, my emphasis)

So, drug-related stimuli both grab attention and cue action *automatically.* Once an action plan or program is activated, it requires effort to inhibit it. As Breslin and colleagues argue, "the action plans are firmly established and require few resources to operate, the 'abstinence plan' is poorly established and demands vigilance (i.e., attention) and effort to maintain." (Breslin et al. 2002:285–286). Successful self-control in this situation will require the agent to override her automatic responses, and this requires controlled cognitive processing capacity (Levy 2011). According to the influential dual-processing model of cognition, controlled processing (or system 2) is a limited resource, and so when we are under cognitive load or our attentional resources are depleted due to fatigue, stress, and so forth, system 1 dominates, and our judgments and behavior will be more strongly influenced by our automatic attitudes. Given that addicts may have reduced system 2 capacity (perhaps as a result of neuroadaptations caused by drug use; Volkow and Li 2005), this would make them especially vulnerable to the hijacking of their attention and subsequent loss of self-control.

Craving and the Wanting/Liking Distinction

Not only can drug-related thoughts automatically grab and monopolize attention, so making it difficult to implement strategies of *redirecting* one's attention, the strength of cravings to use may be largely insensitive to the negative evaluations made by the agent in her attempts at self-control and to her expectations of negative consequences.

How should we understand drug cravings? Are they just very intense desires formed in response to intensely pleasurable experiences? Initially, this is likely the case. But, as Heyman points out, drug rewards typically diminish over time, whereas it appears that drug cravings do not. Consideration of the role of cravings in addiction might appear to fit quite well with Heyman's and Ainslie's monistic account of motivation and the conflict between local and global perspectives on rewards. Perhaps craving makes it the case that drug use *is* the best local option and also causes the addict to apply an overly steep discount rate to future goods. But even if true, the account offers no explanation of the strength of the craving given the diminished rewards of which addicts such as Sartwell are keenly aware. The distinction between wanting and liking first postulated by Robinson and Berridge (1993) provides a more satisfying explanation of the experiences of hardcore addicts such as Sartwell. Although wanting and liking normally go hand in hand, they are neurologically separable. According to Berridge:

> Human drug addiction may be a special illustration of intense "wanting" that results from permanent sensitization of mesocorticolimbic systems (Robinson & Berridge, 1993; Robinson & Berridge, 2003). Sensitized "wanting" may rise to quite irrational levels. That is, the intensity of cue-triggered "wanting" to take drugs for brain-sensitized addicts could outstrip their "liking" even for pleasant drugs, outstrip their expectation of how much they will like the drugs, and outlast any feelings of withdrawal if they stop. Brain-sensitized addicts may be unable to give a reason for their drug taking in such a case. *Indeed, there is no reason, there is only a cause for why they "want" so much.*" (Berridge 2009:384–385, my emphasis)

Crucially, Berridge (2009) points out, this sensitization may last years after an individual stops taking any drugs, rendering them vulnerable to relapse, just as the disease model claims.

Willpower as a Limited Resource

Of course, the addict subject to irrational wanting that she cannot moderate or ignore can attempt to combat it head on by an exercise of willpower. In 1970, Joel Feinberg argued that no desire is literally irresistible; in Feinberg's words, "human endurance puts a severe limit on how long one can stay afloat in an ocean; but there is no comparable limit to our ability to resist temptation" (Feinberg 1970:283). It now appears that Feinberg was wrong. Willpower is hypothesized to be a limited resource, rather more like muscular endurance than Feinberg imagined. The influential work of Baumeister (2003) demonstrates that repeated calls on willpower deplete it (perhaps because, as suggested by Levy [2011], it is a system 2 process). The fact that a desire of a particular strength is resisted on some occasion does not show that it can be resisted on a subsequent occasion. The persistent and intrusive nature of drug cravings may therefore eventually defeat the addicts' best efforts to resist drug use.

Diachronic Self-Control Again

The obstacles to successful self-control for long-term addicts in particular look to be formidable. In sum, there appear to be significant compulsive or involuntary elements motivating drug use in this group that are not responsive to the ordinary social incentives and disincentives Heyman focuses upon and that are resistant to everyday techniques of self-control. Sustained abstinence would appear to rely on the development and deployment of diachronic strategies of self-control, on better pharmaceutical agents, and on explicit retraining of the automatic system to instill new action plans (e.g., Wiers and Stacey 2006). But although techniques of diachronic self-control are familiar and widely available, they are not always available to addicts when needed, and they are not always successful. Self-control requires a degree of social scaffolding, and it is unlikely that alcoholics, in particular, can avoid the many situations that could trigger alcohol cravings without largely withdrawing from social life, given the ubiquity of alcohol use in Western countries. Social pressure to use alcohol can be intense, such use may be expected, and abstinence may be subtly and not so subtly criticized. Moreover, alcohol cues in the form of advertising, alcohol outlets, and depictions of use in film and television are so pervasive as to be impossible to ignore. The business of ridding oneself of the offending urges through therapy, training, or medication is also not guaranteed to be successful and may require financial and social resources that the person does not have.

CONCLUSION: ADDICTION NEUROSCIENCE, RESPONSIBILITY, AND THE LAW

Ordinarily, we think that if some action is voluntary, then the person can be held responsible for it. However, if it can be shown that a person suffers a mental impairment that affects her decision making and control capacities, then we think that the person may have some excuse for her actions. Heyman insists that addicts act voluntarily and that their choices are responsive *in the ordinary way* to their choice situation, subject only to very common biases in reasoning. So the conclusion many will draw from his work is that drug users are fully responsible and thus blameworthy for their actions. I've argued here that some forms of voluntary behavior—that is, behavior that could plausibly be described as intentional or under the executive control of the agent—may nevertheless be the product of a disease or a disorder and are best understood in terms of causes, not reasons. I've also suggested that addictive behavior has significant involuntary or nonvoluntary features that undermine claims by Heyman and others (e.g., Pickard 2012) that it conforms to standard principles of motivation and is thus responsive, in the ordinary way, to sufficient incentives.

What should this mean for the law? Insofar as criminal sanctions against drug use are intended as a deterrent to such use, the evidence I've surveyed suggests that they will be largely ineffective against those hardcore users who constitute the bulk of those seen by clinicians. They may also be unjust. The disconnection of wanting from both liking and valuing arguably constitutes an impairment of the addict's capacities for choice and control that mitigates her responsibility for at least some of her drug-related criminal actions[8] and mandates treatment rather than imprisonment.

Yet the evidence that Heyman relies on also suggests that the bulk of problematic drug users *are* responsive to incentives and may well be induced to change their behavior by threat of punishment. Can courts discriminate between those users whose choices are reasons responsive, and those whose choices are not? Some have argued that this is no easy task. Bernadette McSherry suggests that "it is impossible to devise an objectively verifiable test to determine when an accused could not control him or herself and when he or she merely would not" (2004:188). Robert Sapolsky also points to the difficulties associated with "distinguishing between an irresistible impulse and one that is to any extent resistible but which was not resisted" (2004:1790). Behavioral tests and behavioral history will not reliably discriminate between these possibilities.

In the case of addiction, however, neuroimaging techniques and neuropsychological testing can provide objective measures of such things as hyperresponsiveness to drug-related stimuli and impairments to brain structure and function that may affect the addicts' capacities to either attend to or weigh relevant considerations, or to control their actions in accordance with their values and the law. How should such evidence be weighed? Greely argues for a very strong test: one that "requires specific proof that ties some characteristic of the defendant, some Condition X...that correlates extremely strongly with the criminal Behaviour Y, so strongly that we can say...that *anyone* with those characteristics, including the defendant, 'can't help himself'" (2010:75, my emphasis). Although such evidence should indeed be exculpatory, the kinds of evidence I have been talking about would not usually satisfy such a demanding standard. It is unlikely to be the case that *everyone* with the impairments to brain structure and function observed in long-term addicts is an addict or commits criminal actions—although it remains to be established whether they all display disorders of control. Greely's suggested standard would exclude such evidence. I think the standard is too high. We should not require such evidence to stand alone in order to have any probative value. Neuropsychological and neurobiological evidence, while not decisive on its own, can serve to confirm the existence of specific impairments suggested by behavioral data and clinical history and should be considered by courts.[9]

There is one final point to briefly consider. Although hardcore drug addicts may suffer the kinds of impairments that normally function as an excuse, it might be argued that these impairments are self-inflicted and thus do not mitigate their responsibility. I agree that addicts do, depending on their history (many begin problematic use at a very young age), bear some responsibility for their condition. But it should be noted that they are also the victims of bad moral luck. A huge number of people use drugs and alcohol recreationally; the vast majority do not become addicts. Those who do become addicted may have taken no more moral risks than many of those who would condemn them, including those who sit on the bench and in the jury room. How the law should respond to the issue of hard moral luck is, however, a topic for another time.

NOTES

1. The work on this paper was supported by an Australian Research Council Grant: *Addiction and Moral Identity.* I thank my team members, Craig Fry, Steve Matthews, Anke Snoek, and Doug McConnell, participants at the Addiction and Responsibility Conference in Oslo 2011, and Nicole Vincent for helpful discussions of the material herein.
2. A number of participants in our current study (five) do not identify or describe mental health problems, such as psychosis, depression, or PTSD, as occurring until their middle to late twenties, after commencing drug us in their mid-teens. They trace those problems to their drug use, not the other way round. Of course such first-person reports do not have independent confirmation, but they should be accorded some status.
3. The term *reasons responsive* is originally owed to Fischer and Ravizza (1998); the reward model understands reasons narrowly in terms of incentives and disincentives, which Fischer is not, as I understand him, committed to.
4. See Hursthouse (1991) for a discussion for actions that she terms "arational," that do not fit the standard belief desire model.
5. Heyman might dismiss this challenge by denying that such cases of subconscious motivation for drug use are primary or central cases of addiction. Some other psychological disease or disorder explains the addictive behavior. But this will not do, since on Heyman's account, these other disorders explain the continuation of such behavior via their impact on the other options available to the addict, not via the supply of subconscious motives.
6. This might not require that your attention is focused on what you are doing: the steering of your car is normally under your executive control even when you are not paying explicit attention to it. The evidence for this is in the way you respond to traffic contingencies.

7. Some philosophers (e.g., Feinberg 2007:282) have argued that there is no such thing as an irresistible desire since it will always be true that if the agent had tried harder it could have been resisted. Although this may be strictly true, the ego depletion model of willpower discussed later suggests that it is not always possible to try harder. Others have argued that the fact that the a person *would have* resisted their urge in some quite different circumstances, e.g., if the building was burning and she had to leave to save her life, shows that a person *could have* resisted in her current circumstances. I think that we need to stick to quite close possible worlds—and worlds that are practically available—if such claims are to be plausible. The addict may be able to refrain from using while the policeman is at her elbow, but the permanent provision of such a policeman is not a practical possibility.
8. It would not mitigate responsibility for criminal actions that were not directly related to securing or consuming drugs or that were disproportionate or excessive or easily avoidable—e.g., torturing someone for money. It might mitigate small-time dealing, petty theft, or fraud and of course, use of the illicit substance.
9. Here I am abstracting away from controversies over the interpretation of brain imaging. If and when imaging provides reliable information about relevant volitional impairments, it should be accepted.

REFERENCES

Ainslie, G. (1975). "Specious reward: A behavioral theory of impulsiveness and impulse control." *Psychological Bulletin* 82(4): 463–496.

Ainslie, G., and J. Monterosso (2003). Hyperbolic discounting as a factor in addiction: A critical analysis. In: Choice, Behavioural Economics and Addiction. R. E. Vucinich and N. Heather. Amsterdam, Pergamon/Elsevier Science, 35–69.

Baler, R. D., and N. D. Volkow (2006). "Drug addiction: The neurobiology of disrupted self-control." *Trends in Molecular Medicine* 12(12): 559–566.

Baumeister, R. F. (2003). "Ego depletion and self-regulation failure: A resource model of self-control." *Alcoholism-Clinical and Experimental Research* 27(2): 281–284.

Bechara, A. (2005). "Decision making, impulse control and loss of willpower to resist drugs: A neurocognitive perspective." *Nature Neuroscience* 8(11): 1458–1463.

Bechara, A., S. Dolan, and A. Hindes (2002). "Decision-making and addiction (part II): Myopia for the future or hypersensitivity to reward?" *Neuropsychologia* 40(10): 1690–1705.

Berridge, K. C. (2009). "Wanting and liking: Observations from the neuroscience and psychology laboratory." *Inquiry (Oslo)* 52(4): 378.

Breslin, F. C., M. Zack, and S. McMain (2002). "An information-processing analysis of mindfulness: Implications for relapse prevention in the treatment of substance abuse." *Clinical Psychology: Science and Practice* 9(3): 275–299.

Camerer, C. F. (2006). "Wanting, liking, and learning: Neuroscience and paternalism." *University of Chicago Law Review* 73(1): 87–110.

Feinberg, J. (1970). "What's so special about mental illness?" In: *Doing and Deserving: Essays in the Theory of Responsibility.* Princeton NJ, Princeton University Press, pp, 272–292.

Fischer, J. M., and M. Ravizza (1998). *Responsibility and Control: A Theory of Moral Responsibility.* Cambridge Studies in Philosophy and Law. Cambridge, UK, and New York, Cambridge University Press.

Gallagher, S . (2006) *How the Body Shapes the Mind.* Oxford, UK, Clarendon Press.

Garavan, H., and R. Hester. (2007). "The role of cognitive control in cocaine dependence." *Neuropsychology Review* 17(3): 337–345.

Greely, H. T. (2010). Neuroscience and criminal responsibility: Proving 'can't help himself' as a narrow bar to criminal liability." In: *Law and Neuroscience (Current Legal Issues Volume 13).* M. Freeman. Oxford, UK, Oxford University Press.

Heyman, G. M. (2009). *Addiction: A Disorder of Choice.* Cambridge, MA, Harvard University Press.

Holmwood, C. (2003). *Comorbidity of Mental Disorders and Substance Use. A Brief Guide for the Primary Care Clinician.* Canberra, Australia, Common Wealth Department of Health and Ageing.

Hursthouse, R . (1991). "Arational actions." *Journal of Philosophy* 88(2): 57–68.

Johnson, M. W., and W. K. Bickel (2003). "The behavioral economics of cigarette smoking: The concurrent presence of a substitute and an independent reinforcer." *Behav Pharmacol* 14(2): 137–144.

Kalivas, P. W., and N. D. Volkow (2005). "The neural basis of addiction: a pathology of motivation and choice." *Am J Psychiatry* 162(8): 1403–1413.

Kennett, J . (2001). *Agency and Responsibility: A Common-Sense Moral Psychology.* Originally a thesis (PhD)—Monash University, early 1990s. Oxford, UK, Clarendon Press; Oxford University Press.

Kennett, J. Just say no?. Addiction and the elements of self-control. Forthcoming In: *Addiction and Self-Control.* Neil Levy. Oxford, UK, Oxford University Press.

Kennett, J., and S. Matthews (2006). The moral goal of treatment in cases of dual diagnosis. In: *Ethical Challenges for Intervening in Drug Use: Policy, Research and Treatment Issues.* J. Kleinig and S. Einstein. Huntsville, TX, Office of International Criminal Justice.

Koob, G. F. and N. D. Volkow (2010). "Neurocircuitry of addiction." *Neuropsychopharmacology* 35(1): 217–238.

Leshner, A. I. (1997). "Addiction is a brain disease, and it matters." *Science* 278(5335): 45–47.

Levy, N. (2011). "Resisting 'weakness of the will'" *Philosophy and Phenomenological Research* 82(1): 134–155.

McGowan, K. (2004). "Addiction: Pay attention." *Psychology Today* Retrieved March 16, 2012, from http://www.psychologytoday.com/articles/200411/addiction-pay-attention.

McSherry, B. (2004). "Criminal responsibility, "fleeting" states of mental impairment, and the power of self-control." *International Journal of Law Psychiatry* 27(5): 445–457.

Mischel, W., and N. Baker (1975). "Cognitive appraisals and transformations in delay behavior." *Journal of Personality and Social Psychology* 31(2): 254–261.

Mischel, W., E. B. Ebbesen, and A. R. Zeiss (1972). "Cognitive and attentional mechanisms in delay of gratification." *Journal of Personal and Social Psychology* 21(2): 204–218.

Pickard, H. (2012) "The purpose in chronic addiction." *AJOB Neuroscience* 3(2): 40–49.

Posternak, M. A., D. A. Solomon, A. C. Leon, T. I. Mueller, M. T. Shea, J. Endicott, and M. B. Keller (2006). "The naturalistic course of unipolar major depression in the absence of somatic therapy." *Journal of Nervous and Mental Disease* 194(5): 324–329.

Robinson, T. E., and K. C. Berridge (1993). "The neural basis of drug craving: An incentive-sensitization theory of addiction." *Brain Research Reviews* 18(3): 247–291.

Robinson, T. E., and K. C. Berridge (2003). "Addiction." *Annu Rev Psychol* 54: 25–53.

Samuelson, P. (1938). "A note on the pure theory of consumers' behaviour." *Economica* 5: 61–71.

Sapolsky, R. M. (2004). "The frontal cortex and the criminal justice system." *Philosophical Transactions of the Royal Society of London: B Biological Sciences* 359(1451): 1787–1796.

Sartwell, C. (no date). "Detritus." Retrieved December 3, 2012, from http://www.crispinsartwell.com/addict.htm.

Schadé, A.(2003). "The treatment of alcohol-dependent patients with a comorbid phobic disorder." Amsterdam: Vrije Universiteit.

Schatzberg, A, R. Weiss, K. Brady, and L. Culpepper (2008). "Bridging the clinical gap: Managing patients with co-occurring mood, anxiety, and alcohol use disorders." *CNS Spectrum* 13(4): 1–16.

Stewart, S. H., and P. J. Conrod, Eds. (2008). *Anxiety and Substance Use Disorders: The Vicious Cycle of Comorbidity*. New York, Springer.

Volkow, N., and T. K. Li (2005). "The neuroscience of addiction." *Nature Neuroscience* 8(11): 1429–1430.

Wiers, R. W., and A. W. Stacy (2006). "Implicit cognition and addiction." *Current Directions in Psychological Science* 15(6): 292–296.

12

How May Neuroscience Affect the Way that the Criminal Courts Deal with Addicted Offenders?

WAYNE HALL AND ADRIAN CARTER

INTRODUCTION

A central question for the courts in dealing with addicted offenders is deciding whether they had the capacity to make free choices about the criminal acts in which they engaged (e.g., theft, drug dealing or violence) to fund or support their drug use. We contrast two views that have dominated recent public debates about this issue: the medical model, in which addiction is seen as "a chronic relapsing brain disease" that renders addicted persons not responsible for drug-related crimes; and the commonsense moral view, which is skeptical about addiction and sees both drug use and criminal acts to support it as freely chosen acts for which offenders should be held responsible.

TWO MODELS OF ADDICTION

The Brain Disease Model of Addiction

Advocates of the brain disease model of addiction argue that chronic drug use impairs an individual's ability to control their drug use by making it difficult to refrain from using drugs (Baler and Volkow 2006; Leshner 1997; Volkow and Li 2005). A number of empirical observations are cited in support of the brain disease view.

First, a significant minority of people who use drugs become addicted to them, and the size of this minority depends on the drug, the way in which it is used (e.g., whether swallowed, smoked, or injected) and its pharmacological actions (e.g., whether its effects have a fast or slow onset and offset) (Anthony and Helzer 1991). Generally, drugs that are injected or smoked, that act quickly and for a short period of time (e.g., heroin, cocaine, and nicotine), are more likely to produce addictive behavior than slower onset, longer-acting drugs that are taken orally (e.g., alcohol, codeine, methylphenidate).

Second, there is an identifiable subset of individuals who are more likely to develop an addiction. Twin and adoption studies suggest that genetic factors make a substantial contribution to addiction liability (Ball 2008; Goldman, Oroszi, and Ducci 2005; True et al. 1999). Genetic factors have been estimated to account for between 40% and 60% of addiction liability (Uhl et al. 2004, 2008).

Third, neuroimaging studies suggest that the brains of addicted and non-addicted persons differ in their responses to drugs and drug-related stimuli (Garavan et al. 2007). Addicted individuals also perform worse on neurocognitive tests of decision making (Bechara 2005; Jentsch and Taylor 1999). These differences are correlated with the amount of drugs that have been used and with self-reported cravings for drugs; they often persist for months after abstinence has been achieved; and they are thought to be mediated by increased dopamine activity in the brain's reward pathway (Koob and Volkow 2010). Neuroimaging studies provide a bridge between animal models of addiction and addictive behavior in humans (Koob and Le Moal 2006).

The neurobiological view that addiction arises from repeated effects of drugs on the brain's dopaminergic reward pathway has recently been supported by reports of addictive disorders arising from therapeutic use of drugs. As many as one in six patients with Parkinson's' disease (PD) who are treated with dopamine (DA) agonists (drugs that stimulate dopamine release in the brain) develop problem gambling, compulsive sexual behavior, and overeating (Ambermoon et al. 2011). There is reasonable evidence that DA agonists play a causal role in these instances of problem gambling (PG): its incidence is much higher in PD patients treated with DA agonists than in the general population (Ambermoon et al. 2011; Voon et al. 2006; Weintraub et al. 2006, 2010). PG often develops after the initiation of these drugs or increases in dose, and it usually significantly improves after the drugs are discontinued, the dose is reduced, or a different drug is used. The impact of chronic DA agonist use upon the brain's reward system provides a biologically plausible explanation of the relationship between DA agonists and compulsive behavior (Dagher and Robbins 2009; Merims and Giladi 2008; Zack and Poulos 2009).

Leading addiction clinicians have argued that animal and human neuroimaging studies challenge the assumption that addictive behavior is always a voluntary choice (Dackis and O'Brien 2005; Leshner 1997; McLellan et al. 2000; Volkow and Li 2004). Prolonged drug use, they argue, produces changes in brain function that persist after abstinence and undermine addicted persons' capacity to control their drug use. Advocates of this model also argue that its acceptance will change social policies for dealing with addiction (e.g., Dackis and O'Brien 2005; Leshner 1997; McLellan et al. 2000; Volkow and Li 2004), namely, it will increase medical treatment of addiction and investment in neuroscientific research to develop more effective drug treatments for addiction. Such a view has other, possibly less welcome, implications (Carter and Hall 2012). If, for example, addicted individuals are lacking in autonomy, then an argument can be made for compulsorily treating their addiction to reinstate their autonomy (Caplan 2006, 2008). We examine the evidence for such a claim below.

The Moral Model of Addiction

The moral model of addiction is a folk psychological view in which drug offenders are simply drug users who knowingly and willingly choose to use drugs without regard for the consequences of their actions for themselves or others. On this account, "addiction" is an excuse that these drug users invoke to deny responsibility for their socially deviant behavior (Dalrymple 2006; Davies 1997; Foddy and Savulescu 2006; Szasz 1975). Drug users on this view should be held legally responsible for any harmful effects that their drug use has on others because these harms arise from their freely chosen actions (Dalrymple 2006; Satel 1999; Szasz 1975).

The moral view of drug use makes sense of a number of features of the behavior of drug-using offenders. Drug use is (at least) initially a voluntary choice. "Addicted" drug users often engage in considerable planning in seeking drugs and engage in long sequences of actions that change in response to environmental contingencies such as the likelihood of arrest (Foddy and Savulescu 2006). In this view, the fact that addictive patterns of drug use occur in a minority of drug users is explained by the characteristics of these drug users and the environments in which they live rather than by the effects of the drugs (Peele 2004). Most "addicted" drug users, it is stressed, stop using drugs without assistance (Heyman 2001, 2009; Peele 2004). Even severely "addicted" drug users will cease using drugs if offered small incentives (e.g., shopping vouchers or cash for drug free urine samples) in "contingency management" programs (Higgins 2006; Higgins and Petry 1999) or in response to lifestyle changes (Heyman 2009). Heavy users of alcohol and tobacco also respond to changes in their availability and price (Room 2007).

The moral view of addiction forms the primary basis of the criminal law's response to addicted persons who engage in crimes (e.g., dealing drugs and theft), although medical models of addiction are beginning to influence some of these responses, as we discuss later. The criminal law sees the threat of punishment, such as imprisonment, as the best way to deter drug use and drug-related crime because, it is assumed, addicted offenders will change their behavior when provided with the right incentives to do so (Satel 2006).

Any appraisal of the comparative merits of the moral and medical views of addiction requires an acquaintance with the evidence on the effectiveness of different approaches to treating addiction. We briefly review this evidence and then discuss the ways in which the Australian legal system deals with addicted offenders.

COMMUNITY-BASED ADDICTION TREATMENT: WHAT WORKS?

Detoxification is supervised drug withdrawal that uses pharmacotherapies or psychosocial support to minimize the severity of withdrawal symptoms that occur when an addicted person abruptly ceases using drugs. It is *not* a treatment for addiction, but rather is a precondition for engaging in drug-free or abstinence-based forms of treatment (Mattick and Hall 1996).

Drug-free treatment approaches include residential treatment in therapeutic communities (TCs), outpatient drug counseling (DC), and attendance at self-help groups such as Narcotics Anonymous (NA). These approaches all aim to support abstinence by using group and psychological interventions. There have been very few randomized controlled trials for TCs, and the few that have been conducted have produced equivocal results (Smith, Gates, and Foxcroft 2006). Most of the support for their effectiveness comes from observational studies in the United States (Gerstein and Harwood 1990), the United Kingdom (Gossop et al. 2003), and Australia (Teesson et al. 2008) that have followed up addicted persons who chose these treatments. In these studies, TCs have been less successful than opioid substitution treatment (OST) in attracting and retaining heroin users in treatment, but they substantially reduce heroin use and crime in those who remain in treatment for at least 3 months (Gossop et al. 2003; Teesson et al. 2008).

A drug that can be used in abstinence-oriented addiction treatment is naltrexone (Caplan 2008; Marlowe 2006; Sullivan et al. 2008). Naltrexone is an opioid antagonist that blocks the euphoric effects of opioids. Compliance with oral naltrexone is therefore poor (Minozzi et al. 2006). Long-acting naltrexone implants or depot injections have recently been developed to overcome problems with compliance. These have shown positive results for up to a year in several small trials in selected patients (Comer et al. 2006; Hulse et al. 2009; Krupitsky et al. 2011; Kunoe et al. 2009). A slow-release injectable form of naltrexone was approved by the US Food and Drug Administration (FDA) in 2010, although some researchers have been critical of the limited evidence on which the FDA made this decision (Wolfe et al. 2011). It has not yet been approved in Australia because of uncertainty about its benefits and adverse effects, although a trial of a long-acting naltrexone implant has been conducted there (Hulse et al. 2007, 2009).

The aim of OST is not to achieve immediate abstinence from opioids. It substitutes a long-acting, usually orally administered, opioid (e.g., methadone or buprenorphine) for the shorter-acting injected opioid, heroin (van den Brink and Haasen 2006; Ward, Hall, and Mattick 2009). Longer-acting oral opioids avoid the oscillation between euphoria and withdrawal that occurs in heroin injectors. Ideally these treatments are provided in combination with psychological treatments (such as cognitive behavior therapy) that address the comorbid psychiatric disorders that are common in addicted individuals (e.g., anxiety and depression) and with social support and crisis management (Ward, Mattick, and Hall 1998). Often, these ancillary services are provided in a perfunctory and inadequate way.

Methadone maintenance treatment (MMT) is the oldest and most widely used form of OST. When taken daily in high doses, methadone blocks the effects of heroin. Its effectiveness is supported by randomized controlled trials and observational studies, all of which have found that MMT decreases heroin use and criminal activity (Mattick et al. 2009; Ward, Hall, and Mattick 2009) and reduces human immunodeficiency virus (HIV) infection (Ward, Hall, and Mattick 2009).

Buprenorphine is a mixed agonist-antagonist, that is, it has opiate effects like those of morphine while blocking the effects of heroin, like naltrexone. It has a longer period of action than methadone, thereby enabling doses to be given every 2 to 3 days (rather than daily, as with methadone). Buprenorphine maintenance treatment (BMT) is marginally less effective in retaining patients in treatment than MMT but approximately equivalent in reducing heroin use (Mattick et al. 2008). Buprenorphine is associated with a substantially lower risk for overdose than methadone because of its mixed agonist and antagonist effects. It is also easier to withdraw from than methadone (Mattick et al. 2008) and, because it can be taken every 2 or 3 days, requires less supervision than methadone.

In heroin maintenance treatment (HMT), dependent heroin users are maintained on daily injectable heroin (van den Brink and Haasen 2006).

The safety and effectiveness of HMT in patients who have failed to respond to MMT have been evaluated in randomized controlled trials in Switzerland (Perneger et al. 1998), the Netherlands (van den Brink et al. 2003), Spain (Perea-Milla et al. 2009), Germany (Haasen et al. 2010), Canada (Oviedo-Joekes et al. 2009), and the United Kingdom (Strang et al. 2010). All studies have shown HMT to be more effective than MMT in reducing illicit heroin use and crime in treatment-refractory heroin-addicted individuals (Lintzeris 2009). HMT is much more expensive than MMT because of the need to supervise daily injection of a drug with a short half-life. It is more cost-effective than MMT when its effects on crime are included, but its higher costs mean that it is a second- or third-line treatment. Its scale of implementation has been modest, in part because of its costs and in part because of limited patient demand (Scherbaum and Rist 2010). For these reasons, the public health impact of HMT is likely to be modest compared with that of MMT and BMT (Hall 2005; Lintzeris 2009).

CURRENT LEGAL PRACTICE IN DEALING WITH ADDICTED OFFENDERS

A distinction is often drawn between voluntary (freely chosen) and involuntary (compelled) treatment of addiction. In reality, the decision to enter treatment is rarely completely free. Internal forces (such as withdrawal symptoms) and external pressures from family and friends or workmates may prompt a decision to seek treatment for addiction. Compulsory addiction treatment is rarely imposed in most Western democracies, either to protect the community or, even more rarely, for the addicted person's own good. The most commonly used form of legally coerced addiction treatment is that which occurs within the legal system when addicted offenders enter treatment as an alternative to imprisonment. This approach is described as legally coerced addiction treatment (Hall 1997). Legally coerced treatment has most often been used with offernders who inject opioids, a group that is overrepresented among prisoners in many Western countries (e.g., AIHW 2010; Fazel, Bains, and Doll 2006; Makkai 2002).

The Case for Legally Coerced Addiction Treatment

The major justification for coerced addiction treatment depends on a consequentialist theory of punishment: addiction treatment, it is argued, is more effective in reducing drug offenders' recidivism than imprisonment (Chandler, Fletcher, and Volkow 2009). This argument has particular weight in the case of heroin offenders who, if untreated, are very likely to relapse to heroin use and reoffend after release (Gerstein and Harwood 1990; Hall 1996). Providing addiction treatment under coercion in the community is also more effective and cost-effective than imprisonment in reducing recidivism (Gerstein and Harwood 1990; Moore, Ritter, and Caulkins 2007). The advent of HIV and acquired immunodeficiency syndrome (AIDS) among injecting heroin and other drug users has added a potent public health argument to the case for coerced addiction treatment of drug offenders, namely that they are at high risk for transmitting HIV and hepatitis to other inmates by needle sharing in prison (Dolan 1991; Wodak et al. 1992).

The argument for coerced addiction treatment raises the following questions: What types of addiction treatment should be provided (e.g., drug-free only or opioid substitution treatment)? Where is such treatment best provided (e.g., in the community or in prison)? What degree of coercion is ethically permissible to encourage treatment entry? What resources are required to provide coerced addiction treatment in ways that do not undermine access to or the effectiveness of community-based addiction treatment?

Legally coerced addiction treatment also raises important ethical issues that we have discussed elsewhere (Carter and Hall 2012; Hall 1997). Here we simply agree with a 1986 World Health Organization (WHO) consensus view (Porter, Arif, and Curran 1986) that *legally coerced* drug treatment is legally and ethically justified if (1) the rights of the individuals are protected by "due process" and (2) effective and humane treatment is provided. We would add that legally coerced drug treatment should be provided in ways that allow offenders (1) a choice of whether or not they participate in treatment, with those who decline being processed in the usual way by the criminal justice system; and (2) a choice as to the type of treatment used, should they agree to be treated (Fox 1992; Gerstein and Harwood 1990).

How Effective Is Legally Coerced Drug Treatment in the Community?

The effectiveness of coerced community-based addiction treatment was assessed in observational studies of heroin- and cocaine-dependent offenders in the United States in the 1970s and 1980s (e.g., Hubbard 1989; Hubbard et al. 1988; Simpson and Friend 1988). In these studies, drug-dependent individuals who entered TCs and drug-free out-patient counseling under "legal pressure" (i.e., either on probation or parole) did as well as those who entered these treatments without such pressure (Hubbard 1989; Simpson 1981). There were too few individuals entering MMT under coercion to assess its outcome because judges preferred to coerce offenders into treatments that aimed at abstinence (Leukefeld and Tims 1988).

MMT was found to reduce recidivism in two small studies that randomly assigned addicted prisoners to parole with or without MMT (Dole et al. 1969;

Gordon et al. 2008). Dole and colleagues (1969) found a greater reduction in heroin use and rates of imprisonment in a small group of offenders randomly assigned to MMT in the year after their release. These results were supported in a larger, more recent study by Gordon and colleagues (2008). Both results were consistent with earlier observational studies of MMT under coercion in California (Anglin, Brecht, and Maddahian 1989; Brecht, Anglin, and Wang 1993), which found that offenders who entered MMT under legal coercion substantially reduced their heroin use and criminal behavior.

Drug Courts: Therapeutic Jurisprudence

Drug courts were introduced in the United States in the late 1980s when cocaine offenders swamped US prisons (Belenko 2002; US Government Accountability Office 1995). Drug courts quickly spread throughout the US and to other countries. By the turn of the 21st century, drug courts had been established in Australia (Makkai 2002), the United Kingdom (Bean 2002), and Canada (Fischer 2003). In drug courts, judges supervise addiction treatment, monitor offenders' progress in regular hearings, use urinalyses to monitor their drug use, and impose custodial penalties (e.g., short periods of imprisonment) if drug use is detected or the offender fails to comply with treatment. Typically, US drug courts only order drug-free treatment over a period of 6 to 12 months, and success is defined as the offender demonstrating abstinence for some specified minimum period such as 6 months.

There have been major challenges in evaluating the effectiveness of drug courts (Belenko 2002; Covington 200;1Manski, Pepper, and Petrie 2001). Much of the US research has used quasi-experimental designs with poorly constructed comparison groups and often only measured treatment retention and drug use while in treatment (Belenko 2002; Covington 2001; Fischer 2003; Harvey et al. 2007; Klag, O'Callaghan, and Creed 2005; Wild, Roberts, and Cooper 2002). The degrees of legal coercion involved are often not well defined (Klag, O'Callaghan, and Creed 2005; Wild, Roberts, and Cooper 2002). Rates of successful completion of programs have typically been near 40% to 50% (Belenko 2002).

An early randomized controlled trial (RCT) of a US Drug Court by the RAND Corporation also failed to find any impact on recidivism (Deschenes, Turner, and Greenwood 1995). Similarly, the early findings from a randomized controlled evaluation of the NSW Drug Court showed very marginal benefits in recidivism compared with routine imprisonment (Lind et al. 2002). A later quasi-experimental evaluation of the NSW Drug Court found lower recidivism after the court reduced its use of incarceration as a penalty for drug use and included OST among its treatment options (Weatherburn et al. 2008).

Observational studies of the effectiveness of Drug Courts in Australia (Lind et al. 2002; Weatherburn et al. 2008), Canada (Fischer 2003), and the United Kingdom (Scottish Government Community Justice Services 2010) have reported modest impacts on recidivism. This has prompted some critics to question the exportability of the US drug court model (Fischer 2003). Recent evaluation of drug courts in Glasgow and Fife, for example, were unable to detect any effects on recidivism or offending after 3 years in opioid offenders (Scottish Government Community Justice Services 2010).

On balance, quasi-experimental studies suggest that drug courts reduce recidivism over 1 to 3 years (Belenko 2002; Krebs et al. 2007; Turner et al. 2002; Wilson, Mitchell, and MacKenzie 2006). However, these effects (if causal) are modest, as one would expect with the recidivist offenders typically dealt with by drug courts. A meta-analysis of 57 quasi-experimental studies of drug courts (primarily in the United States) found that they reduced the recidivism rate from 50% to 42% (compared with 41% in community-based drug treatment) (Aos et al. 2006).

Very few evaluations have assessed the cost-effectiveness of drug courts (Covington 2001; Fischer 2003) compared with imprisonment or community-based addiction treatment. An evaluation of the NSW Drug Court compared its cost-effectiveness with that of conventional imprisonment using the results of an RCT (Lind et al. 2002). It found no difference in the cost-effectiveness of the drug court and imprisonment because the court often used imprisonment as a penalty for noncompliance with treatment. The Scottish study of drug courts found that they were much more expensive than community addiction treatment supervised by criminal courts (Scottish Government Community Justice Services 2010).

Despite their modest benefits, drug courts have enjoyed considerable support among policy makers in developed countries. Pilot programs have often been declared a "success" and widely implemented before their effectiveness has been evaluated (Clancey and Howard 2006; Fischer 2003). As a consequence, they have often become part of the system, and their roles have expanded in the absence of evidence that they are effective and often in ways that preclude more rigorous evaluation (Clancey and Howard 2006; Fischer 2003).

HOW PLAUSIBLE IS THE BRAIN DISEASE MODEL OF ADDICTION?

The brain disease model of addiction is being increasingly invoked to justify further legal intervention in the treatment of addiction (Caplan 2008). The argument is that if addicted persons are suffering from a disease of the brain that hijacks their ability to choose not to use drugs, then the government should intervene for the addicted person's own good.

A key question in assessing how addiction neuroscience should affect legal practice in this way is the plausibility of the claim that addiction is a "chronic relapsing brain disease". The primary evidence for a brain disease model of addiction comes from laboratory animal models in which animals trained to self-administer addictive drugs at a high rate are shown to be resistant to extinction after the drug is withdrawn (Koob and Le Moal 2006; Olds and Milner 1954). These animal models appear to have good face validity, but their relevance to human compulsive drug use and the contexts in which humans typically use addictive drugs has been questioned (Ahmed 2010; Cantin et al. 2010; Epstein et al. 2006; Foddy and Savulescu 2006; Kalant 2010). Getting rodents "hooked" on drugs is surprisingly difficult unless scientists use animals that are selectively bred to be more easily addicted. Moreover, many of these "addictive behaviors" (e.g., self-administration, conditioned-place preference, cue-induced reinstatement) disappear when animals are placed in more ecologically realistic environments that are often misleadingly referred to as "enriched environments" (Ahmed 2010; Alexander et al. 1981; Alexander, Coambs, and Hadaway 1978; Xu et al. 2007).

Neuroimaging and neurocognitive studies of addicted humans provide bridges between animal and human studies of addiction. These studies have identified changes in brain regions that are believed to be involved in addictive behavior (e.g., motivation, reward learning, and executive control)—changes that are often associated with self-reported difficulties in controlling drug use and with poorer performance on decision-making tasks. Such findings have two major limitations. First, these findings only reflect *average group* differences between addicted and nonaddicted populations. It is not clear that the observed changes in brain and cognition undermine autonomous decision making in addicted individuals. These studies find that addicted individuals, as a group, have *diminished* neurocognitive capacity compared with nonaddicted individuals, but not all addicted persons show these deficits, whereas some nonaddicted controls do so (Bechara 2001, 2005). Second, these studies are typically cross-sectional, so it is not clear whether the observed changes in brain function are causes or consequences of addictive drug use (Schumann 2007) or some combination of the two.

As critics of the brain disease model of addiction argue (Foddy and Savulescu 2006; Heyman 2009), the claim that addictive drug use is compulsive is at odds with behavioral evidence that addicted individuals retain some degree of control over their drug use. This includes the types of evidence outlined previously, namely, that many, if not most, addicted individuals in the general population overcome their addiction without treatment (Granfield and Cloud 1996; Heyman 2009) and that many addicted persons cease using drugs without assistance for varying periods (e.g., to reduce their tolerance, to take time out from a drug-using lifestyle) (Granfield and Cloud 1996; Heyman 2009) or in response to major life events (e.g., the birth of a child, ultimatums from friends and family, or threats from employers) (Dalrymple 2006; Heyman 2009). The fact that small incentives such as vouchers can reduce drug use in addicted persons (Elster and Skog 1999; Fingarette 1988; Higgins and Petry 1999; Marteau, Ashcroft, and Oliver 2009; Neale 2002; Stitzer et al. 1993) is especially difficult to reconcile with the claim that addicted persons' drug use is the result of compulsions that are very difficult to resist.

We are persuaded by these observations that "addicts" are not, by definition, unable to make autonomous, rational decisions whether to use a drug or not, as per the strong form of the brain disease model. We therefore do not accept that addicted persons are not legally responsible for criminal behavior committed to facilitate their drug use. However, we do accept neuroscience evidence that chronic drug use affects brain function and cognition, and therefore find it difficult to accept the moral model. A similar view has been expressed by the leading addiction neuroscientist and Harvard Provost, Steven Hyman, that "addicted individuals have substantial impairments in cognitive control of behavior, but this 'loss of control' *is not complete* or simple" (Hyman 2007:8).

The scientific, ethical, and legal challenge we have yet to meet is to provide a plausible alternative to these sharply distinguished and, arguably, equally implausible moral and medical models of addiction. The brain disease model, in our view, overstates the degree of compulsion that is experienced by *most* addicted individuals, as indicated by the natural history of untreated addiction and the efficacy of contingency management in treating addiction. The moral view, by contrast, ignores the effects that chronic drug use may have on brain function and the biological and social obstacles that many addicted individuals face in becoming and remaining abstinent. It also is unresponsive to the evidence on the effectiveness of OST in treating heroin dependence, reviewed earlier.

There is considerable variability among addicted persons in the severity of their addiction and in the degree to which they can control their drug use. The challenge for the courts is in specifying offenders' degree of impaired control over their drug use in ways that are independent of their self-reported difficulty in refraining from using drugs. As disease model skeptics note, self-reported difficulties in controlling drug use lack evidentiary value because addicted individuals often have significant incentives to dissemble, such as evading a conviction in a court of law, or in seeking to excuse the adverse effects of their drug use on their family and friends.

One of the hopes of addiction neuroscience research is that neuroimaging or cognitive tests will objectively measure impaired control in ways that predicts either risk of relapse (Paulus, Tapert, and Schuckit 2005; Schutz 2008) or responses to different types of treatment (Cox et al. 2002; Goudriaan, Grekin, and Sher 2008; Goudriaan et al. 2008). Such tests, should they prove reliable, could potentially be used by the courts to assess addicted offenders' degree of legal responsibility, or their likelihood of recidivism and relapse to drug use upon release, when making decisions about sentences or parole (Nadelhoffer et al. 2010). Neuroimaging and cognitive tests, however, are a long way from being able to do any of these things at present.

There is a much simpler alternative approach to assessing addicted offenders' capacity to control their drug use—"behavioral triage." This approach uses offenders' responses to contingency management in setting penalties (Kleiman 2009) and is discussed later.

HOW COULD WE IMPROVE LEGAL PRACTICE IN DEALING WITH ADDICTED OFFENDERS?

As outlined earlier, legal practice in Australia and other developed countries implicitly acknowledges that addicted offenders are not fully in command of themselves, even if their degree of impairment is not adjudged to be sufficient to eliminate legal responsibility for their criminal acts. The criminal courts have not allowed addiction to exculpate drug users who engage in criminal acts to fund their drug use (Hall and Lucke 2010). They often, however, accepted addiction treatment as an alternative to punishment when setting penalties for addicted offenders who commit nonviolent offences. (Hall 1997; Pritchard, Mugavin, and Swan 2007; Wild, Roberts, and Cooper 2002).

This approach is arguably a reasonable compromise between the brain disease and moral models, and an approach that is cognizant of the neuroscientific and observational evidence reviewed previously. The suspension or reduction of sentences for crimes that would otherwise receive custodial sentences signals the courts' recognition that addiction impairs offenders' decision making in ways that may make them less deserving of custodial sentences than nonaddicted persons who commit the same offences. From the perspective of a consequentialist theory of punishment, legally coerced addiction treatment also potentially produces greater public order benefits by reducing recidivism more than imprisonment. But these benefits are not likely to satisfy politicians, social commentators, and citizens whose theories of punishment place a greater emphasis on retribution and deterrence.

Current US legal practice represents an uneasy compromise between the medical and moral models of addiction. While implicitly acknowledging that the responsibility of addicted offenders is diminished, US drug court practice remains captive to the moral model in restricting the choice of treatments offered as an alternative to imprisonment to drug-free treatment. The evidence suggests that these treatments are the least likely to be effective in recidivist addicted offenders with serious cocaine and heroin problems. US drug courts, for example, have eschewed the use of OST for heroin dependence, despite the evidence of its effectiveness (Chandler, Fletcher, and Volkow 2009). US courts often paradoxically compel attendance at self-help groups such as NA, whose treatment philosophy places a heavy emphasis on addicted persons accepting responsibility for their drug use and their actions by voluntarily attending the groups.

Wider acceptance by courts of research evidence on the effectiveness of addiction treatment could change the types of treatment that are provided in coerced addiction treatment programs. Such research could, for example, provide a stronger rationale for the inclusion of OST among the treatment options from which legally coerced heroin-addicted offenders can choose. This seems to have occurred in some Australian drug courts (Weatherburn et al. 2008). However, in the United States, evidence of the effectiveness of OST has failed to sway judges, who object on moral grounds to agonist maintenance treatment. US drug courts may be more positively disposed to coerced addiction treatment using sustained-release naltrexone to block the effects of opioids. Some US bioethicists have already argued that heroin-addicted offenders should be *compelled* to use implantable forms of naltrexone (Caplan 2008). We have criticized this proposal elsewhere (Hall, Capps, and Carter 2008). It would, however, be acceptable to include this form of treatment among a menu of options from which offenders could choose.

The courts should also be prepared to experiment with less coercive forms of treatment for addicted or drug-using offenders. They could include, for example, the use of incentives (e.g., vouchers or money) for refraining from drug use (Higgins 2006; Higgins and Petry 1999) or requiring offenders to provide urinalysis evidence of abstinence from drugs to remain in the community (Kleiman 2009). There is reasonable evidence for the effectiveness of the former approach (Flodgren et al. 2011). As we have argued above, the strong form of the brain disease model provides a poor fit to most addicted offenders, many of whom may respond to less coercive forms of management. These will also be cheaper than drug court–supervised addiction treatment.

Kleiman (2009) has argued that we do not need to routinely use the expensive infrastructure of judges, lawyers, drug courts, and specialist addiction treatment in order to reduce recidivism among addicted offenders. He advocates for "behavioral triage" to match the intensity of the addiction intervention to the offender's drug use. According to Kleiman, the response of addicted offenders to modest incentives (avoiding short periods in jail) should be used to select the level of coerced treatment that the offender requires to avoid reoffending. He reports observational evidence on the effectiveness of "coerced abstinence" as implemented in a court in Hawaii. This trial showed that drug use and recidivism in methamphetamine-using offenders could be substantially reduced by requiring them to undergo random weekly urinalyses and penalizing evidence of recent drug use with 24 hours of immediate and certain imprisonment. Kleiman proposes that court-supervised addiction treatment should be reserved for offenders who fail to respond to the coerced abstinence program described previously. That is, offenders who failed to remain abstinent in such a program would be referred to court-supervised addiction treatment to assist them to remain abstinent. Drug courts, the most intensive and expensive form of coerced treatment, would be reserved for addicted offenders who failed in court-supervised addiction treatment.

This approach of coerced abstinence is consistent with the use of incentives to assist addicted persons to control their drug use. It is also consistent with neuroeconomic theories of addiction (e.g., Ainslie 2001; Ross et al. 2008), which predict that addicted individuals will be insensitive to large disincentives that uncertainly occur in the distant future (e.g., a potential long prison sentence) because such individuals heavily discount future punishment against the small but immediate benefits of drug use. On this theory (Glimcher 2008), addicted individuals will be more responsive to smaller but immediate disincentives for drug use (e.g., avoidance of immediate 24 hours incarceration) and rewards for abstinence. It is an approach that we believe is deserving of careful trial and evaluation.

CONCLUSION

The Australian courts have not been persuaded that addicted offenders suffer from a chronic brain disease that renders them not legally responsible for their criminal actions. The courts do appear, however, to have accepted that addicted offenders are better dealt with by offering them addiction treatment rather than imprisonment. The most plausible argument for this approach is a consequentialist one—that it is a more effective and less expensive than imprisonment in reducing drug use and crime. The most ethically defensible form of such coercion is the use of imprisonment as an incentive for treatment entry, with the threat of return to prison as a reason for complying with drug treatment in the community. Offenders should still have a choice as to whether they take up the treatment offer, and, if they choose to do so, they should have a choice of treatment rather than being compelled to enter a particular form of treatment.

The courts should pay more attention to research on the effectiveness of addiction treatment. If they did, they would make greater use of agonist maintenance treatment in coerced treatment of heroin-addicted offenders. They should also be prepared to experiment with less intensive forms of coerced treatment, such as the behavioral triage approach advocated by Kleiman, an approach that research in behavioral neuroscience and neuroeconomics provides a strong warrant for trialing.

ACKNOWLEDGMENTS

We would like to thank Sarah Yeates for her assistance in searching the literature and preparing the paper for publication and Jayne Lucke, Stephanie Bell, and Nicole Vincent for their helpful comments on an earlier draft of this paper.

REFERENCES

Ahmed, S. H. (2010). "Validation crisis in animal models of drug addiction: Beyond non-disordered drug use toward drug addiction." *Neuroscience & Biobehavioral Reviews* 35(2): 172–184.

AIHW (2010). *The Health of Australia's Prisoners 2009*. Canberra, Australian Institute of Health and Welfare.

Ainslie, G. (2001). *Breakdown of Will*. New York, Cambridge University Press.

Alexander, B. K., B. L. Beyerstein, P. F. Hadaway, and R. B. Coambs (1981). "Effect of early and later colony housing on oral ingestion of morphine in rats." *Pharmacology, Biochemistry and Behavior* 15(4): 571–576.

Alexander, B. K., R. B. Coambs, and P. F. Hadaway (1978). "The effect of housing and gender on morphine self-administration in rats." *Psychopharmacology* 58(2): 175–179.

Ambermoon, P., A. Carter, W. Hall, N. N. W. Dissanayaka, and J. D. O'Sullivan (2011). "Impulse control disorders in patients with Parkinson's disease receiving dopamine replacement therapy: Evidence and implications for the addictions field." *Addiction* 106(2): 283–293.

Anglin, M., M. Brecht, and E. Maddahian (1989). "Pretreatment characteristics and treatment performance of legally coerced versus voluntary methadone maintenance admissions." *Criminology* 27(3): 21.

Anthony, J. C., and J. Helzer (1991). Syndromes of drug abuse and dependence. In: *Psychiatric disorders in America*. L. N. Robins and D. A. Regier. New York, Academic Press, pp. 116–154.

Aos, S., M. Miller, E. Drake, and Washington State Institute for Public Policy (2006). *Evidence-Based Public Policy Options to Reduce Future Prison Construction, Criminal Justice Costs, and Crime Rates*. Olympia, WA, Washington State Institute for Public Policy.

Baler, R. D., and N. D. Volkow (2006). "Drug addiction: The neurobiology of disrupted self-control." *Trends in Molecular Medicine* 12(12): 559–566.

Ball, D. (2008). "Addiction science and its genetics." *Addiction* 103(3): 360–367.

Bean, P. (2002). "Drug treatment courts, British style: The drug treatment court movement in Britain." *Substance Use and Misuse* 37(12–13): 1595–1614.

Bechara, A. (2001). "Neurobiology of decision-making: Risk and reward." *Seminars in Clinical Neuropsychiatry* 6(3): 205–216.

Bechara, A. (2005). "Decision making, impulse control and loss of willpower to resist drugs: A neurocognitive perspective." *Nature Neuroscience* 8(11): 1458–1463.

Belenko, S. (2002). "The challenges of conducting research in drug treatment court settings." *Substance Use & Misuse* 37(12–13): 1635–1664.

Brecht, M. L., M. D. Anglin, and J. C. Wang (1993). "Treatment effectiveness for legally coerced versus voluntary methadone maintenance clients." *American Journal of Drug and Alcohol Abuse* 19(1): 89–106.

Cantin, L., M. Lenoir, E. Augier, N. Vanhille, S. Dubreucq, F. Serre, C. Vouillac, and S. H. Ahmed (2010). "Cocaine is low on the value ladder of rats: Possible evidence for resilience to addiction." *PLoS ONE* 5(7): e11592.

Caplan, A. (2006). "Ethical issues surrounding forced, mandated, or coerced treatment." *Journal of Substance Abuse Treatment* 31(2): 117–120.

Caplan, A. (2008). "Denying autonomy in order to create it: The paradox of forcing treatment upon addicts." *Addiction* 103(12): 1919–1921.

Carter, A., and W. Hall (2012). *Addiction Neuroethics: The Promises and Perils of Neuroscience Research on Addiction*. London, Cambridge University Press.

Chandler, R. K., B. W. Fletcher, and N. D. Volkow (2009). "Treating drug abuse and addiction in the criminal justice system: Improving public health and safety." *JAMA* 301(2): 183–190.

Clancey, G., and J. Howard (2006). "Diversion and criminal justice drug treatment: Mechanism of emancipation or social control?" *Drug and Alcohol Review* 25(4): 377–385.

Comer, S. D., M. A. Sullivan, E. Yu, J. L. Rothenberg, H. D. Kleber, K. Kampman, C. Dackis, and C. P. O'Brien (2006). "Injectable, sustained-release naltrexone for the treatment of opioid dependence: A randomized, placebo-controlled trial." *Archives of General Psychiatry* 63(2): 210–218.

Covington, J. (2001). Linking treatment to punishment: An evaluation of drug treatment in the criminal justice system. In: *Informing America's Policy on Illegal Drugs: What We Don't Know Keeps Hurting Us*. C. Manski, J. Pepper, and C. Petrie. Washington, DC, National Academy Press, pp. 349–381.

Cox, W. M., L. M. Hogan, M. R. Kristian, and J. H. Race (2002). "Alcohol attentional bias as a predictor of alcohol abusers' treatment outcome." *Drug and Alcohol Dependence* 68(3): 237–243.

Dackis, C., and C. O'Brien (2005). "Neurobiology of addiction: Treatment and public policy ramifications." *Nature Neuroscience* 8(11): 1431–1436.

Dagher, A., and T. W. Robbins (2009). "Personality, addiction, dopamine: Insights from Parkinson's disease." *Neuron* 61(4): 502–510.

Dalrymple, T. (2006). *Romancing Opiates: Pharmacological Lies and the Addiction Bureaucracy*. New York, Encounter Books.

Davies, J. B. (1997). *The Myth of Addiction*. 2nd ed. Amsterdam, Harwood Academic.

Deschenes, E. P., S. Turner, and P. W. Greenwood (1995). "Drug Court or probation: An experimental evaluation of Maricopa County's Drug Court." *Justice System Journal* 18:55.

Dolan, K. (1991). "Prisons and AIDS: A review of the VIIth International Conference on AIDS." *International Journal of Drug Policy* 2:23–26.

Dole, V. P., J. W. Robinson, J. Orraca, E. Towns, P. Searcy, and E. Caine (1969). "Methadone treatment of randomly selected criminal addicts." *New England Journal of Medicine* 280(25): 1372–1375.

Elster, J., and O. J. Skog (1999). *Getting Hooked: Rationality and Addiction*. Cambridge, UK, Cambridge University Press.

Epstein, D. H., K. L. Preston, J. Stewart, and Y. Shaham (2006). "Toward a model of drug relapse: An assessment of the validity of the reinstatement procedure." *Psychopharmacology* 189(1): 1–16.

Fazel, S., P. Bains, and H. Doll (2006). "Substance abuse and dependence in prisoners: A systematic review." *Addiction* 101(2): 181–191.

Fingarette, H. (1988). *Heavy Drinking: The Myth of Alcoholism as a Disease*. Berkeley, CA, University of California Press.

Fischer, B. (2003). "Doing good with a vengeance: A critical assessment of the practices, effects and implications of drug treatment courts in North America." *Criminal Justice* 3(3): 22.

Flodgren, G., M. Eccles, S. Shepperd, A. Scott, E. Parmelli, and F. Beyer (2011). "An overview of reviews evaluating the effectiveness of financial incentives in changing healthcare professional behaviours and patient outcomes." *Cochrane Database of Systematic Reviews* 7: CD009255.

Foddy, B., and J. Savulescu (2006). "Addiction and autonomy: Can addicted people consent to the prescription of their drug of addiction?" *Bioethics* 20(1): 1–15.

Fox, R. G. (1992). "The compulsion of voluntary treatment in sentencing." *Criminal Law Journal* 16: 37–54.

Garavan, H., A. Lingford-Hughes, T. Jones, P. Morris, J. Rothwell, and S. Williams (2007). Neuroimaging. In: *Drugs and the Future: Brain Science, Addiction and Society*. D. Nutt, T. Robbins, G. Stimson, M. Ince, and A. Jackson. London, Academic Press, pp. 285–314.

Gerstein, D. R., and H. J. Harwood (1990). *Treating Drug Problems (Volume 1). A Study of Effectiveness and Financing of Public and Private Drug Treatment Systems*. Washington, DC, Institute of Medicine, National Academy Press.

Glimcher, P. W. (2008). *Neuroeconomics: Decision Making and the Brain*. London, Academic.

Goldman, D., G. Oroszi, and F. Ducci (2005). "The genetics of addictions: Uncovering the genes." *Nature Reviews. Genetics* 6(7): 521–532.

Gordon, M. S., T. W. Kinlock, R. P. Schwartz, and K. E. O'Grady (2008). "A randomized clinical trial of methadone maintenance for prisoners: Findings at 6 months post-release." *Addiction* 103(8):1333–1342.

Gossop, M., J. Marsden, D. Stewart, and T. Kidd (2003). "The National Treatment Outcome Research Study (NTORS): 4–5 year follow-up results." *Addiction* 98(3): 291–303.

Goudriaan, A., E. R. Grekin, and K. J. Sher (2008). "The predictive value of neurocognitive decision-making for heavy alcohol use in a longitudinal, prospective study." *European Neuropsychopharmacology* 18: S69–S70.

Goudriaan, A., J. Oosterlaan, E. De Beurs, and W. Van Den Brink (2008). "The role of self-reported impulsivity and reward sensitivity versus neurocognitive measures of disinhibition and decision-making in the prediction of relapse in pathological gamblers." *Psychological Medicine* 38(1): 41–50.

Granfield, R., and W. Cloud (1996). "The elephant that no one sees: Natural recovery among middle-class addicts." *Journal of Drug Issues* 26(1): 45–61.

Haasen, C., U. Verthein, F. J. Eiroa-Orosa, I. Schafer, and J. Reimer (2010). "Is heroin-assisted treatment effective for patients with no previous maintenance treatment? Results from a German randomised controlled trial." *European Addiction Research* 16(3): 124–130.

Hall, W. (1996). *Methadone Maintenance Treatment as a Crime Control Measure. Crime and Justice Bulletin (No. 29).* Sydney, NSW Bureau of Crime Statistics and Research.

Hall, W. (1997). "The role of legal coercion in the treatment of offenders with alcohol and heroin problems." *Australian and New Zealand Journal of Criminology* 30(2): 103–120.

Hall, W. (2005). Assessing the population level impact of the Swiss model of heroin prescription. In: *Heroin-Assisted Treatment: Work in Progress.* M. Rihs-Middel, R. Hammig, and N. Jacobshagen. Bern, Verlag Hans Huber, pp. 49–61.

Hall, W., B. Capps, and A. Carter (2008). "The use of depot naltrexone under legal coercion: The case for caution." *Addiction* 103: 1922–1924.

Hall, W., and J. Lucke (2010). "Legally coerced treatment for drug using offenders: Ethical and policy issues." *Crime and Justice Bulletin* 144: 1–12.

Harvey, E., A. Shakeshaft, K. Hetherington, C. Sannibale, and R. P. Mattick (2007). "The efficacy of diversion and aftercare strategies for adult drug-involved offenders: A summary and methodological review of the outcome literature." *Drug and Alcohol Review* 26(4): 379–387.

Heyman, G. (2001). Is addiction a chronic, relapsing disease? In: *Drug Addiction and Drug policy. The Struggle to Control Dependence.* Cambridge, MA, Harvard University Press, pp. 81–117.

Heyman, G. (2009). *Addiction: A Disorder of Choice.* Cambridge, MA, Harvard University Press.

Higgins, S. T. (2006). "Extending contingency management to the treatment of methamphetamine use disorders." *American Journal of Psychiatry* 163(11): 1870–1872.

Higgins, S. T., and N. M. Petry (1999). "Contingency management—Incentives for sobriety." *Alcohol Research & Health* 23(2): 122–127.

Hubbard, R. L. (1989). "Drug abuse treatment: A national study of effectiveness." London, University of North Carolina Press.

Hubbard, R. L., J. J. Collins, J. H. Rachal, and E. R. Cavanaugh (1988). The criminal justice client in drug abuse treatment. In: *Compulsory Treatment of Drug Abuse: Research and Clinical Practice.* C. Leukefeld and F. M. Tims. Rockville, MD, National Institute of Drug Abuse.

Hulse, G. K., V. H. Low, V. Stalenberg, N. Morris, R. I. Thompson, R. J. Tait, C. T. Phan, H. T. Ngo, and D. E. Arnold-Reed (2007). "Biodegradability of naltrexone-poly(DL) lactide implants in vivo assessed under ultrasound in humans." *Addiction Biology* 13(3–4): 364–372.

Hulse, G. K., N. Morris, D. Arnold-Reed, and R. J. Tait (2009). "Improving clinical outcomes in treating heroin dependence: Randomized, controlled trial of oral or implant naltrexone." *Archives of General Psychiatry* 66(10): 1108–1115.

Hyman, S. E. (2007). "The neurobiology of addiction: Implications for voluntary control of behavior." *American Journal of Bioethics* 7(1): 8–11.

Jentsch, J. D., and J. R. Taylor (1999). "Impulsivity resulting from frontostriatal dysfunction in drug abuse: Implications for the control of behavior by reward-related stimuli." *Psychopharmacology* 146(4): 373–390.

Kalant, H. (2010). "What neurobiology cannot tell us about addiction." *Addiction* 105(5): 780–789.

Klag, S., F. O'Callaghan, and P. Creed (2005). "The use of legal coercion in the treatment of substance abusers: An overview and critical analysis of thirty years of research." *Substance Use & Misuse* 40(12): 1777–1795.

Kleiman, M. (2009). *When Brute Force Fails: How to Have Less Crime and Less Punishment.* Princeton, NJ, Princeton University Press.

Koob, G. F., and M. Le Moal (2006). *Neurobiology of Addiction.* New York, Academic Press.

Koob, G. F., and N. D. Volkow (2010). "Neurocircuitry of addiction." *Neuropsychopharmacology* 35(1): 217–238.

Krebs, C. P., C. H. Lindquist, W. Koetse, and P. K. Lattimore (2007). "Assessing the long-term impact of drug court participation on recidivism with generalized estimating equations." *Drug and Alcohol Dependence* 91(1): 57–68.

Krupitsky, E., E. V. Nunes, W. Ling, A. Illeperuma, D. R. Gastfriend, and B. L. Silverman (2011). "Injectable extended-release naltrexone for opioid dependence: A double-blind, placebo-controlled, multicentre randomised trial." *Lancet* 377(9776): 1506–1513.

Kunoe, N., P. Lobmaier, J. K. Vederhus, B. Hjerkinn, S. Hegstad, M. Gossop, O. Kristensen, and H. Waal (2009). "Naltrexone implants after in-patient treatment for opioid dependence: randomised controlled trial." *British Journal of Psychiatry* 194(6): 541–546.

Leshner, A. I. (1997). "Addiction is a brain disease, and it matters." *Science* 278(5335): 45–47.

Leukefeld, C. G., and F. M. Tims (1988). *Compulsory Treatment of Drug Abuse: Research and Clinical Practice.* Rockville, MD, National Institute on Drug Abuse.

Lind, B., D. Weatherburn, S. Chen, M. Shanahan, E. Lancsar, M. Haas, and R. De Abreu Lourenco (2002). *New South Wales Drug Court Evaluation: Cost-Effectiveness, Legislative Evaluation Series (No. 15).* Sydney, NSW Bureau of Crime Statistics and Research.

Lintzeris, N. (2009). "Prescription of heroin for the management of heroin dependence: Current status." *CNS Drugs* 23(6): 463–476.

Makkai, T. (2002). "The emergence of drug treatment courts in Australia." *Substance Use and Misuse* 37(12–13): 1567–1594.

Manski, C. F., J. V. Pepper, and C. V. Petrie (2001). *Informing America's Policy on Illegal Drugs: What We Don't Know Keeps Hurting Us.* Washington, DC, National Academy Press.

Marlowe, D. B. (2006). "Depot naltrexone in lieu of incarceration: A behavioral analysis of coerced treatment for addicted offenders." *Journal of Substance Abuse Treatment* 31(2): 131–139.

Marteau, T., R. E. Ashcroft, and A. Oliver (2009). "Using financial incentives to achieve healthy behaviour." *British Medical Journal* 338: 983–985.

Mattick, R. P., C. Breen, J. Kimber, and M. Davoli (2009). "Methadone maintenance therapy versus no opioid replacement therapy for opioid dependence." *Cochrane Database of Systematic Reviews* 3: CD002209.

Mattick, R. P., J. Kimber, C. Breen, and M. Davoli (2008). "Buprenorphine maintenance versus placebo or methadone maintenance for opioid dependence." *Cochrane Database of Systematic Reviews* 2: CD002207.

Mattick, R. P., and W. Hall (1996). "Are detoxification programmes effective?" *Lancet* 347(8994): 97–100.

McLellan, A. T., D. C. Lewis, C. P. O'Brien, and H. D. Kleber (2000). "Drug dependence, a chronic medical illness: Implications for treatment, insurance, and outcomes evaluation." *JAMA* 284(13): 1689–1695.

Merims, D., and N. Giladi (2008). "Dopamine dysregulation syndrome, addiction and behavioral changes in Parkinson's disease." *Parkinsonism and Related Disorders* 14(4): 273–280.

Minozzi, S., L. Amato, S. Vecchi, M. Davoli, U. Kirchmayer, and A. Verster (2006). "Oral naltrexone maintenance treatment for opioid dependence." *Cochrane Database of Systematic Reviews* 1: CD001333.

Moore, T. J., A. Ritter, and J. P. Caulkins (2007). "The costs and consequences of three policy options for reducing heroin dependency." *Drug and Alcohol Review* 26(4): 369–378.

Nadelhoffer, T., S. Bibas, S. Grafton, K. A. Kiehl, A. Mansfield, W. Sinnott-Armstrong, and M. Gazzaniga (2012). "Neuroprediction, violence, and the law: Setting the stage." *Neuroethics* 5:67–99.

Neale, J. (2002). *Drug Users in Society.* New York, Palgrave.

Olds, J., and P. Milner (1954). "Positive reinforcement produced by electrical stimulation of septal area and other regions of rat brain." *Journal of Comparative and Physiological Psychology* 47(6): 419–427.

Oviedo-Joekes, E., B. Nosyk, D. Marsh, D. Guh, S. Brissette, C. Gartry, M. Krausz, A. Anis, and M. Schechter (2009). "Scientific and political challenges in North America's first randomized controlled trial of heroin-assisted treatment for severe heroin addiction: Rationale and design of the NAOMI study." *Clinical Trials* 6(3): 261.

Paulus, M. P., S. F. Tapert, and M. A. Schuckit (2005). "Neural activation patterns of methamphetamine-dependent subjects during decision making predict relapse." *Archives of General Psychiatry* 62(7): 761–778.

Peele, S. (2004). "The surprising truth about addiction." *Psychology Today* 37(3): 43.

Perea-Milla, E., L. C. Aycaguer, J. C. Cerda, F. G. Saiz, F. Rivas-Ruiz, A. Danet, M. R. Vallecillo, and E. Oviedo-Joekes (2009). "Efficacy of prescribed injectable diacetylmorphine in the Andalusian trial: Bayesian analysis of responders and non-responders according to a multi domain outcome index." *Trials* 10: 70.

Perneger, T. V., F. Giner, M. del Rio, and A. Mino (1998). "Randomised trial of heroin maintenance programme for addicts who fail in conventional drug treatments." *British Medical Journal* 317(7150): 13–18.

Porter, L., A. Arif, and W. J. Curran (1986). *The Law and the Treatment of Drug- and Alcohol-Dependent Persons: A Comparative Study of Existing Legislation*. Geneva, World Health Organization.

Pritchard, E., J. Mugavin, and A. Swan (2007). *Compulsory Treatment in Australia: A Discussion Paper on the Compulsory Treatment of Individuals Dependent on Drugs and/or Alcohol*. Canberra, Australian National Council on Drugs.

Room, R. (2007). Social policy and psychoactive substances. In: *Drugs and the Future: Brain Science, Addiction and Society*. D. Nutt, T. Robbins, G. Stimson, M. Ince, and A. Jackson. London, Academic Press, pp. 337–358.

Ross, D., C. Sharp, R. Vuchinich, and D. Spurrett (2008). *Midbrain Mutiny: The Picoeconomics and Neuroeconomics of Disordered Gambling*. Cambridge, MA, MIT Press.

Satel, S. (1999). "The fallacies of no-fault addiction." *Public Interest* 134: 52–67.

Satel, S. (2006). "For addicts, firm hand can be the best medicine." *New York Times*. Retrieved January 24, 2009, from http://query.nytimes.com/gst/fullpage.html?res=9D07E4DB173EF936A2575BC0A9609C8B63&scp=1&sq=For%20addicts,%20firm%20hand%20can%20be%20the%20best%20medicine&st=cse.

Scherbaum, N., and F. Rist (2010). "Opiate addicts' attitudes towards heroin prescription." *Open Addiction Journal* 3: 109–116.

Schumann, G. (2007). "Okey Lecture 2006: Identifying the neurobiological mechanisms of addictive behaviour." *Addiction* 102(11): 1689–1695.

Schutz, C. (2008). "Using neuroimaging to predict relapse to smoking: Role of possible moderators and mediators." *International Journal of Methods in Psychiatric Research* 17(S1):S78–S82.

Scottish Government Community Justice Services (2010). *Review of the Glasgow and Fife Drug Courts. Report, 2009*. Edinburgh, Scottish Government.

Simpson, D., and H. Friend (1988). Legal status and long-term outcomes for addicts in the DARP Followup Project. In: *Compulsory Treatment of Drug Abuse: Research and Clinical Practice*. C. Leukefeld and F. M. Tims. Rockville, MD, NIDA, pp. 81–98.

Simpson, D. D. (1981). "Treatment for drug abuse: Follow-up outcomes and length of time spent." *Archives of General Psychiatry* 38(8):875–880.

Smith, L. A., S. Gates, and D. Foxcroft (2006). "Therapeutic communities for substance related disorder." *Cochrane Database of Systematic Reviews* 1: CD005338.

Stitzer, M., M. Iguchi, M. Kidorf, and G. Bigelow (1993). Contingency management in methadone treatment: The case for positive incentives. In: *Behavioral Treatments for Drug Abuse and Dependence*. L. Onken, J. Blaine, and J. Boren. Rockville, MD, National Indicate of Drug Abuse, pp. 19–36.

Strang, J., N. Metrebian, N. Lintzeris, L. Potts, T. Carnwath, S. Mayet, H. Williams, D. Zador, R. Evers, T. Groshkova, V. Charles, A. Martin, and L. Forzisi (2010). "Supervised injectable heroin or injectable methadone versus optimised oral methadone as treatment for chronic heroin addicts in England after persistent failure in orthodox treatment (RIOTT): A randomised trial." *Lancet* 375(9729): 1885–1895.

Sullivan, M., F. Birkmayer, B. Boyarsky, R. Frances, J. Fromson, M. Galanter, F. Levin, C. Lewis, E. Nace, and R. Suchinsky (2008). "Uses of coercion in addiction treatment: Clinical aspects." *American Journal on Addictions* 17(1): 36–47.

Szasz, T. S. (1975). *Ceremonial Chemistry: The Ritual Persecution of Drugs, Addicts, and Pushers*. London, Routledge.

Teesson, M., K. Mills, J. Ross, S. Darke, A. Williamson, and A. Havard (2008). "The impact of treatment on 3 years' outcome for heroin dependence: Findings from the Australian Treatment Outcome Study (ATOS)." *Addiction* 103(1): 80–88.

True, W. R., H. Xian, J. F. Scherrer, P. A. F. Madden, K. K. Bucholz, A. C. Heath, S. A. Eisen, M. J. Lyons, J. Goldberg, and M. Tsuang (1999). "Common genetic vulnerability for nicotine and alcohol dependence in men." *Archives of General Psychiatry* 56(7): 655–661.

Turner, S., D. Longshore, S. Wenzel, E. Deschenes, P. Greenwood, T. Fain, A. Harrell, A. Morral, F. Taxman, and M. Iguchi (2002). "A decade of drug treatment court research." *Substance Use & Misuse* 37(12–13): 1489–1527.

Uhl, G. R., M. D. Li, J. Gelertner, W. Berrettini, and J. Pollock (2004). "Molecular genetics of addiction vulnerability and treatment responses." *Neuropsychopharmacology* 29: S26–S26.

Uhl, G. R., T. Drgon, C. Johnson, C. Y. Li, C. Contoreggi, J. Hess, D. Naiman, and Q. R. Liu (2008). "Molecular genetics of addiction and related heritable phenotypes." *Annals of the New York Academy of Sciences* 1141(1): 318–381.

US Government Accountability Office (1995). *Drug Courts: Information on a new approach to address drug-related crime, briefing report to the Committee on the Judiciary, U.S. Senate, and the Committee on the Judiciary, House of Representatives*. Washington, DC, United States General Accounting Office.

Van den Brink, W., and C. Haasen (2006). "Evidenced-based treatment of opioid-dependent patients." *Canadian Journal of Psychiatry* 51(10): 635–646.

Van den Brink, W., V. M. Hendriks, P. Blanken, M. W. Koeter, B. J. van Zwieten, and J. M. van Ree (2003). "Medical prescription of heroin to treatment resistant heroin addicts: Two randomised controlled trials." *British Medical Journal* 327(7410): 310.

Volkow, N. D., and T. K. Li (2004). "Drug addiction: The neurobiology of behaviour gone awry." *Nature Reviews. Neuroscience* 5(12): 963–970.

Volkow, N. D., and T. K. Li (2005). "Drugs and alcohol: Treating and preventing abuse, addiction and their medical consequences." *Pharmacology & Therapeutics* 108(1): 3–17.

Voon, V., K. Hassan, M. Zurowski, S. Duff-Canning, M. de Souza, S. Fox, A. E. Lang, and J. Miyasaki (2006). "Prospective prevalence of pathologic gambling and medication association in Parkinson disease." *Neurology* 66(11): 1750–1752.

Ward, J., W. Hall, and R. Mattick (2009). Methadone maintenance treatment. In: *Pharmacotherapies for the Treatment of Opioid Dependence: Efficacy, Cost-Effectiveness and Implementation Guidelines*. R. Mattick, R. Ali and N. Linzteris. New York, Informa Health Care, pp. 107–141.

Ward, J., R. P. Mattick, and W. Hall (1998). *Methadone Maintenance Treatment and Other Opioid Replacement Therapies*. Sydney, Harwood Academic Press.

Weatherburn, D., C. Jones, L. Snowball, and J. Hua (2008). *The NSW Drug Court: A Re-evaluation of its Effectiveness. Crime and Justice Bulletin (No. 121)*. Sydney, NSW Bureau of Crime Statistics and Research.

Weintraub, D., A. D. Siderowf, M. N. Potenza, J. Goveas, K. H. Morales, J. E. Duda, P. J. Moberg, and M. B. Stern (2006). "Association of dopamine agonist use with impulse control disorders in Parkinson disease." *Archives of Neurology* 63(7): 969–973.

Weintraub, D., J. Koester, M. N. Potenza, A. D. Siderowf, M. Stacy, V. Voon, J. Whetteckey, G. R. Wunderlich, and A. E. Lang (2010). "Impulse control disorders in Parkinson disease: A cross-sectional study of 3090 patients." *Archives of Neurology* 67(5): 589–595.

Wild, T. C., A. B. Roberts, and E. L. Cooper (2002). "Compulsory substance abuse treatment: An overview of recent findings and issues." *European Addiction Research* 8(2): 84–93.

Wilson, D. B., O. Mitchell, and D. L. MacKenzie (2006). "A systematic review of drug court effects on recidivism." *Journal of Experimental Criminology* 2(4): 459–487.

Wodak, A., J. Shaw, M. Gaughwin, M. Ross, M. Miller, and J. Gold (1992). Behind bars: HIV risk-taking behaviour of Sydney male drug injectors while in prison. In: *HIV/AIDS and Prisons: Proceedings of a conference held November 19–21, 1990*. J. Norberry, M. Gaughwin, and S. Gerull. Canberra, Australian Institute of Criminology, pp. 239–244.

Wolfe, D., M. P. Carrieri, N. Dasgupta, A. Wodak, R. Newman, and R. D. Bruce (2011). "Concerns about injectable naltrexone for opioid dependence." *Lancet* 377(9776): 1468–1470.

Xu, Z., B. Hou, Y. Gao, F. He, and C. Zhang (2007). "Effects of enriched environment on morphine-induced reward in mice." *Experimental Neurology* 204(2): 714–719.

Zack, M., and C. X. Poulos (2009). "Parallel roles for dopamine in pathological gambling and psychostimulant addiction." *Current Drug Abuse Reviews* 2(1): 11–25.

PART 5

MODIFICATION

13

Enhancing Responsibility

NICOLE A VINCENT

We normally think that the degree of a person's responsibility co-varies with her mental capacity—or put another way, that *responsibility tracks capacity*.[1] This is an important part of the reason that children, senile adults, and those suffering from certain kinds of mental illness or retardation are often thought to be less than fully responsible for what they do—because they have significant deficits in the mental capacities that are required for responsible moral agency. Some plausible mental capacity candidates that come to mind include the abilities to perceive the world without delusion, to think clearly, to guide actions by the light of our judgments, and to resist acting on mere impulse. Or, for the legal context, H. L. A. Hart suggests that "[t]he capacities in question are those of understanding, reasoning, and control of conduct: the ability to understand what conduct legal rules or morality require, to deliberate and reach decisions concerning these requirements, and to conform to decisions when made" (1968:227). This is also why children's responsibility increases as their mental capacities mature, and why responsibility is reestablished as people recover from mental illness or brain injury (e.g., Chatfield, Heffernan et al. 2002)—because we think that responsibility is acquired or restored as mental capacities are developed or regained. Finally, this is also the operative assumption behind a significant portion of current "neurolaw" research, which aims to help the law to assess and to restore people's responsibility by using modern neuroscientific techniques to detect and to treat mental disorders.[2] Responsibility depends on *other* things, too—for instance, on how the particular (in)capacities came about, and on norms of reasonableness that influence what we think it is legitimate to expect of one another in a range of different situations (see Vincent 2010:93–95)—but mental capacities certainly play an important role.

But if responsibility indeed tracks capacity—that is, if it diminishes when mental capacities are lost and is restored when they are subsequently regained—then what would happen if a person's mental capacities were increased through cognitive enhancement *beyond the normal range* accessible to most humans? Would such a person become "hyperresponsible," and if so, then in what sense? For instance, would cognitively enhanced people acquire new responsibilities that they otherwise wouldn't have had? Might they, as a consequence, be legitimately blamed when they fail to discharge those greater responsibilities? And would that increase the likelihood that they would subsequently be held responsible (i.e., liable) when things went wrong? Would hyperresponsible people necessarily be less irresponsible, or might this depend more on their character than on their mental capacities? On a different note, if cognitive enhancers indeed improved mental performance, then might it eventually become morally and legally obligatory for some people to cognitively enhance themselves under some circumstances? For instance, given how much is at stake in an operating room, on a military battlefield, in long-haul flights, and in courtrooms, it could be argued that surgeons, soldiers, airplane pilots, and judges and jurors, respectively, have a moral duty to take cognitive enhancers to ensure the highest performance possible (Sandberg, Sinnott-Armstrong et al. 2011) and that they would be negligent or even reckless if they didn't enhance themselves.[3]

Various authors have reported a sharp increase in off-label use of prescription medications such as donepezil, modafinil, guanfacine, and methylphenidate (Ritalin) for cognitive enhancement purposes, and this has prompted discussions in the current literature of issues such as safety, effectiveness, coercion, and distributive justice (e.g., Appel 2008; Chatterjee 2009; Forlini and Racine 2009; Larriviere, Williams et al. 2009; Racine and Forlini 2009; Persson and Savulescu 2011; Schermer, Bolt et al. 2009; though Hall and Lucke 2010 present a deflationary argument). But next to nothing is written about these drugs' effects on people's responsibility.[4] This leaves pharmaceutical companies, medical practitioners, people in socially important roles, the legal sector, and the general public in a morally and legally uncertain position. This paper addresses this absence of theory on this important topic by investigating the effects that cognitive enhancement might have on moral and legal responsibility.

In section 1, I offer four familiar examples suggesting that responsibility tracks capacity even once capacity rises above the normal level—I refer to this as the claim that *responsibility tracks hypercapacity*—and in section 2, I develop eight objections to this claim. In section 3, these objections are tackled, in the process revealing important nuances to the claim that responsibility tracks (hyper)capacity. Finally, in section 4, I summarize my argument and draw out some moral and legal implications.

Before proceeding, however, the reader is asked to note one caveat. To avoid weighing down the early discussion with too many technical distinctions, in the first half of this paper, I use the word "responsibility" and its cognates in an *intuitive* sense. However, as the discussion progresses, it will become increasingly clear that "responsibility" has many subtly different meanings, and hence that the claim that *responsibility tracks (hyper)capacity* itself needs to be disambiguated. A natural opportunity will present itself to do this in the section entitled "More Responsible *In What Sense*" starting on page 318, but readers who would like this clarification sooner are welcome to read that section first.

RESPONSIBILITY TRACKS *HYPER*CAPACITY—FOUR EXAMPLES

That responsibility tracks hypercapacity, and not just capacity, is suggested by intuitions in the following four examples.

First, we often express our disappointment when a particularly bright young child (a "precocious developer") fails to meet the higher standards that we expected them to meet, and we may blame him for many more things than we might otherwise have blamed him (or that we might blame others of a similar age) precisely because we think that it is legitimate to expect more of the precocious developer on account of his greater than normal mental capacities. It is not uncommon to reprimand a precocious developer who disappoints us by acting in a way that fails to meet the higher standards that we had set for him by saying something like, "I expected more of you," even though we would not expect more of others in his age group, and the reason why we do this is precisely because, ceteris paribus, it seems reasonable to expect more of someone who has above-average mental capacities. His mental capacity is like that of someone several years his senior, and this is why we might expect more of him—because we would expect more of a person that many years his senior.

Second, a person who becomes more highly educated and trained often acquires new responsibilities. Sometimes such responsibilities are *taken on voluntarily* by the said party, whereas other times they are *imposed* upon her by others (e.g., by management); and some of these responsibilities might be made *explicit* (for instance, a new duty statement might be drawn up), whereas others might be left *implicit* (for instance, the said person might be expected to show more initiative). And when such a person subsequently fails to discharge her new responsibilities, then we may eventually hold her responsible for things for which previously (i.e., before her education or training) we would not have held her responsible. For instance, a trained and equipped medical expert who knowingly stands by and does nothing while a person whom she could have saved dies of anaphylactic shock (a severe allergic reaction) will likely be judged more severely (at least from a moral if not from a legal perspective) than the rest of the onlookers in the crowd who also did nothing but who lacked their training and equipment, precisely because she possessed the capacity to save that person, whereas the others in the crowd did not. Thus, because greater capacity acquired through education affects responsibility, hypercapacity from cognitive enhancement might also affect responsibility.

Third, as Stephen Morse (2006:40) has argued and as I have also endorsed (2008:200–201), a person who suffers from hypomania may be more rather than less responsible at some stages of his illness—we might think of him as being "hyperresponsible"—precisely because during his hypomanic episodes, he may possess greater capacities than others (e.g., increased speed and clarity of thought), and his possession of those capacities might make it reasonable to expect more of him on those occasions.

Finally, consider popular intuitions about the responsibility of superheroes like the DC Comics character Superman (DC 2010) and Marvel Comics character Spider-Man (Marvel 2010). In the third Superman movie (1983) starring the late Christopher Reeve, under the effect of a synthetic kryptonite, Superman took to hanging out in bars and getting drunk, he stopped caring for society, and he no longer tried to protect the world from villains. Those who didn't know what motivated this dramatic change of character thought ill of him for becoming so indifferent to humanity's plight. After all, he was the only one who could protect the world from villains, but yet he did nothing. Naturally, he had a good excuse—he was under the influence of the synthetic kryptonite and so he was not himself—but the point remains that if he had chosen to be like that without the influence of the synthetic kryptonite, then we *would* have been justified in thinking worse of him. In the second Superman movie (1980), after initially casting aside his super powers so that he could live a normal life with Lois Lane, to save the world from super villains General Zod, Ursa, and Non, who threatened to wreak havoc on the earth, Superman again restored his super powers after realizing how reckless he had been in previously casting them aside. Similarly, as depicted in the last comic book panel of the first Spider-Man story, a young Peter Parker (the true teenage identity of Spider-Man) comes to realize that "with great power comes great responsibility." When peril strikes, we expect our superheroes to save the world, and we expect them to not cast their super powers aside. Admittedly, these are just science fiction examples, and so they are not necessarily representative of widely held moral intuitions. But, they do provide at least a glimpse of nonacademic thinking about how hypercapacity might affect responsibility by helping us to reach reflective equilibrium, and my present suggestion is simply that this example (like the three examples before it) also suggests that at least prima facie, hypercapacity does result in hyperresponsibility.

In various ways, our intuitions already suggest that we expect more of people with greater capacities. Furthermore, they also suggest that when the stakes are sufficiently high and the costs of raising our capacities are sufficiently low, we expect people to do just that—to enhance themselves. Thus, because the ability to cognitively enhance ourselves is now coming into our grasp, and because (by definition) the capacities of cognitively enhanced people are greater than the capacities of nonenhanced people, prima facie some people (e.g., those mentioned earlier) may under some circumstances (i.e., when the stakes are sufficiently high and the costs are sufficiently low) have a moral responsibility to enhance themselves, and once enhanced, they may acquire further responsibilities. Put another way, it would indeed seem that responsibility tracks *hyper*capacity as well as capacity.

OBJECTIONS TO THE CLAIM THAT RESPONSIBILITY TRACKS *HYPER*CAPACITY

In this section, I briefly develop the following eight objections to the claim that responsibility tracks hypercapacity: (i) it is wrong to expect more of precocious developers; (ii) consent is required for the expansion of responsibility; (iii) hypomanic individuals are not really more responsible; (iv) what happened to supererogation for superheroes? (v) responsibility is a threshold concept, so hypercapacity is irrelevant; (vi) smart people can be evil, too, so enhanced people need not be more responsible; (vii) only blame but not praise is affected by increased capacity, and so not all responsibility judgments are affected by hypercapacity; and (viii) *can* does not imply *ought*. The first four objections directly engage with the four examples that were just discussed, whereas the remaining four objections raise more general worries. Because I offer replies to these objections in the following section, I will refrain from commenting on their adequacy here.

It Is Wrong to Expect More of Precocious Developers

Although I argued above that the practice of holding precocious developers to higher standards of expectation provides some support for the claim that responsibility tracks hypercapacity, a foreseeable objection is that it is actually wrong to have such higher expectations of precocious developers in the first place, and thus that this practice can't really support the claim that responsibility tracks hypercapacity.

One reason that such a practice might be wrong is its potential harmfulness. Although in some ways their mental capacities might indeed be better developed than those of their age-wise peers, often this greater mental development is localized to only a group of capacities—for instance, academic talent—but in other ways, precocious developers still lack the life experience and the emotional strength to cope with greater expectations. Precocious developers already have a tendency to suffer more as a group from high-expectation-induced stress—much of this expectation comes from within themselves—and the last thing that adults should do is to exacerbate the situation by saddling them with yet more high expectations.

Another reason why such a practice might be wrong is because it seems patently unfair. After all, these kids did not ask to be more intelligent, more capable, or just more mature than their age-wise peers—that is just how they turned out—and it seems unfair to rob them of their childhood by expecting them to carry adult burdens that their age-wise peers are not expected to endure. After all, why should they have to work harder, to perform better, and to satisfy these higher expectations, just because they were burdened with their supposed "gifts"? They did not ask for or earn those gifts, and so why should they now be burdened with the "talent slavery" that we wish to impose onto them?

Thus, the first objection is that because it is wrong to expect more of precocious developers (i.e., it is harmful and unfair), this practice can hardly justify the claim that responsibility tracks hypercapacity.

Consent Is Required for the Expansion of Responsibility

I argued previously that a person who becomes more highly educated and trained often acquires new responsibilities, and hence that for similar reasons a person who acquires new capacities through cognitive enhancement might also acquire new responsibilities. However, an opponent might retort that it is only fair to expect more of a person who acquires new capacities—that is, to impose upon them and to expect them to deliver upon new responsibilities—when they have consented to taking on those responsibilities. After all, if a petrol station attendant studies accounting in his spare time, the mere fact that he is now capable of doing accounting will not entail that his employer may legitimately expect him to do the petrol station's bookkeeping. The petrol station attendant must first consent to this before his employer may expect him to discharge such responsibilities, and so likewise it might also be argued that hypercapacity will only entail increased responsibility if the person consents to taking on those greater responsibilities.

Hypomanic Individuals Are Not Really More Responsible

I suggested earlier that because people suffering from hypomania are hyperresponsible on account of their greater than average mental capacities, cognitively enhanced people may also be hyperresponsible as a consequence of their greater than average mental capacities. But as Andrew Turner points out, there are good grounds to resist the intuition that hypomanic individuals are indeed hyperresponsible.

First, Turner points out that hypomanic individuals lack the ability to accurately gauge their own capacities, and consequently he conjectures that they might either bite off more than they can chew, or (continuing the metaphor) they might fail to realize that they could have bitten off more than what they did. As Turner argues, "the hypomanic individual is not an objective or capable judge of their own abilities. Their capacity to recognize the presence, or absence, of capacities required to carry out some task is flawed, such that we would reduce [rather than increase] their responsibility for [failing to] carry...it out, even when they possess the required capacity" (2010:5).

Second, Turner also points out that hypomanic individuals lack *manifestation control*—that is, they lack the ability to determine *when* their greater mental capacities will manifest themselves (2010:5–7). After all, given the many negative aspects of hypomania, if they possessed *that* capacity, then there would be no need for them to take medications. Because they lack the ability to determine precisely when they will experience an episode of hypomania, Turner argues that hypomanic individuals should not be expected to perform at a higher standard than others because when the time comes to deliver on those expectations, they may lack the capacity to do so.

Thus, because by Turner's account hypomanic individuals are not really hyperresponsible despite their sometimes greater mental capacities, this example also can't be cited to support the claim that responsibility tracks hypercapacity.

What Happened to Supererogation for Superheroes?

Although I suggested earlier that our intuitions about the hyperresponsibility of superheroes support the claim that responsibility tracks hypercapacity, it could be argued that my analysis fails to leave sufficient room for supererogation.

On a different analysis, although we might privately think worse of Superman when he casts aside his greater capacities to live a life of urban mediocrity with Lois Lane, or when he fritters away his time and talents in public bars getting intoxicated rather than saving the world from peril, we would not be *justified* in thinking these things—let alone in publicly expressing such thoughts or imposing sanctions upon him—because in the end we are not justified in expecting him to do anything more than what we expect of anybody else. Likewise, although Peter Parker may feel guilt ridden by the idea that with his greater powers comes greater responsibility, perhaps we should put this down to his teenage angst or megalomania because (as per my comments about Superman), in fact, nobody is entitled to expect him to use his super powers for the greater good. In other words, perhaps my analysis of the responsibility of superheroes is tacitly infused with too great a measure of Utilitarian intuition—something that has led me to leave no room for supererogation, and thus for the idea that superheroes might be praiseworthy if they use their talents for the greater good, but that they would not be blameworthy if they failed to do this.

If superheroes are not hyperresponsible on account of their greater capacities, however, then again this example cannot be cited to support my claim that responsibility tracks hypercapacity.

Responsibility Is a Threshold Concept, so Hypercapacity is Irrelevant

Another objection to my suggestion that responsibility tracks hypercapacity stems from the observation that in law—which at least partly reflects commonsense morality, and certainly it does this no worse than my superheroes example—responsibility is a *threshold* concept.

For instance, in law, people are expected to take *sufficient* care to avoid causing harm to others—that is, as long as the amount of care that they take passes the threshold of sufficient care, then they will not be deemed responsible for bad consequences should things go wrong—and what is deemed to be sufficient is pegged to what a *reasonable person* would have done. Put a different way, to avoid responsibility for bad accidental outcomes, we must merely observe the same threshold level of care as a reasonable person would have observed rather than a level of care pegged to our individual capacities, and the law uniformly expects everyone to observe this *objective* standard of care.[5] In a much-quoted passage, Oliver Wendell Holmes Jr. argued that:

> If for instance, a man is born hasty and awkward, is always having accidents and hurting himself or his neighbours, no doubt his congenital defects will be allowed for in the courts of Heaven, but his slips are no less troublesome to his neighbours than if they [had] sprang from guilty neglect. His neighbours accordingly require him, at his proper peril, to come up to their standard, and the courts which they establish decline to take his personal equation into account. (Holmes 2000:108)

Given that *that* is all the care that we are entitled to expect of other people—that is, given that the law imposes an objective rather than a subjective standard of care onto everyone, and that once your actions reach that threshold of care you can't be blamed for things going wrong—an enhanced individual's "personal equation" (i.e., their greater mental capacities) should make no difference to their responsibility.

Smart People Can Be Evil, Too, so Enhanced People Need Not Be More Responsible

A different objection stems from the observation that smart people can be evil, too, and that unintelligent people need not be bad—that is, that intelligence does not necessarily correlate with how responsible a person might be. If responsibility increased as mental capacities increased, then we should expect that those people who have greater mental capacities would invariably be more responsible than those with lower mental capacities. However, quite the opposite is sometimes the case: sometimes a person of greater intelligence might use her superior intellect to take unfair advantage of others, and a dim-witted person may nevertheless be kind and make every effort to do the responsible thing. Mental capacity does not necessarily correlate with how responsible a person will be because this also seems to be in part a matter of one's character, and so this observation also seems to undermine my claim that responsibility tracks hypercapacity—indeed, it might even undermine the claim that responsibility tracks capacity, let alone hypercapacity.

Only Blame, but Not Praise, Is Affected by Increased Capacity

Even if we set aside the prior objections, another problem with the claim that responsibility tracks hypercapacity is that it seems *too broad*. As Susan Wolf has argued, "[w]hen we ask whether an agent's action is deserving of praise, it seems [that] we do not require that he could have done otherwise" (Wolf 1980:156). Although ascriptions of blame seem to require the ability to do otherwise—or put another way, ascriptions of blame seem to be governed by the *ought implies can* principle—a person can still be praised even if they couldn't have helped to do the praiseworthy thing. John Fischer and Mark Ravizza summarize Wolf's point thus:

> Wolf concludes that the freedom requirement for responsibility is *asymmetrical*: to be morally responsible for bad actions one must be able to do otherwise, but this ability is not required, in order to be responsible for good actions. (1998:57, emphasis added)

If this is so—that is, if only attributions of blame but not attributions of praise are sensitive to (or track) a person's actual capacities (let alone hypercapacities)—then at most, only those responsibility judgments that attribute blame might be influenced (and thus track) (hyper)capacity, but those responsibility judgments that attribute praise will not track (hyper) capacity.

Can Does Not Imply *Ought*

The final objection to my argument is that it seems to presuppose the claim that *can implies ought*, and this is surely an implausible idea. After all, what motivates my claim that the respective parties' responsibility is greater when their capacities are expanded, in the four examples that I offer above, is the idea that because each of them *can* do more, they therefore *ought* to do more. But this is surely a non sequitur, a fallacy. The fact that I can right now jump out of the window of my fourth floor study and plummet to my death surely does not yet imply that I ought to do so. The fact that a shopkeeper can make greater profit by giving me the wrong change in no way implies that he ought to do this either. Although the opposite relation might indeed be a condition of just ascriptions of moral responsibility—that is, *ought implies can*—*can implies ought* is an embarrassing mistake. Hence, the final reason that my opponents might offer to reject my arguments in support of the claim that responsibility tracks hypercapacity is that those arguments presuppose the fallacious idea that can implies ought.

DEFENSE OF THE CLAIM THAT RESPONSIBILITY TRACKS *HYPER*CAPACITY

The previous section described eight objections to the suggestion that responsibility tracks hypercapacity, and this section offers rejoinders to those objections.

Precocious Developers

The previous section argued that because it is wrong to expect more of precocious developers on account of their greater mental capacities, that this example therefore can't support my claim that responsibility tracks hypercapacity. Two reasons were given to support the claim that expecting more of precocious developers would be wrong: first, that it might harm their health (after all, they still lack the emotional strength to psychologically cope with higher expectations); and second, that it is unfair to impose greater burdens on them than on others of the same age. However, I do not believe that these reasons present a problem for my position.

Regarding the second point about unfairness, given that ex hypothesi the capacities of precocious developers are higher than those of their age-wise peers, it is not true that they would necessarily be disproportionately burdened by these higher expectations. If two people with different physical strength are asked to lift two boxes of different weights, then as long as the weights are equally matched to their respective strengths, the mere fact that the stronger person will be expected to lift the heavier box is no reason to suppose that the stronger person will be burdened more than the weaker person—relative to their strengths, each will be burdened equally. By analogy, the mere fact that precocious developers might be expected to perform at a higher standard than their age-wise peers is not yet a reason to suppose that this would burden them more because that would only occur if, *relative to their abilities,* they were asked to do more than their age-wise peers, or if the expectations that were imposed onto them were badly matched to their actual capacities.

Regarding the first point about harmfulness, I concede that if precocious developers lack the mental capacity to deal with higher expectations (e.g., because of their lack of emotional strength or whatever else), then that might indeed be a good reason to not expect more of them than their age-wise peers. However, note that the reason we would refrain from imposing higher expectations onto them is *precisely because* they would lack some capacity that is required to justify the imposition of those greater expectations (in this case, the emotional coping capacity). Had they possessed sufficient emotional strength or whatever other capacities might be required, then it would no longer necessarily be unreasonable to expect more of them.

Thus, in a way, considering these objections provides an opportunity to underscore the very point I am making—namely, that an important determinant of what responsibilities it is reasonable to impose upon a person is that person's capacities. Naturally, this means that before we set out to impose responsibilities onto people, we should first establish *what* capacities are required to justify the imposition of *which* responsibilities, and this, it seems, will be in part an empirical task. For instance, I suggested earlier that to be fully responsible, a person needs the ability to perceive the world without delusion, to think clearly, to guide his actions by the light of his judgments, and to resist acting on mere impulse. However, it is also plausible that future psychological and neuroscientific research may reveal that other (and perhaps very different kinds of) capacities—for instance, certain affective or emotional capacities—are needed for responsible moral agency.

Consent

In response to the second objection, I agree that consent—its presence, or its absence—often plays an important role in our ruminations about what responsibilities it is reasonable to expect a person to discharge. However, the idea that people can *only* be expected to discharge those responsibilities to which they have given their prior consent is surely far-fetched and not reflective of ordinary moral intuitions.

For instance, although some of our responsibilities are indeed related to the formal agreements that we make with others, many others are imposed upon us by society in virtue of the roles that we occupy whether we like it or not (e.g., parents have a responsibility to see to it that their children attend school), sometimes we occupy a role and are burdened with its responsibilities by accident (e.g., when a couple unintentionally conceive a child who is then carried to full term), and as Garrath Williams points out, other responsibilities are ours simply because of "the imperatives of basic human decency" (2008:467). Furthermore, the fact that "good Samaritan laws" generate so much heated debate—that is, laws that hold people responsible when they stand by and do nothing to save a person's life even though they could have done so at little cost or risk to themselves—also shows that it is not a far-fetched idea that some responsibilities might be acquired not through explicit consent but rather through coming to possess a particular capacity.

Consent is important—the fact that someone gave her consent is undoubtedly a ground for expecting her to discharge the relevant responsibility, and the fact that she withheld her consent is also undoubtedly a good prima facie reason to not expect her to discharge that responsibility. But consent is not everything—it is not the touchstone of responsibility.[6] What matters in the final analysis is *whether it is reasonable to expect* someone to do something or not, and my contention is that this in turn depends on *many* factors, only one of which is consent—but among the other factors are surely also the person's capacities.

Hypomania

Turner's comments about the responsibility of hypomanic individuals are sensible and important. The fact that such individuals lack a reliable *self-assessment capacity* and *manifestation control* are both good reasons to retract the blanket claim that hypomanic people are hyperresponsible. A person who can't accurately assess what capacities they have may either overestimate or underestimate his abilities, and consequently he might either get himself into deep water by taking on more responsibilities than what he can realistically be expected to discharge, or his self-doubt may prevent him from carrying out others' otherwise-legitimate expectations ("otherwise" because he *actually* possesses the required capacities, but just *thinks* that he does not possess them). Furthermore, we can't reasonably expect a person who lacks manifestation control to perform at a higher level, even though when the time comes, his capacities may indeed be so high that he could have (perhaps even easily) discharged those greater responsibilities, because there is insufficient certainty that at the crucial moment he will indeed be able to perform at that higher level. However, although these points are indeed important, I again do not believe that they are damaging to my claim that responsibility tracks hypercapacity.

Firstly, Turner's paper yet again draws attention to the point made while discussing precocious developers—namely, that it is crucially important to figure out *precisely which capacities* are required for the imposition of *which responsibilities*. After all, what Turner's conceptual analysis reveals is that in addition to any content-related capacities that a person might require to perform a given task—for instance, that the capacity to lift a 50-kilogram box is a prerequisite (though not the *only* prerequisite) of having a responsibility to lift a 50-kilogram box—the person must also possess the *meta* capacities of manifestation control and self-assessment ability. A person who often suffers great lethargy over which she has little control cannot be *expected* to lift a 50-kilogram box—if she does so, then that's a bonus, but we shouldn't expect this of her. Likewise, we shouldn't base our decisions about the reasonableness of imposing certain responsibilities onto her on her self-reports because that strategy is bound to eventually land us in deep water. Thus, Turner's discussion is important because it highlights the following point: I said earlier that it is partly an *empirical* task to figure out which capacities are required for which responsibilities, but I now add that this is also partly a *conceptual* task.

Second, Turner's discussion highlights yet again that what matters in the final analysis is *what it is reasonable to expect* of someone, and that this depends partly on the person's capacities. Whether people with hypomania may be expected to discharge certain responsibilities depends in part on their capacities, partly on the situation—that is, on what is at stake and who can do what about it—and partly also on our norms concerning reasonableness. For instance, suppose that something very important needs doing *right now*, and that a hypomanic individual with just the right capacities (both their *kind* and their *extent*) to do that thing is standing by our side, and that nobody else can do it but him. If what needs doing is indeed sufficiently important and urgent, if he knows that he has the requisite capacities to do it, and if in the immediately foreseeable future his capacities are not likely to wane, then mightn't *that* be precisely the sort of situation in which that individual's hypercapacity might warrant expecting him to help? I take it that the answer to this question is not an immediate "no"—rather, that it depends on the precise things that are at stake and on the sort of imposition that this would be on the hypomanic person's liberty[7]—and hence that the deciding factor is ultimately whether we think such an imposition to be reasonable or unreasonable.

Thus, although I now agree with Turner that people with hypomania should not be thought of as hyperresponsible, as long as cognitively enhanced people can control when their greater capacities will manifest themselves, and as long as they can accurately assess their own capacities, there will be no reason to deny that *they* would be hyperresponsible on account of their hypercapacities.

Superheroes and Supererogation

Given what I have said previously about the centrality of reasonableness assessments to assessments of responsibility, I will keep this discussion brief.

Basically, it is no part of my claim that there is only one thing that superheroes may do in any particular situation. A Utilitarian may claim that superheroes must always perform the optimific action, and that is indeed why Utilitarian analysis leaves no room for supererogation. But *my* claim is only that Superman's and Spider-Man's super powers make it reasonable to expect at least a little more of them than what can be expected of everybody else—I leave it up to the reader to decide *how much more* would be reasonable, but I assume that the reader will share my intuition that at least a little more can legitimately be expected of superheroes. And in light of the fact that it is reasonable to expect somewhat more of them than of others, when superheroes fail to meet these reasonable expectations—and again, I stress that I am not suggesting that we should expect *more* than what is justified and reasonable, but only what *is* justified and reasonable—we may reasonably blame them for failing to do what they ought to have done, or put another way, we may blame them for their failure to discharge their responsibilities.

Thresholds and the Reasonable Person Standard

It was argued earlier that because responsibility is a threshold concept—that is, because the law expects everybody's actions to come up to the same threshold standard of care, a standard that is defined by reference to the reasonable person—neither the availability of cognitive enhancement technologies, nor the fact that cognitively enhanced people would ex hypothesi have greater mental capacities, would have any bearing on their responsibility. However, I will now argue that what this analysis overlooks is that the objective standard of care that the law imposes on everyone is itself sensitive to (i.e., tracks) people's capacities. To see this, consider an example.

When various medical diagnostic techniques that are common today (e.g., x-ray imaging) were originally developed, their diagnostic value was not yet established, their risks were unknown, they were available only in select research laboratories, and all of these factors coupled with their expense meant that a medical practitioner who failed to use them to diagnose what was wrong with their patient could not have been considered negligent or reckless for failing to use them. But today, now that the clinical value of these diagnostic techniques is widely recognized, and they are relatively inexpensive, largely free of risk, and ubiquitously available, a medical practitioner who fails to request that her patient be tested using these techniques (when such a test could be diagnostically useful) *could* be deemed negligent and maybe even reckless.

As time went by, the applicable standard of care changed—it got updated—such that a medical practitioner who still only uses 19th-century diagnostic techniques today could be justifiably accused of gross negligence, recklessness, and maybe even stupidity. Put another way, what has happened in the intervening time is that as circumstances have changed—that is, as new enabling technologies have been developed that extend our capacity to diagnose a range of previously undiagnosable conditions—so, too, have the applicable standards of care been updated. What was reasonable a century ago was tied to our capacities back then. But now that our capacities have been extended through the progress of science, medicine, and new technologies, what is reasonable today is tied to our present extended capacities.

The significance of this example for the present discussion is that although Holmes' comments about the law's imposition of the same objective standard of care on everyone may still hold true, *that standard is pegged to today's capacities*. But, by analogy, once the ratio of costs and benefits associated with enhancing our cognitive powers is deemed to reach the right level—something that will inevitably happen as we develop safer and more effective cognitive enhancement techniques—the new objective standard may well include the expectation that under some circumstances (e.g., surgeons performing operations during extended night shifts in hospital, or perhaps particularly challenging kinds of surgery that require impeccable concentration, perceptiveness, and mental stamina), people should avail themselves of cognitive enhancement techniques. Furthermore, once people come to be expected to use these techniques in certain situations (e.g., in the operating room), the new standard of care that people will be expected to reach in those contexts will also be pegged to the standard that an *enhanced reasonable person* can be expected to attain.

Responsibility is indeed a threshold concept—for instance, to avoid being deemed negligent, a reasonable person must find our actions to be unobjectionable. But what a reasonable person will think is unobjectionable is not something that stands still over time, and it is certainly not something that is unaffected by our continually expanding capacities. Rather, capacities—both *to do things* and *to develop the capacity* to do previously unattainable but valuable things—play a crucial role in defining the objective reasonable person standard—the threshold of reasonableness is itself sensitive to capacities. Thus, the fact that responsibility is a threshold concept in no way undermines my claim that responsibility tracks hypercapacity, but rather it yet again helps to underscore and to emphasize it.

More Responsible *in What Sense*?

The observation that sometimes an intelligent person might be less rather than more responsible, and that an unintelligent person may be more rather than less responsible, reveals an important ambiguity (flagged in the introduction), which I will now discuss.

As I have argued elsewhere, the word "responsibility" is not really a single concept, but is rather a syndrome or a collection of interrelated concepts (2009b:44–45, 2010:82). Sometimes, "responsible" describes a kind of person, either in terms of their moral-agency-relevant capacities ("a fully responsible person") or in terms of their character virtues or lack thereof ("an irresponsible person"). At other times, it describes relations between events, either in causal terms ("his depression was responsible for his behavior") or in moral terms ("he is responsible for that accident"). And finally, it can also refer to a person's duties ("these are your responsibilities") or to how that person should be treated ("you will be held responsible for that accident"). Because the word "responsibility" has so many subtly different meanings, the claim that responsibility tracks hypercapacity (and equally capacity) is inherently ambiguous—that is, it is not clear in *which* of these senses responsibility is meant to track capacity—and what the present objection identifies is precisely this ambiguity.

However, once the different senses of "responsibility" are clearly distinguished from one another and the relations between them are spelled out,[8] what emerges is that only *some* senses of responsibility track (hyper)capacity. For instance, in regard to the first sense mentioned previously, which I call *capacity responsibility*—that is, the sense in which someone (e.g., a young adult) can be (or fail to be) a fully responsible person—responsibility clearly tracks capacity; in fact, it is almost an analytic truth that in this sense, responsibility tracks capacity because this sense of responsibility is all about what capacities a person has (Vincent 2008). Second, for the reason that has already been mentioned numerous times—that is, that what it is reasonable to expect of someone depends in part upon what capacities she possesses—I also think that claims about what responsibilit*ies* people have (i.e., the fifth sense mentioned earlier, which I call *role responsibility*) will also track claims about what capacities they possess. Third, because to be deemed responsible for a bad outcome (in the fourth sense mentioned earlier, which I call *outcome responsibility*), one must not only cause that outcome to occur but also be at fault in doing so—that is, one must act counter to one's just-mentioned role responsibilities—it is plausible that the more (or more demanding) role responsibilities a person has (something that depends in part on their capacities), the more likely he will be to breach one of them[9] and thus that he will be blamed for some state of affairs, and so for this reason it also seems plausible to me that, in this sense, responsibility also tracks capacity. Fourth, if outcome responsibility (the sense just mentioned in the previous sentence) tracks capacity—that is, if people who have more or greater responsibilities are more likely to encounter situations in which they are blamed for having breached their responsibilities—it also seems plausible that *liability responsibility* (the sixth sense mentioned earlier) tracks capacity—that is, because people are usually *held* responsible for those things for which they *are* responsible. Finally, given the meaning of claims that use the third sense of "responsibility" mentioned earlier, or what I call *causal responsibility*—that is, these are claims about how a person acted, and what effects those actions had in the world—it is plausible that a person who gains greater perceptive capacities and greater dexterity may indeed be less often causally responsible for various bad outcomes (because they will notice dangers, and because they will more skillfully control their actions), and so perhaps it might even be said that causal responsibility also tracks capacity. In summary, in the following five senses of "responsibility," responsibility indeed seems to track capacity: *capacity responsibility*, *role responsibility*, *outcome responsibility*, *liability responsibility*, and *causal responsibility*.

However, it is not clear that, in the remaining sense, responsibility also tracks capacity. In the second sense mentioned earlier—a sense I call *virtue responsibility*, which is exemplified by the italicized word in the sentence "John is such an *irresponsible* young man, even though he is a fully responsible person"—for precisely the reasons that were outlined in the objection that is currently under consideration, it is indeed doubtful that responsibility tracks capacity. Put a different way, I indeed see no prima facie reason to suppose that those people who are better at figuring out for themselves what they ought to do and who can better control their actions (those whose capacities are enhanced) will necessarily act more responsibly; rather, their superior knowledge may simply help them to find more ways to exploit people, it may alert them to the need to cover up what they did, or it may help them to think up more convincing excuses should they get caught. Similarly, I also see no prima facie reason to suppose that those people who are worse at figuring out for themselves what they ought to do will necessarily act more irresponsibly; rather, they may hold steadfastly to tried and true rules of thumb and as a result be model citizens. As I argue elsewhere, a fully responsible person (in the capacity sense) can be very irresponsible (in the virtue sense), and a person who is not yet fully responsible (in the capacity sense) may never the less be very responsible (in the virtue sense) (2009a).

Thus, in summary, my response to this objection is essentially a nod of agreement because it is indeed not true that in every sense of "responsibility," responsibility tracks capacity. Nevertheless, this objection has provided a very useful opportunity to disambiguate the main claim defended by this paper—that responsibility tracks hypercapacity.

On the Alleged Asymmetry of Praise and Blame

I will now argue that the problem with the seventh objection—that is, that due to the asymmetry of praise and blame, at most *only some* responsibility ascriptions (i.e., those that attribute blame) might track (hyper)capacity—is that it trades on an ambiguity between two different kinds of responsibility claims. However, rather than initially using my own terminology, let me instead first quote Gary Watson at length because in my view he has already addressed this issue:

> Wolf takes her [asymmetry] thesis to be supported by common judgments. If someone acts well because of a moral clarity and commitment so strong that she could not have done otherwise, then we still think her praiseworthy. But if she acts badly because her deprived childhood has rendered her unable to care about the moral considerations in question, then she is not thought to be blameworthy.
>
> [But t]he appeal of Wolf's thesis depends, I think, on a shift between the perspectives of *accountability* and *aretaic* appraisal. It is from the aretaic perspective that the agent in Wolf's example is plainly praiseworthy despite her supposed inability to do otherwise. For she conducts herself well, and that is to *be* praiseworthy. However, whether she deserves praise in the sense of some further favourable treatment in response to her virtuous *conduct* is more doubtful. Similarly, if we remain within the same perspective, the victim of the deprived childhood is blameworthy as well, since his conduct reflects badly upon him as a moral agent. From within this outlook, there is no difference. [Thus, t]he sense of asymmetry results from a shift to the perspective... (2004:283, emphasis altered)

Put in terms of the terminology and responsibility concepts I introduced in the previous section, what I take Watson to be saying is that the appearance of there being an asymmetry between praise and blame judgments can only be sustained when we equivocate between assessments of a person's *virtue* responsibility and assessments of his *outcome* responsibility (the latter of which rely on assessments of his capacity responsibility). However, when *the same* perspective or sense of responsibility is retained—specifically, when we stick to attributions of what I called *outcome responsibility*—there is no asymmetry between judgments of praise and blame. If neither party had the required capacity, then it is only a fluke that the person who did the good thing succeeded, and it is not the other person's fault that she couldn't have avoided the bad outcome—that is, in neither case is the person responsible (praiseworthy or blameworthy) for the state of affairs that obtains. On the other hand, if both parties possessed the required capacities, then both would be responsible: first, the person who does the good thing wouldn't have just done it through a *fluke*, and so the good outcome could be attributed to her as something of her doing; second, the person who does the bad thing would have done it despite the fact that he could have avoided doing it, and so the bad outcome could also be attributed to him as something of his doing. Given that there is no asymmetry between praise and blame when the *kinds* of responsibility assessments that we make stay the same (i.e., when we do *not* equivocate between assessments of *outcome* and *virtue* responsibility), this asymmetry can't support the objection that at best only blame judgments, but not praise judgments, would track (hyper)capacity.

Nevertheless, it is worth noticing that even if my own and Gary Watson's distinctions were cast aside, and thus even if it *were* true that only blame-focused assessments of responsibility would track hypercapacity, this would still be no small discovery. After all, it would entail that under certain circumstances, people *could* be legitimately expected to enhance themselves, that once enhanced they *could* legitimately be expected to perform at a higher standard of care than nonenhanced individuals, that they *could* be blamed when things go wrong (either for not enhancing themselves, or for performing at a lower than reasonable level for an enhanced person), and that they *could* perhaps even be held responsible (i.e., liable) when things go wrong. In other words, even if most of this section's argument were cast aside, there would still be very serious practical ramifications to blame judgments tracking hypercapacity.

From *Can* to *Ought*

Do my arguments in support of the claim that responsibility tracks hypercapacity commit me to the objectionable claim that *can implies ought*? I do not believe that they do.

The simple idea that my analysis rests on—an idea to which I have returned throughout the foregoing discussion—is that what responsibilities a person has, is something that is *in part* determined by what capacities they possess.[10] Put another way, the idea is not that we can read off a person's responsibilities simply from the assessment of their capacities (this would indeed be an instance of *can implies ought*), but it is rather that in determining what responsibilities a person has, we should, among other things, consider what capacities they possesses—that is, that *can*, taken together with a range of other considerations, implies *ought*. But perhaps even this is not a sufficiently clear way of stating my position, so let me now elaborate on how I conceive of the relationship between *can* and *ought* a bit further.[11]

There are actually two ways of conceiving the relationship between capacity (can), duty (ought), and blame (responsibility). On one account, capacity plays a *positive* role by *generating* duties: we ought to do what we have most reason to do, and what we can and can't do (presumably along with a range of many other things) generates the reasons that we have to do various things.[12] On this first account, if I cannot save a child from drowning—perhaps because I do not know that she is drowning, or because I can not swim, or because I do not have a rope to throw to her—then in the final analysis, it is simply not true that I ought to save her (unless I am responsible for the fact that I cannot do this); thus, the reason I would not be blameworthy for not saving her is that I did not, in the final analysis, actually have that saving duty. On the other account, though, capacity plays a *negative* role by *regulating* duties: regardless of the source of our duties, on this second view our *in*capacities can justify or perhaps excuse departures from those duties. On this latter account, the three cited considerations—that is, they don't know that the child is drowning, they can't swim, or they have no rope—do not outweigh the saving duty, but rather they justify or provide an excuse for departing from it—that is, the reason I would not be blameworthy is because although I did have the saving duty, my incapacity provided a justification or an excuse for departing from it. Two advantages of the second view is that only it has an explicit place for justifications and excuses (which occupy an important place in legal discourse), and it arguably also more adequately captures the rich structure of practical reasoning in which some considerations *discount*, *undermine*, and *invalidate* (rather than just *outweigh*) other considerations. Nevertheless, I suspect that both views of the relationship between capacity, duty, and blame generate the same conclusions about when a person is to blame and when they are not to blame (and thus are responsible or not responsible for some state of affairs), and because I find it simpler to explain ideas within the conceptual framework of the first account, that is the account of the relationship between *can*, *ought*, and *blame* reflected in this paper's foregoing discussion.

Hence, although I acknowledge that my position may give off the impression that I endorse *can implies ought*, I hope that this section's discussion sets aside this worry.

SUMMARY, IMPLICATIONS, AND CONCLUSION

The point of this paper was to investigate whether cognitive enhancement might result in enhanced responsibility. The kernel for this idea came from two observations: first, that responsibility tracks capacity—that is, that responsibility diminishes when mental capacities are lost, and that it is restored when they are subsequently regained; and second, that cognitive enhancement increases our capacities even further, beyond the normal range. Taken together, these observations prompted the idea that if reductions in capacity result in reduction of responsibility, and restorations of capacity result in restoration of responsibility, then perhaps enhancements of capacity result in the enhancement of responsibility. To motivate the discussion further, I presented four examples that at least prima facie suggest that responsibility does indeed track hypercapacity and not just capacity—precocious developers, education and training, hypomania, and superheroes—and in the following two sections, I investigated the resilience of our intuitions about those examples. Although I have made one important concession along the way—namely, I have admitted that there is one sense of "responsibility" (what I call *virtue responsibility*, which matches up with Gary Watson's *aretaic* face of responsibility) in which responsibility tracks neither capacity nor hypercapacity—I believe that as far as the other senses of responsibility are concerned, the foregoing arguments have confirmed the initial intuition that responsibility does indeed track hypercapacity.

But what of it? Who cares?

From the perspective of a chapter in a book about neuroscience and legal responsibility, I take it to be a novel observation that cognitive enhancement might result in hyperresponsibility. Quite a few chapters in this volume, and indeed in the wider current literature, debate whether neuroscientific discoveries about the brain-based foundations of human cognition and volition spell doom for the very notion of responsibility—that is, whether neuroscientific findings show that responsibility does not exist (and most, though not all, deny that this is the case). Many papers also explore the question of whether, and if so how, neuroscientific techniques and technologies might be used to help the law to *assess* people's responsibility (for instance, by detecting deception, or by measuring a person's mental capacities). Finally, some papers also touch on the idea that perhaps the techniques and technologies of neuroscience might be used to *restore* people's responsibility (for instance, so that they can stand trial, be punished, or maybe even returned to society once the condition that causes their irresponsibility has been "treated," or by helping to rid people of addictions that diminish their volitional capacities). However, few people have considered the prospect that neuroscience might be used to *enhance* responsibility, rather than just to *deny* that it exists, to *assess* it, or to *restore* it, but yet this seems to me like the next natural question to ask in a *deny-assess-restore-enhance* progression. Thus, I hope that this paper not only offers a first serious exploration of this idea but also provides a framework (via the distinctions between the different senses of the term "responsibility" and the associated discussion) for thinking about this topic—a framework that others can also use in their own analyses.

However, I also think that this finding—that responsibility tracks hypercapacity—has many serious practical ramifications, especially when applied to the legal context, and to draw these out I shall now return to the questions I raised at the start of this paper.

First, I asked whether a person who becomes cognitively enhanced might become "hyperresponsible," and if so, then in what sense? My answer to this question is that in some senses he would become hyperresponsible, but in others he wouldn't, or at least he wouldn't necessarily become hyperresponsible. For instance, as long as the capacities that were gained through cognitive enhancement were relevant to that person's status as a moral agent—for instance, if they increased his ability to reason about what he ought to do, and to be moved to action through that reasoning—then in a very straightforward sense such a cognitive enhancement would make the said person more *capacity responsible.*

Second, I asked whether cognitively enhanced people might acquire new responsibilities that they otherwise wouldn't have had? In response, if the cognitive enhancement indeed made the person more capacity responsible (see above paragraph), then in virtue of this, it might indeed be reasonable to expect more of her—that is, it might be reasonable to expect her to discharge *role responsibilities* that previously she couldn't have been expected to discharge, or maybe the strength or weightiness of those responsibilities (i.e., how important it is that they be discharged, and how blameworthy the person would be for failing to discharge them) would also increase. Whether it would or wouldn't be reasonable to impose new (or weightier) role responsibilities onto enhanced people depends, as I argued in the last sub-section of the previous section, on a great many things, but at least one of these things is what the person's greater capacities make it possible for her to do.

Third, I also asked whether, as a consequence of becoming cognitively enhanced, a person might be legitimately blamed when he fails to discharge those greater responsibilities—in my own terminology, whether the person might become more prone to being judged *outcome responsible* for various states of affairs. Here my answer is a guarded affirmative one because the more responsibilities that we have (or the weightier those responsibilities become), the more likely it might be that we will infringe one of them (or not discharge them properly). However, the reason this answer is guarded rather than unqualified is because, as I explained earlier, we can only impose responsibilities on people if it is reasonable to expect them to discharge those responsibilities in the first place. But if, ex hypothesi, it is just as reasonable to expect enhanced people to discharge their responsibilities as it is to expect us nonenhanced folk to discharge our responsibilities, then it had better not turn out that it is, after all, more likely that a person with more responsibilities (or with weightier responsibilities) will fail to discharge one of them (properly) on account that her greater responsibilities would burden her more, because that would violate the original assumption that it was, all things considered, reasonable to impose those responsibilities onto her in the first place (see my discussion at the start of the previous in section). Nevertheless, I leave open the possibility that a person with more responsibilities (or with weightier responsibilities) might be more prone to accidentally failing do discharge (properly) one of them, if only because with more (or with weightier) responsibilities, there is greater opportunity to not discharge one of them (properly).

Fourth, would that—that is, their being more prone to assessments of outcome responsible for various states of affairs—increase the likelihood that they will subsequently be held responsible (i.e., liable) when things go wrong? If our answer to the previous question is affirmative, then our answer to this question should also be affirmative. On the other hand, if our answer to the previous question is negative, then our answer to this question should also be negative. Given that *liability responsibility* is usually imposed onto those people who are deemed to be outcome responsible for the thing in question, our answer to this question should track our answer to the previous question.

Fifth, however, as I already explained above, I do not believe that hyper-responsible people will necessarily be less irresponsible. As I hinted at earlier, I suspect that whether someone is responsible or irresponsible (in the *virtue responsibility* sense) depends more on their character than on their mental capacities. But this position is not unproblematic. For instance, as I acknowledge elsewhere (2010:90–91, 94), I am not even sure how we are supposed to navigate the conceptual distinction between people's character and their capacities—after all, does the fact that I have a brain-based inability to control my angry outbursts show that I have an *incapacity* and thus that I should be excused, or does it show that I have a serious and maybe even a condemnable *character* flaw? On the face of it, both interpretations seem plausible (Maibom 2008; Reimer 2008), and I am not sure how to even begin to address this conceptual difficulty (though see Vincent 2011a). However, although I am not personally fond of the idea that people might be condemned merely for *who they are* rather than just for *what they do*, I do believe that ordinary language and legal practice recognize both of these kinds of responsibility assessments. As I argue elsewhere:

> Our language must make it possible to *criticize* someone for the fact that they are irresponsible (that they lack *virtue* responsibility) but at the same time to also attribute responsibility to them for the things that they do (on account that they possess *capacity* responsibility), or to *praise* someone for the fact that they are so responsible (that they possess *virtue* responsibility) but without this necessarily having to entail that they are legitimate targets for attributions of responsibility for the things that they do (since they may lack *capacity* responsibility). Put another way, our language must make it possible to criticize someone's character without this entailing that they are not responsible moral agents, and it must allow us to praise someone's character without this entailing that they are responsible moral agents. However, these things can only be done if we clearly distinguish the concepts of capacity responsibility and virtue responsibility... (2009a:124–5)

Ordinary language and the law's responsibility practices reveal that assessments of responsibility sometimes attach to character, and as long as the distinction between character and capacity remains within our thinking, there will be good reason to doubt that enhancements of capacity must necessarily result in enhancements of responsibility in the character-relevant sense of responsibility. Some people might disagree with me on this point. For instance, in his paper "Moral Enhancement," Thomas Douglas argues that certain brain-based interventions might make us morally better people (2008). However, I suspect that Douglas' argument conflates two issues—the question of whether cognitive enhancement might result in greater *capacity responsibility* (I said earlier that it almost certainly will, as long as the capacities that are enhanced bear on our abilities as moral agents) and the idea that cognitive enhancement might result in greater *virtue responsibility* (I suggested previously that this need not necessarily occur). But rather than attempting to resolve this question here and now, I would prefer to acknowledge that this is indeed an important issue and then set it aside to see what others say about it.

Sixth, and on a very practical note, I asked "if cognitive enhancers indeed improve mental performance, then might it eventually become morally and legally obligatory for some people to cognitively enhance themselves under some circumstances?" In response, I now endorse my original comments, namely, that given how much is at stake in an operating room, on a military battlefield, and in long-haul flights—and undoubtedly in many other contexts—it could indeed be argued that surgeons, soldiers, and airplane pilots, respectively, at least sometimes have a moral duty to take cognitive enhancers to ensure the highest performance possible, and that they would be negligent or even reckless if they didn't enhance themselves when the stakes are particularly high. Legal, and arguably moral, determinations of responsibility hinge critically on assumptions about what it is reasonable to expect of people in different circumstances (Vincent 2010:80–85, 87–88). But as circumstances change—for instance, as new enabling technology is introduced that extends what is possible without imposing significant risks or costs—so do our expectations. Put another way, people are blamed when they act counter to how it is reasonable to expect them to act, and how it is reasonable to expect people to act depends in part on what it is possible for them to do, on the benefits that doing it would produce, and on the costs of doing it (e.g., financial expense, risks imposed, pain endured). Thus, as I explained above, once the benefits of cognitive enhancement are deemed to outweigh the risks of their unwanted side effects, people in socially important roles may indeed come to be morally and perhaps even legally required to enhance themselves in certain situations, cognitively enhanced people (and those who *ought* to have cognitively enhanced themselves) may be expected to satisfy higher standards of care than nonenhanced people, and they may even be deemed negligent or reckless for failure or refusal to do so, and possibly even sanctioned.

****

This paper has discussed one effect that cognitive enhancement might have on moral and legal responsibility. My chief contention is that for precisely the same reason as responsibility tracks capacity, so too responsibility tracks hypercapacity. Put another way, just as responsibility is reduced when mental capacities are diminished, and just as responsibility is restored when those capacities are regained, so, too, responsibility can be enhanced when mental capacities are enhanced. I have said nothing about precisely which enhancements would enhance responsibility because this depends on two things. First, much research still needs to be done on precisely what capacities are required for responsible moral agency, and until we know this, we will hardly be in a position to enhance the right capacities. Second, responsibility is not a thing—a single unified concept—but is rather a syndrome of concepts, and so before we can comment on how a particular enhancement will affect responsibility, we must first determine precisely what *kind* of responsibility is at issue. I have tried to make headway on this latter task, and I hope that others may build upon this work to further investigate how cognitive enhancement affects moral and legal responsibility.

NOTES

1. The idea that responsibility tracks capacity appears throughout the compatibilist literature (Dennett 1984; Fischer and Ravizza 1998; Glannon 2002; Wallace 1994) and in philosophy of law (Hart 1968; Honoré 1999), and it also seems intuitively plausible, and so I shall not defend it here. If interested, see Vincent 2008, 2011b, 2010.
2. For instance, Aharoni et al. (2008) and Glannon (2005) suggest some ways in which neuroscientific techniques might help us to *assess* people's responsibility, though Eastman and Campbell (2006) and Morse (2006) are critical about whether neuroscience can be used for this purpose. In regard to *restoration*, Latzer (2003) discusses the use of psychopharmaceuticals to restore convicts' capacity to be executed, Tancredi's (2005) discussion of the dubious (in his opinion) distinction between "madness" and "badness" suggests a position from which irresponsibility is seen as something to be treated rather than punished, and Lekovic (2008) discusses (without endorsing) various neurosurgical interventions such as deep brain stimulation and surgical ablative treatments used in China to treat addicts, which may be adopted as a means to rehabilitate criminals rather than merely to punish or isolate them.
3. These examples are meant to be *demonstrative* rather than *exclusive* because a lot can also be at stake in other contexts—e.g., maybe teachers owe it to society to enhance themselves.
4. One exception is Oliver Warren and colleagues' (2009) paper, which considers it possible that surgeons may one day be expected to enhance themselves. In a related context, James Blair also raises the question that if "we consider moral responsibility and blame proportional to the level of [emotional development, then] what should we do about individuals who show heightened emotional responses, for example, patients with specific anxiety disorders? There are already some data that inhibited children show heightened conscience development. If such children commit moral transgressions, should we regard them as more morally responsible and more blameworthy than regular folks?" (2007:150, internal citations omitted). Although Blair seems to endorse a negative answer, I cite him to demonstrate that others have pondered similar questions.
5. Everyone, that is, within the same category, such as *all surgeons*, *all pilots*, or *all soldiers*.
6. Libertarians are bound to disagree, but libertarianism is an extreme position (Nozick 1974).
7. Judy Jarvis Thomson's (1971) *violinist* example comes to mind.
8. In the ensuing discussion, I presuppose an analysis of responsibility that I develop elsewhere (2010:80–85, 2011c), and the reader is directed there for clarifications.
9. However, see my comments on this in section 4 below at point three.
10. ...and by what capacities they *ought* to possess—see Vincent 2010:95.
11. The following paragraph is borrowed, with modifications, from my (2011a) publication.
12. An inference is first made from what *capacities* I have to what I have *reason* to do, and then another inference is made from what I have *reason* to do to what I *ought*

to do—i.e., the reasoning moves from *capacity* claims via *reasons* claims to *ought* claims. In both inferences, there is room to take a range of other considerations into account—my particular capacities (e.g., to swim) may give me reason to do a range of things (e.g., to laugh at the drowning child, or to save the drowning child), and other reasons (e.g., that I am running late for work) may compete with the capacity-based reasons for determination of what I ought to do—and this is why my reasoning does not presuppose that *can* implies *ought*. I owe this analysis of the relationship between *can* and *ought* (via *reasons*) to Rosemary Lowry who discusses this topic in much greater depth and precision (2011). Peter Vranas (2007) also develops a similar nuanced account of the relationship between capacity and obligation.

REFERENCES

Aharoni, E., C. Funk, et al. (2008). "Can neurological evidence help courts assess criminal responsibility? Lessons from law and neuroscience." *Annals of the New York Academy of Sciences* 1124: 145–160.

Appel, J. M. (2008). "When the boss turns pusher: A proposal for employee protections in the age of cosmetic neurology." *Journal of Medical Ethics* 34: 616–618.

Blair, R. J. R. (2007). "What emotional responding is to blame it might not be to responsibility." *Philosophy, Psychiatry & Psychology* 14(2): 149–151.

Chatfield, G., K. Heffernan, et al. (2002). Competence to stand trial: An eleven-year case study. In: *Social Work and the Law: Proceedings of the National Organization of Forensic Social Work, 2000*. I. Neighbors, A. Chambers, E. Levin, G. Nordman and C. Tutrone. Binghamton, NY, The Haworth Press, pp. 13–19.

Chatterjee, A. (2009). "Is it acceptable for people to take methylphenidate to enhance performance? No." *British Medical Journal* 338: 1532–1533.

DC. (2010). "DC Comics: Superman." Retrieved January 28, 2010, from http://www.dccomics.com/sites/superman/.

Dennett, D. C. (1984). *Elbow Room: The Varieties of Free Will Worth Wanting*. Cambridge, MA, MIT Press.

Douglas, T. (2008). "Moral enhancement." *Journal of Applied Philosophy* 25(3): 228–245.

Eastman, N., and C. Campbell (2006). "Neuroscience and legal determination of criminal responsibility." *Nature Reviews Neuroscience* 7(April): 311–318.

Fischer, J. M., and M. Ravizza (1998). *Responsibility and Control: A Theory of Moral Responsibility*. Cambridge, UK, Cambridge University Press.

Forlini, C., and E. Racine (2009). "Disagreements with implications: Diverging discourses on the ethics of non-medical use of methylphenidate for performance enhancement." *BMC Medical Ethics* 10(9): 13 pages.

Glannon, W. (2002). *The Mental Basis of Responsibility*. Aldershot, UK, Ashgate Publishing Limited.

Glannon, W. (2005). "Neurobiology, neuroimaging, and free will." *Midwest Studies in Philosophy* 29: 68–82.

Hall, W. D., and J. C. Lucke (2010). "The enhancement use of neuropharmaceuticals, more scepticism and caution needed." *Addiction* 105: 2041–2043.

Hart, H. L. A. (1968). IX. Postscript: Responsibility and retribution. In: *Punishment and Responsibility.* Oxford, UK, Clarendon Press, pp. 210–237.

Holmes, O. W. (2000). "The common law." Retrieved May 12, 2004, 2004, from http://www.gutenberg.net/etext/2449.

Honoré, T. (1999). *Responsibility and Fault.* Portland, OR, Hart Publishing.

Larriviere, D., M. A. Williams, et al. (2009). "Responding to requests from adult patients for neuroenhancements: Guidance of the Ethics, Law and Humanities Committee." *Neurology* 73(17): 1406–1412.

Latzer, B. (2003). "Between madness and death: The medicate-to-execute controversy." *Criminal Justice Ethics* 22(2): 3–14.

Lekovic, G. P. (2008). "Neuroscience and the law." *Surgical Neurology* 69: 99–101.

Lowry, R. (2011). Blame, reasons and capacities. In: *Moral Responsibility: Beyond Free Will and Determinism.* N. Vincent, I. van de Poel, and J. van den Hoven. New York, Springer, pp. 71–81.

Maibom, H. L. (2008). "The mad, the bad, and the psychopath." *Neuroethics* 1(3): 167–184.

Marvel. (2010). "Spider-man (Peter Parker)." Retrieved January 28, 2010, from http://marvel.com/universe/Spider-Man_%28Peter_Parker%29.

Morse, S. J. (2006). Moral and legal responsibility and the new neuroscience. In: *Neuroethics: Defining the Issues in Theory, Practice, and Policy.* J. Illes. Oxford, UK, Oxford University Press, pp. 33–50.

Nozick, R. (1974). *Anarchy, State and Utopia.* New York, Basic Books.

Persson, I., and J. Savulescu (2011). "The perils of cognitive enhancement and the urgent imperative to enhance the moral character of humanity." *Journal of Applied Philosophy* 25(3): 228–245.

Racine, E., and C. Forlini (2009). "Expectations regarding cognitive enhancement create substantial challenges." *Journal of Medical Ethics* 35(8): 469–470.

Reimer, M. (2008). "Psychopathy without (the language of) disorder." *Neuroethics* 1(3): 185–198.

Sandberg, A., W. Sinnott-Armstrong, et al. (2011). Cognitive enhancement in courts. In: *Oxford Handbook of Neuroethics.* J. Illes and B. J. Sahakian. Oxford, UK, Oxford University Press.

Schermer, M., I. Bolt, et al. (2009). "The future of psychopharmacological enhancements: expectations and policies." *Neuroethics* 2(2): 75–87.

Tancredi, L. R. (2005). The bad and the mad. In: *Hardwired Behavior: What Neuroscience Reveals About Morality.* New York, Cambridge University Press, pp. 143–161.

Thomson, J. J. (1971). "A defence of abortion." *Philosophy & Public Affairs* 1(1): 47–66.

Turner, A. J. (2010). "Are disorders sufficient for reduced responsibility?" *Neuroethics* 3(2): 151–160.

Vincent, N. (2008). "Responsibility, dysfunction and capacity." *Neuroethics* 1(3): 199–204.

Vincent, N. (2009a). "Responsibility: Distinguishing virtue from capacity." *Polish Journal of Philosophy* 3(1): 111–126.

Vincent, N. (2009b). "What do you mean I should take responsibility for my own ill health?" *Journal of Applied Ethics and Philosophy* 1: 39–51.

Vincent, N. (2010). "On the relevance of neuroscience to criminal responsibility." *Criminal Law and Philosophy* 4(1): 77–98.

Vincent, N. (2011a). Madness, badness and neuroimaging-based responsibility assessments. In: *Law and Neuroscience, Current Legal Issues Volume 13*. M. Freeman. Oxford, UK, Oxford University Press, pp. 79–95.

Vincent, N. (2011b). "Neuroimaging and responsibility assessments." *Neuroethics* 4(1): 35–49.

Vincent, N. (2011c). A structured taxonomy of responsibility concepts. In: *Moral Responsibility: Beyond Free Will and Determinism*. N. Vincent, I. van der Poel, and J. van der Hoven New York, Springer, pp. 15–35.

Vranas, P. B. M. (2007). "I ought, therefore I can." *Philosophical Studies* 136(2): 167–216.

Wallace, R. J. (1994). *Responsibility and the Moral Sentiments*. Cambridge, MA, Harvard University Press.

Warren, O. J., D. R. Leff, et al. (2009). "The neurocognitive enhancement of surgeons: an ethical perspective." *Journal of Surgical Research* 152(1): 167–172.

Watson, G. (2004). Two faces of responsibility. *Agency and Answerability*. G. Watson. Oxford, UK, Oxford University Press, pp. 260–288.

Williams, G. (2008). "Responsibility as a virtue." *Ethical Theory and Moral Practice* 11(4): 455–470.

Wolf, S. (1980). "Asymmetrical freedom." *Journal of Philosophy* 77(3): 151–166.

14

Guilty Minds in Washed Brains?

Manipulation Cases and the Limits of Neuroscientific Excuses in Liberal Legal Orders

CHRISTOPH BUBLITZ AND REINHARD MERKEL

INTRODUCTION

Novel means to intervene into minds and to modulate thoughts, emotions, and behavioral dispositions raise at least two questions for the law: What kinds of interventions into other peoples' minds are permissible? And how do impermissible interventions in the inner workings of a person's mind affect her responsibility for actions springing from manipulated preferences?

An illustration: Rumors have it that some Casinos secretly spray the odorless substance oxytocin in gambling halls to increase players' trust in their luck. Suppose oxytocin disposes gamblers to be overconfident and to take higher risks—would this be permissible, or could players claim their stakes back?[1] What if they commit illicit acts under oxytocin's influence—are they responsible? And what is at stake for society at large if the growing body of neuroscientific and psychological knowledge of decision making and behavior is increasingly used for manipulative purposes and private gain?

Legal scholarship will have to address these questions, which branch out into constitutional, civil, and criminal law. In this chapter, we deal with what we take to be the fundamental question bearing on all the others: Are manipulated persons responsible? We want to approach it from two angles: moral philosophy as well as legal doctrine. Both provide strikingly different answers. In the rich philosophical debate, it is widely taken for granted that manipulated agents are not responsible, whereas the law rarely addresses the issue and even more rarely excuses defendants merely because they were unduly influenced.

Contrasting these largely unconnected debates reveals that they may benefit from taking note of each other. Legal thinking may cast some doubts on philosophical premises and positions centering too narrowly on mental states and their history, undervaluing the functions of responsibility and its interdependences with freedom in various contexts. Conversely, the philosophical concerns help to unveil principles of liberal legal orders that often remain implicit: liberal states seek to govern conduct of citizens primarily not by preventive police measures but through norms. A societal setup based on the freedom of the individual is viable only if everyone can be expected to abide by the law, and this presupposes corresponding capacities of its citizens. This foundational liberal premise sets limits to the relevance of neuroscience in matters of legal excuses. And of course, it stands to be argued whether this is the only type of society desirable, but this is ultimately a political matter to which neuroscience cannot contribute much.

We shall consider manipulation cases in comparison to regular coercion cases, in regard to civil law responsibility as well as in light of the idea that freedom and responsibility are inextricably related. Conclusively, we present a theory of punishment built on the rationale of securing normative expectations and the validity of a broken norm. It provides good reasons against granting excuses merely because of the manipulative history of preferences, even though this might seems to contravene moral intuitions. At any rate, manipulation cases provide interesting insights for a—yet to be fully developed—compatibilist theory of excuses. The more neuroscientific findings point to the susceptibility of the mind to manipulative interferences, philosophical worries about fairness and legitimacy of responsibility will, in one way or another, come up in court cases and make judges, jurors, and jurists wonder whether the position of the law is correct. It's time to get the matter straight.

MANIPULATION CASES IN PHILOSOPHICAL PERSPECTIVE

To begin with, here is a typical manipulation case as discussed in moral philosophy (Pereboom 2003:112):

Case 1: P was created by neuroscientists who can manipulate him directly, but he is as much like an ordinary human being as is possible, given this history. Suppose the neuroscientists manipulate him to undertake the process of reasoning by which his desires are brought about and modified, directly producing his every state from moment to moment. The neuroscientists manipulate him by pushing a series of buttons just before he begins to reason about his situation, thereby causing his reasoning process to be rationally egoistic.

P is not constrained to act in the sense that he experiences irresistible desires—the neuroscientists do not provide him with irresistible desires—and he does not think and act contrary to character because he is often manipulated to be rationally egoistic. He forms the desire to kill V, and this first-order desire conforms to his second-order desires. Also, his reasoning process is moderately reason-responsive P kills V.

Is P responsible for killing V? First, it is important to note that in regular manipulation cases, perpetrators fulfill all conditions usually considered necessary for responsibility. In the moment of action, P is a rational and reason-responsive agent, not suffering from any incapacity and endorsing his volition to kill V. Nevertheless, he appears to be a victim rather than a perpetrator and not a suitable target for negative sanctions. This result is widely agreed upon in philosophical circles, but the reasons for it are fiercely contested. To understand why so much attention is given to these cases, let us take a look at the broader argument in which they are embedded.

Background: Manipulation Cases and Free Will

Manipulation cases play a pivotal role in the free-will debate. Roughly, authors affirming that responsibility can be reconciled with determinism are compatibilists, and this is denied by their incompatibilist opponents. Incompatibilists draw on manipulation cases to expose weaknesses and counterintuitive consequences of compatibilist views. They claim that acting from preferences shaped by a deterministic universe is not relevantly different from acting from manipulated preferences. In both cases, agents do not have "ultimate control" over the processes forming preferences and volitions. In this vein, incompatibilists create an array of cases unfolding in the following fashion: After a scenario like case 1 in which agents are (intuitively) not responsible for their actions, further variations are introduced in which the powers of manipulators gradually vanish. In the end, there is no manipulator, only a world of deterministic causal processes. This challenges compatibilists to pinpoint the step at which a poor victim of manipulative forces turns into an autonomous and fully accountable agent. Here are some variations to which we will resort subsequently.

Case 2: From his early days, P suffers from a brain disorder and has to live with an implanted deep brain stimulator (DBS). For the sake of research, neuroscientists have secretly installed one electrode of the DBS in proximity to the reward pathway in the nucleus accumbens. They can condition and fashion P's tastes and preferences by triggering his reward center, just like in the famous rat experiments in the 1950s.[2] Over time, they condition P into a rational egoistic, strongly pleasure-seeking person, not shying away from violence against others. Acting from these traits, P kills V.

Case 3: Through control of DBS devices, neuroscientists can elicit volitions, the inner feeling of wanting to commit particular acts. One day, they instill in P the volition to kill V. Because he does not have any countervailing preferences or moral reservations, he does so.[3]

Case 4: P wears an arm prosthesis controllable through a brain-computer interface (BCI). The BCI monitors P's neuronal activity; when it detects motor cortex signals resulting from P thinking about moving his arm, it triggers movement of the prosthesis. To avoid malfunctioning and optimize detection, the BCI relays the recorded signals to a computer where scientists analyze the data. Through this, they can observe P's brain almost in real time. One day, by pressing a button, scientists instruct the prosthesis to move, killing V.

Variation 4a: P is contemplating killing V. Neuroscientists monitor his brain signals and are prepared to intervene to make P kill V. However, P decides to do so on his own. The neuroscientists remain idle bystanders.

Case 5: Hackers have found a way to remote-control P's DBS.[4] When they turn it on (of which P is not aware), P experiences rapid changes in mood, energy levels, and preferences, which he attributes to other situational factors. One day, this induces P to become rationally egoistic, aggressive, and hostile toward V, who dies from P's attack.

Case 6: Psychologists abduct P. Through pharmaceuticals and sophisticated psychological persuasion, they extensively transform P's preference structure into one of rational egoism. P kills V.

Variation 6b: P is an orphan child. Roaming around the streets, he is picked up by N, a war veteran full of hatred against Muslims. N takes care of P and indoctrinates him with anti-Muslim beliefs and an aggressive and egoistic preference structure. One day, P's preferences make him kill Muslim V.

Remarkably, neuroscience might be about to turn into reality some mind interventions that were a couple of years ago only conceivable in philosophical thought experiments. In all of these cases, manipulators (M) severely influence perpetrators (P) to kill victims (V), raising questions about P's responsibility. To legal scholars, these stories may confer the false impression that the normative problem only concerns rare and extreme scenarios and, as the proverb has it, that such hard cases make bad law. However, these cases are of great illustrative value and let the salient normative features of manipulation cases obtrude. In the end, nothing in our argument hinges on the means employed. DBS may well be replaced by other electrochemical interventions into brains (e.g., TMS, pharmaceuticals) or even by intensive psychological interferences. If one continues these cases with gradually decreasing strength of manipulation, one may end up here:

Case X: P grows up as an ordinary human being. Because of his genetic dispositions (monoamine oxidase A [MAOA]), poor upbringing, and a myriad of ordinary influences, he ends up recklessly rationally egoistic and kills V.[5]

The Incompatibilist Argument

Manipulated persons, incompatibilists suggest, may act according to their preferences in cases 1 to 6, but because these preferences have been brought about through manipulation, they are too flimsy to ground responsibility. This leads to their first premise:

(Inc 1) Manipulated agents are not responsible.

Incompatibilists then go to great lengths to argue that there is *no relevant difference* between acquiring preferences through manipulation or through living in a deterministic world (Kane 1996). After all, so they contend, if determinism is true, every action can be fully inferred from the state of the universe at the moment of the agent's conception and deterministic laws of nature (and perhaps psychological dynamics). At any rate, agents do not have ultimate control over the forces shaping their preferences; they are neither prime movers unmoved nor ultimate creators of their destiny. Thus:

(Inc 2) There is no relevant difference between a deterministic world and manipulation as sources of human action.

From which follows:

(Inc 3) In a deterministic world, agents are not responsible (incompatibilism).

Those incompatibilists believing in free will (libertarians) suggest that what is necessary is the possibility to act otherwise (Frankfurt's "principle of alternative possibilities", PAP, 1969). Truly speaking of freedom and responsibility, in their view, requires an indeterministic element somewhere in the process from preference formation and decision making to action.[6] Hence, libertarians add:

(Inc 4) Somewhere along the way, responsibility conferring indeterminism comes in.

A Compatibilist Answer: History

By definition, compatibilists reject (inc 3), so they must deny either (inc 1) or (inc 2). Most accept (inc 1) and deny (inc 2). There seems widespread agreement that if there is any plausible candidate for a categorically nonresponsible agent, it is a manipulated person. Few authors are prepared to bite the bullet and argue for a so-called hard-line reply, rejecting (inc 1) and holding manipulated agents responsible (McKenna 2008). The prima facie soundness of the incompatibilist reasoning has been acknowledged even by such outspoken compatibilists as John Fischer, who concedes that manipulation cases are compatibilists' "dirty secret" (2000:390).

For rebuttal, many compatibilists introduce further conditions of responsibility, claiming that responsibility is a *historical* phenomenon requiring that agents have acquired their preferences in certain ways. Otherwise, they do not act from their own "authentic" preferences and cannot be held responsible. As a consequence of historical views, responsibility cannot be assessed by time-sliced, "snapshot" properties of brains and minds at the moment of action, but only in hindsight by reflecting on the formation of respective preferences. The historical conditions necessary for responsibility are spelled out differently.[7] One of the best proposals comes from Alfred Mele, to whom preferences must not have been brought about by causal routes *bypassing* agents' control over their mental lives (1995). Fischer and Ravizza hold that agents have to act from their "own mechanisms", which cannot be formed by pills, electronic stimulation of the brain, or brainwashing (1998:236). To Fischer, it is "surely one of the boundary conditions on a successful general analysis of moral responsibility" that it excludes severe manipulations of the brain, hypnoses, and the like (2006:53). Thus, many authors of both camps concur in denying the responsibility of manipulated agents. Without rehearsing the skillful arguments advanced on both sides, we take the acceptance of (inc 1) and the nonresponsibility of P in cases 1 to 6 as the mainstream position in moral philosophy.

We shall argue, however, that compatibilists need not worry too much. Manipulation cases are in principle insufficient to provide for the incompatibilist conclusions. The whole argument gets off the ground only if manipulated persons are, in fact and simply for the reason of being manipulated, not responsible (inc 1). At least from a legal perspective this seems false. Furthermore, if there are any "secrets" in these cases, they lie in principles of distributing responsibility among several agents, transferring the bulk of blame from the manipulated onto the manipulator. These principles are arguable themselves and may not fully exculpate the former, but both incompatibilists and compatiblists can subscribe to them without substantial concessions to their respective positions. Even more, we claim that both already do so, perhaps unknowingly, in ordinary coercion cases. Thus, *if* there is a reason for not holding manipulated persons responsible, it is grounded in their relation to manipulators, not in any psychological fact or (in)determinate events in their brains. So even if (inc 1) were true, there are good reasons why (inc 2) through (inc 4) would not follow.

LEGAL PERSPECTIVE

No "Brainwashing Defense"

How does the law deal with manipulation cases? First, we wish to recall that there is not *the* law, but many different and diverging legal systems. Just as we cannot do justice to each of the philosophical views, we cannot to every legal system.[8] Keeping this caveat in mind, for the law, the issue of severely manipulated offenders is not entirely new. A good point to start is a former controversy in US criminal law over a "brainwashing defense", spurred by the trial of Patricia Hearst in the 1970s. Hearst, heiress to an influential publisher, was kidnapped by a leftist guerilla group. After several weeks of abuse, mistreatment, and indoctrination, Hearst apparently turned sides, became a revolutionary herself, and participated in a bank robbery. In the trial, her defense argued that she could not be held responsible because she was involuntarily indoctrinated and had to be excused for everything she did subsequently.

This was not the first time a "brainwashing defense" was raised. Similar arguments were made in trials against soldiers who defected to Communism and committed treason after having been incarcerated and "brainwashed" in Maoist thought-reform camps during the Korean War. And quite recently, a similar defense was brought forward in the case against the "D.C. snipers." One perpetrator, the 17-year-old orphan boy Malvo, was raised and indoctrinated by an anti-Muslim fundamentalist who, in the words of the Washington Post, "created what Malvo became just as surely as potter molds clay" (Nolan 2004:452, see case 6b).

In various aspects, these cases are sufficiently similar to the examples discussed in philosophy. A manipulator has transformed preferences of another person through intensive, illegitimate, and control-bypassing interventions, and the newly acquired preferences provide the motivation for committing crimes. So, how does the law deal with them?

If incapacitated, brainwashed agents may fall under the ordinary defenses of duress or insanity. But these are exceptions. Manipulations do not necessarily lead to incapacities in the legal sense, that is, the inability to discern right from wrong or direct one's actions accordingly. The point in manipulation is not to create illness or incapacity but to change preferences of mentally intact persons. When such persons act willfully or knowingly and without severe distress or disorder, the law does not consider the history of preference acquisition an exculpatory factor per se. Because of this, brainwashing defenses are regularly unsuccessful in court practice. At best, they mitigate punishment. Malvo was spared the death sentence and convicted to life in prison. Many American POWs were convicted (Nolan 2004; Robinson 2011), and so was Hearst. Despite her ordeal and the well-documented psychological processes from which she suffered (which Anna Freud called "identification with the aggressor" and is today known as the "Stockholm syndrome"), she was sentenced to 7 years.[9] Public opinion sympathized with her, and after 22 months, Hearst was commuted by President Carter because "she would not have participated" in the crimes "but for the extraordinarily criminal and degrading experiences" that she had to endure (Nolan 2004:443, quoting White House press release). The public support for Hearst is good evidence that (inc 1) tracks widely shared moral intuitions.

The apparent unfairness in the law's treatment of manipulated agents has received surprisingly little scholarly attention. A notable exception comes from Richard Delgado, who, in the wake of the Hearst trial and wisely anticipating today's neurointerventions, proposed to introduce a "brainwashing defense" (1978/1979). Reminiscent of the philosophical views, he argues that under specific conditions, coercively persuaded persons should not be held responsible because "their choice is not freely made, [it is] not their choice at all," brainwashed persons do not act from their "own will". They are "by ordinary moral intuitions, more victims than perpetrators". As the law excuses "one who chooses to perform a criminal act when forced to do so", Delgado suggests, "it would constitute no great extension of the doctrine to excuse those who have been forced to choose to commit a crime consistent with their new, coercively induced beliefs and desires."[10] Since then, the issue has not received much attention. Some authors report an increasing tendency of courts to recognize such a defense (Nolan 2004), so the issue is not settled yet.

Responsibility in Multi-Actor Cases

Rules of Imputation

Before turning to the merits of the defense, we wish to draw attention to legal rules of imputation not often taken notice of in philosophical arguments.

Typically, the elements of a criminal offense proscribe bringing about an event and thus a (new) state of affairs, such as harm to bodily integrity ("result" or "effect" crimes). To *attribute* responsibility for an event to an agent, a specific relation between both has to be established. First, an agent's conduct (act or omission) has to causally contribute to the occurrence of the event—often, an evident matter. But because physical causation (*causation in fact*) is a transitive relation and hence far-reaching in space and time (after all, Adam and Eve are a *condicio sine qua non* of every evil brought about by human beings), legal systems have developed narrowing corrective rules to reasonably demarcate the realm of personal responsibility—so-called rules of objective imputation (*causation in law*). These rules take into account such factors as foreseeability of the resulting effect in the moment of action, deviating causal routes and contributory behavior by others including victims themselves.

Rules of imputation distribute, inter alia, risks and responsibilities for actions and their consequences. As a general principle, agents are liable only for their own deeds, not for those of others. Therefore, whenever a third party intervenes into a "causal chain" set in motion by an agent and displaces the original chain by a new one, ascription of responsibility to the former agent is usually forestalled. Consequently, one can distinguish actors who immediately bring about prohibited events through proximal causal contributions—let us call them *principals*—from more remotely involved persons. Responsibility of the latter has to be established by even other rules and doctrines expanding imputation beyond principals (e.g., complicity, aiding and abetting), which seem to be widely shared across jurisdictions (Rehaag 2009).

In addition to these objective conditions, actors have to have a subjective relation to the event: they have to be in the right state of mind (mens rea). Particular offenses have stronger (intent) and others have lesser (negligence) subjective requirements. Only when actors meet these conditions are they suitable targets for negative sanctions.

Most manipulated or brainwashed agents satisfy these demands (Dressler 1979). It is hard to deny that in our cases, P acts intentionally and hence at least descriptively exhibits the necessary subjective elements. Thus, a prima facie responsibility of P in relation to the victim (P–V) can be established: that of having violated V's rights unlawfully. The only question is whether P can invoke specific exculpatory reasons pertaining, not to the objective and normative quality of his deed, but to the individual situation of his mind.[11]

Voluntary Act Requirement

In some manipulation cases, however, agents do not even *act*. Actions in the legal sense differ from mere bodily movements insofar as they have to be under agential (potentially conscious) control (Fletcher 2007:274).

Reflexes, sleepwalking, and so forth are not actions,[12] and hence automatisms, too, are apt to generate problems for jurisdictions. If we were to generalize findings in the line of Benjamin Libet (2004) and Daniel Wegner (2002) to mean that many actions are initiated and governed by automatic processes responding to external cues rather than conscious reasons, these difficulties might exacerbate (Sie and Wouters 2010). However, we do not think these findings seriously call into question the general ability of persons to act for reasons (Levy and Bayne 2004). At most, they might prompt some revisions of the act requirement in cases of sudden, but not of more complex, behavior.

More troublesome for the act requirement are BCIs enabling persons to control movements or send signals "by thoughts" (case 4). BCIs record neuronal correlates of thoughts and transmit them to other devices. If initiating causal chains in the physical world does not require bodily movement anymore, *mental* actions may have to be legally understood as actions (cf. Metzinger in press). As of yet, however, we are reluctant to expand the scope of criminal law to mental phenomena, as suggested by Douglas Husak (2010:40). This would seriously interfere with the human right to freedom of thought.[13] But the matter certainly deserves further attention by legal theorists.

BCI and other means to directly stimulate brain areas eliciting bodily movements may confer sinister powers of controlling humans like marionettes on a string. Jens Clausen (2009:1080) asks: "Who is responsible for involuntary actions caused by brain/mind machines?" Well, the answer is already given: involuntary movements are no actions proper and do not give rise to criminal liability. Perhaps, law makers may consider civil law regimes of strict tort liability for these devices.

Thus, if neuroscientists in case 1 produce P's *bodily* states moment to moment by tampering with his motor cortex through DBS, his movements are not actions and, hence, the killing of V cannot be imputed to him (but to the neuroscientists). The same applies to remotely controlled BCI (case 4). If interventions, however, target mental states anteceding bodily movements, such as processes of reasoning and forming motivations, ensuing actions would satisfy the act requirement.

Responsibility of Manipulators

What about the responsibility of manipulators? As said, certain interventions of third parties usually forestall attributing responsibility. For this reason, manipulators (M) absent from the scene of the crime and not proximally causal for the offense do not automatically bear responsibility for deeds of principals. Such deeds must be apt to be imputed to the manipulator. A good defense would attack such imputation, for example, on grounds that the action of the principal was not foreseeable for the manipulator, that the manipulation itself did not create an unlawful risk for producing that particular action. This strategy might prove successful in cases of broad value manipulation such as the indoctrination in cases 2 and 6b.

The relation between several offenders (here, M–P) plays an important role, especially in sentencing. Sometimes, it is a same-level relationship. When Bonny and Clyde (or Hearst and her comrades) jointly rob a bank, they are co-principals both bearing full responsibility. In other cases, the relation is hierarchical, between a (dominant) principal and a (subordinate) assistant (accomplice, abettor), with the principal bearing prime responsibility and receiving harsher sentencing. On occasion, however, the culpability of the principal is considerably lesser than of other contributors, although his action directly caused the offense. For instance, East German soldiers found guilty of shooting refugees at the Berlin Wall received comparably lenient sentences.[14] The "real culprits", so to speak, were those who set up the border regime, the political leaders of the former GDR. Similar reasoning applies to other power structures, such as Mafia or organized white-collar crimes. The bulk of blame is not assigned to accountants (knowingly) executing fraudulent financial transactions but to CEOs who ordered them to do so. That culpability is assessed on an individual basis does not imply that social contexts of offenses and offenders are disregarded.[15] Therefore, in exceptional cases involving several agents, their relation may lead to what we loosely call a transfer of responsibility from the principal to the more powerful background participants. In very extraordinary cases, principals might even be fully alleviated from blame. Still, in principle, they remain suitable targets for punitive sanctions. This "blame distribution" surely partially accounts for and corresponds with the moral intuition that manipulated agents are not responsible.

Victims or Perpetrators?

Another reason for intuitive support of (inc 1) is that manipulated agents appear to be, as Delgado says, more victims than perpetrators. Indeed, they are victims of illegitimate manipulations. While this is evident when electrodes are secretly implanted in another's brain or DBS devices are hacked, it is largely unclear in regard to practically more prevalent but less severe interventions (such as the childhood indoctrination in 6b). Philosophy and law seem to neglect the question of where to draw the line between permissible and impermissible ways of changing other peoples' minds, between permissible skillful rhetorical persuasion, undue influence, and illegitimate manipulative indoctrination. There are no "crimes against minds", and the US tort of infliction of mental distress is not recognized by many jurisdictions, nor does it seem to stand on sufficiently stable theoretical grounds (Grey 2011).

Elsewhere, we propose to introduce a criminal offense protecting mental integrity against severe manipulative interferences (Bublitz and Merkel 2012). Leaving details aside here, we will simply assume that all manipulations (M–P) are illegitimate.

Therefore, by intervening into P's mind, manipulators (M) commit a crime against P. The crime P commits in turn against V (P–V) can therefore be imputed to M (M–V), at least if it was foreseeable. Then, the bulk of the blame for P–V can be shifted onto M. However, the fact that M has wronged P does not exculpate P *tout court* from what he did to V. P is not "more a victim than a perpetrator", he is both victim *and* perpetrator. Apparently, this dual status is not easily acceptable to many. Intuitions, so it seems, only allow for truly guilty perpetrators or innocent victims but not for finer graduations. So the question is whether the law should follow philosophical and commonsensical intuitions and excuse manipulated agents (P) for their misdeeds toward V in virtue of them being victims of M.

DIFFERENCES AND SIMILARITIES

Let us quickly note some differences and commonalities between law and philosophy. For the law, even severely manipulated agents are responsible, but sentences may be mitigated. To historicists and advocates of a "brainwashing defense", this is unsatisfying. They insist on (inc 1) demanding outright exculpation. What often remains unclear in philosophical statements is the kind of responsibility in question (Vincent 2011). Are manipulated persons not suitable targets for any negative sanction, should they compensate victims, are they "answerable" for their deeds? This relates to general differences between moral and legal responsibility. While discussions about the latter are often quite unreceptive to arguments beyond positive law, the former concentrate on ideal situations; sometimes authors even remark that they are not interested in practical responsibility at all.[16] The free will debate is primarily concerned with finding at least *one* instance of a responsible agent in an (in)deterministic world, whereas the law, by contrast, regulates common and daily social interactions. These differences presumably account for some disagreements, but for the sake of dialogue, we will assume henceforth that everyone talks about the same subject matter.[17]

Two main differences emerge. First, for the law, the involvement of a second agent does not exculpate the principal but rather inculpates the former, whereas it seems to be the other way around in philosophical reasoning. Second, the philosophical as well as Delgado's proposals heavily draw on *internal defects* in actors' mental states or in the history of their formation. They suggest that manipulated agents do not act from their "own true" will, "authentic preferences", or their "own (psychological or neuronal) mechanisms", but rather from a "superimposed mens rea" that blocks responsibility. We consider these views as *internalist* because, for them, agential responsibility exclusively depends on actual or former facts internal to agents' minds. It is helpful to recall that the philosophical debate very plausibly centers on internal properties because the larger point in dispute is whether responsibility requires indeterminism somewhere along the line between preference formation and decision making.

For the law, by contrast, a will superimposed on a perpetrator generally remains the will of that perpetrator. If sentencing is mitigated, it is not for intrapersonal but for interpersonal reasons: the relation to other agents. So the difference is primarily in perspective: *internalism or externalism*.[18] If externalism is correct, the significance of manipulation cases for the philosophical dispute drastically diminishes as not (meta-)physical but social facts matter. And this, we think, generalizes over many questions concerning neuroscience and responsibility: the more important internal states and their genesis, the greater the relevance of neuroscience—and reversely.

We contend that a purely internalist perspective is insufficient to adequately solve questions of responsibility. It is misleading to suggest that manipulated persons are not responsible in virtue of facts internal to their psyche. A prime example of this imprecise *façon de parler* can be found in regular coercion cases, which are often regarded as partially analogous to manipulation cases and which provide a good argument against across-the-board exculpation of manipulated perpetrators.

COERCION CASES

Let us begin with ordinary two-party coercion cases (in which the victim is coerced into harming herself).

Two-Party Coercion Cases

> *As P walks down the street, C approaches him, holds a gun to his head, and asks for a "donation of money". P, out of fear, complies and "donates" the money.*

Obviously, the pseudo-contractual agreement to donate the money is legally invalid. But why? A common answer says that P did not act freely or voluntarily; his will was not of a kind to be recognized by the law. And, in a sense, this is correct. But if taken too literally, this answer is prone to fall into the internalist trap and to obscure relevant *objective* aspects. Various internal accounts of autonomy have been put forward, but none of them, it seems, can handle these cases satisfactorily. For instance, Harry Frankfurt's prominent autonomy account relies on the structure of internal elements. Volitions have to be endorsed by higher-order preferences. To him, "it seems that a threat is only coercive when the motive from which it causes its victim to act is a motive from which he would prefer not to act." And he further specifies: "A person who submits to a threat necessarily does so in order to avoid a penalty. That is, his motive is not to improve his condition but to keep it from becoming worse. This seems sufficient to account for the fact that he would prefer to have a different motive for acting" (Frankfurt 1988:43).

This is a rather superficial take on the problem. Agents coerced into compliance do not necessarily exhibit a disharmony or incoherence among their desires. Describing their psychological state, Irving Thalberg notes that agents who give in to threats "would at the time and later, give second order endorsement to their cautious motives. They are unlikely to yearn, from their elevated tribune, for more defiant ground-floor urges" (1989:123). And he is right. Every rational agent endorses the first-order desire to hand over the money, sparing his skin. It is a rational and often the best decision.[19]

Surely, Frankfurt correctly notes that the victim does not really want to be moved by the desire to hand over the money. We may say he has a *relative* preference not to be moved by the threat because he'd prefer to be rid of it. Yet, we all, time and again, would prefer to act in circumstances different from those we actually find ourselves in. From trivial ones such as going to work, visiting stepparents, or stopping at red lights to more dramatic ones such as consenting to perilous medical treatments under life-threatening health conditions, we are confronted with obligations we meet and decisions we make only to evade a "penalty or the worsening of the situation"; and even more often we wish to be in very different situations. However, this does not rule out our autonomy. We must not confuse hard decisions devoid of any reasonable alternative with nonautonomous decisions. On the contrary, we can be held responsible for most actions in those not-what-we-wished-for circumstances, even if we carry them out reluctantly and involuntarily in the same sense and psychological state as victims of coercion do. The criterion to separate situations in which a relative preference for being in a different situation undermines autonomy from those in which it is simply unreasonable reverie cannot be found in an agent's psyche. In fact, compulsion is compatible with most mental states; victims may even experience a masochistic thrill from being robbed at gunpoint. Nevertheless, without prior and recognizable consent, it is illegal. Involuntariness in a psychological sense, then, is only an approximation for what is really at stake here.

Complaints about the world not being the way a person prefers it to be is of legal significance only if she is *entitled to be in a different situation*. And an agent's available options can only be short of what assures their being sufficiently free if someone else has illegitimately interfered with them. "Money or life!" is an illegal threat by a robber, but, in more moderate a tone, an *invitatio ad offerendum* by a pharmaceutical company in a capitalist economy.[20] Even presuming the inner unwillingness of the addressees to be identical in both cases, you can claim your money back only in the former. Neither the "pressure" on nor the "overpowering" of the will is decisive. Roughly, the legal reasoning is this: contractual agreements are binding even if one party would have preferred to be in very different circumstances. They are voidable only if one party has induced the other party to enter it by illegal means (threats, exploitation, deception, undue influence).

Three-Party Coercion Cases

The insufficiencies of pure internalist accounts become even more evident in three-party cases in which the victim of coercion harms a third party. They also illustrate another important condition of legal responsibility: fair expectance of norm compliance. Suppose C threatens P with disclosure of photographs revealing P's extramarital affairs unless he severely harms V. If P does so, it is anything but self-evident that he is justified or excused, despite his acting strongly involuntarily and under a massive and illegitimate threat, which, of course, makes the coercer, too, criminally responsible (for duress).

Things change if the threat consists in, say, killing P's wife. Then, P's harming V might be justified under a rationale that allows one to weigh the harm inflicted (on V) against the harm avoided (on P's wife). Deontological and utilitarian ethics and legal systems may diverge in the way of balancing competing interests and in the extent to which they accept a justification of "choosing the lesser of two evils."[21] Still, all agree that grave threats may rule out responsibility of the coerced for considerably less grave harms done to uninvolved others. For the law, the deeper reason is that normative compliance *cannot reasonably be expected in face of severe threats*. But it is important to scrutinize the relations between the agents involved: the coercer (C) illegitimately compels the coerced (P) to harm an innocent third party (V). Despite acting involuntarily, the law holds P prima facie accountable for harms he inflicts on V. Isn't that unfair to P? After all, "by ordinary moral standards," he may, just as much as manipulated agents, appear "more a victim than a perpetrator." In the eyes of the law, he is, once again, both: victim and perpetrator. The reason is that the legally protected interests of V must not be ignored. That C infringes on the rights of P does not entitle P to equally intrude into the protected realm of V.

The legal order guarantees V full protection, which cannot simply vanish in virtue of whatever goes on between C–P. Therefore, the law regularly demands P to withstand the threat and refrain from harming V, or else, take it onto himself and not simply pass it on to others. Why, one may ask, should V be legally obliged to suffer harms only because P suffers such harms with no corresponding legal obligation either? Only in exceptions, the law—in accordance with the "considerably lesser evil" rationale—allows one to divert an impending harm from oneself to others.

The basic principle underlying such justifications is not primarily a concession to the pressured mental state of the coerced, but rather a kind of compulsory solidarity, legally forced on the previously uninvolved third party to alleviate the hardship of an acutely endangered fellow citizen. It is quite obvious that a legal order bound to safeguard equal rights for all—and hence by sharp contrast to the much more demanding utilitarian principle—can allow for such a compulsory solidarity with third persons within very narrow limits only. That accounts for the conspicuously wide asymmetry between the colliding interests that legally guides the process of weighing between them. Thus, the law has to take very seriously "the distinction between persons" (Rawls 1971:27). And that is why in any other than those few cases of legitimately compelled solidarity, P's harming V remains illegal. However, some room for a (personal) exculpation with regard to an (objectively) unlawful deed remains: if and when the threat to the coerced (say, her life, bodily integrity, or freedom of movement) is so severe that it becomes unreasonable to expect her to resist it, she may be excused, the objective unlawfulness of the (excused) action notwithstanding.

Yet again, it would be misleading to say that the reason for the exculpation lies in the fact that the coerced (P) was not in the proper state of mind to be responsible or that his will was overpowered or that he lacked the appropriate mens rea in a psychological sense. Regarding exculpation, too, it is *not* primarily an internal deficit, a psychological state, or its history that may exculpate perpetrators, but rather a normative consideration, namely whether the legal order can safely relinquish its claim to punish unlawful acts without endangering its primary task in the field of criminal law—that of credibly defending the validity of a broken norm and assuring its future efficacy (an issue to which we will return).

Because mental aspects do not contribute much to the solution of these cases, free will theories are largely irrelevant here. The reasoning sketched previously obtains under any account of freedom: add indeterminism, agent causation, libertarian, historical, or any other internal conditions of freedom and run the story again—*libertarian free P walks down the street*... In whichever way internal conditions of responsibility are crafted, they never sufficiently explain the invalidity of the donation. Therefore, every account of responsibility has to be supplemented with rules ascribing responsibility that take account of social context, of relations between agents, and of the harms avoided and inflicted.

In his seminal study on coercion, Alan Wertheimer comes to much the same conclusion. He points out that "we use the *language* of voluntariness to talk about cases of mens rea and coercion"; "judges *refer* to overborne wills" (1987:164, his emphasis), but no empirical inquiries into subjective states can answer the real question. "While the court's language suggests that duress turns on the volitional quality of defendant's act (his freedom of will), the real issue, it seems, is not [about] pressure, but whether the defendant should be held responsible. And that is clearly a moral matter" (162).

The important insight to gain from all of this is that pure internalist accounts are inadequate to solve even standard coercion cases. The final judgment always involves recognition of the normatively shaped relations between persons. And if such external principles are accepted here—and we do not see any chance for internalists to escape them—it seems very likely that they also apply to and explain manipulation cases. So let us now compare cases of coercion to coercive persuasion.

From Coercion to Coercive Persuasion

Two-party manipulation cases (e.g., oxytocin in a casino) are structurally similar to two-party coercion cases, and tripartite manipulation cases (M–P–V) are to three-party coercion cases. Let us begin with the latter. P harms V; but instead of being coerced, P is manipulated by M into harming V. In coercion cases, P could be justified if harms avoided outweigh harms inflicted,[22] or excused if it is normatively unreasonable to expect him to withstand the threat. In manipulation cases, by contrast, P harms V out of his own (though previously manipulated) preferences. He is not threatened, nor are there any interests to be balanced or drawbacks he'd face if he refrained from harming V. Manipulated persons may even, as Delgado acknowledges, act "wholeheartedly." This is the paradox of manipulation, vividly illustrated in Skinner's "Walden Two" (1976): manipulated persons might be free although—perhaps even because—they have been conditioned into wanting what they pursue.

Manipulated P enjoys more freedoms than his coerced counterpart, at least vis-à-vis victim (V). If, despite his inner reservations against harming V, the coerced is not exempted (unless the previous exceptions apply), why should the manipulated be excused *tout court*? Both find themselves equally faultlessly in an unlucky situation—but unlike the coerced, manipulated agents have no relative countervailing preferences and no further argument for exculpation on their side. Comparing these types of cases speaks clearly against a manipulation excuse.

In regard to the psychological side of agents, coercion and coercive persuasion are similar by name only. The former do experience inner turmoil, the latter don't. Excusing the latter to a larger extent would, pace Delgado, not merely constitute a small expansion of well-accepted principles, but would have to be grounded in quite different principles. What both have in common is that agents have been illegitimately harmed by a manipulator/coercer, and, therefore, a principle of responsibility attribution kicks in: the scope of liability of interveners (manipulator/coercer) for their illegitimate acts against others (manipulated/coerced) may expand to harms that the latter foreseeably inflict on third parties (V), thus partially transferring responsibility onto manipulators. The reason for this is the *external* hierarchical relation between offenders (M–P). Pure internalist strategies should appear dubious by now; but as our argument has yet been comparative only, some room for objections might remain. In the following, we shall criticize internal views directly.

NOT THEIR OWN WILL?

The recurring theme in internalist arguments is that the will of manipulated persons is not "their own." We surely understand the picture, but again, this metaphor is too imprecise to generate good law. If the law were to take this line of argument seriously and regard the factually present (previously implanted) will not as the agent's own, dramatic negative consequences for the agent would ensue, inevitably even calling his legal status into question. What about persons without a properly formed and autonomy-manifesting will—do they not have any will at all?

Whenever the law disregards the will of persons, they may not only gain impunity but also lose liberty. Paternalism and legal guardianship dangle over anyone unable to form wills in a manner that demands legal recognition. In practice, how should agents who want to X, identify with X'ing, and do not suffer from incapacities be treated if their respective preferences have been implanted? Should we really disregard them and deny X'ing? Affected agents themselves would obviously strongly object to interventions impeding X'ing because it curbs their freedom. This point is far from academic concern only; structurally similar cases are litigated in court.

Suppose a spiritual community H+ with transhumanist beliefs has lured the young law student S into their group by deception and changed his preference structure with sinister techniques of manipulation and control-bypassing neurotools. "Enlightened" by the futuristic visions, S replaces his democratic with technocratic beliefs. H+ reinforces his new value system through group discussions, readings of prophetic philosophy, and promises of eternal life in virtual space.

This may sound funny or bizarre, but S's parents are not amused. Substitute H+ with a dubious spiritual sect of your choice, and we are back in court-reality. Parents of (allegedly) brainwashed followers of such groups seek custody orders for their adult children, arguing that their will is not their own anymore and hence they are incompetent to make legally binding decisions. Assuming that bypassing interventions took place, such parental arguments are pretty much on Delgado's and the historicists' track. Yet, these orders are regularly not granted, and for a good reason: as long as persons have the capacity to take care of their own affairs, the history of their preferences must be irrelevant for the law. Suggesting otherwise presupposes the *normative priority of a former preference structure*, an idea inherent to historical accounts. But what is the point in preserving and protecting bygone versions of an agent's self? Let us press this point a little further.

Additionally, the parents of S seek an order to "de-program" him, that is, to change his preference structure by intensive psychological and pharmaceutical treatment aiming to reverse his transformation and restore his old "self." These techniques make use of inner conflicts and unconscious forces and directly intervene into his brain, bypassing mental control. Should the order be granted? And should interveners be allowed to invoke a defense of necessity if they save "brainwashed" souls without consent?

These are real questions at courts.[23] If S's current will is not his "real" or his "own," treatment is a morally and legally laudable endeavor. From an opposing view, these are highly dubious and profound paternalistic interventions into the life of persons possessing all necessary capacities to make legally binding decisions for themselves. In a variety of situations, preferences are deeply altered by dubious control-bypassing stimuli; in none of them, upholding the normative priority of a former preference structure over a later one appears to be a reasonable idea. Think of Ulysses contracts in which persons explicitly express that a former decision should prevail and be binding in the future. Even then, it is—and cannot—be binding if future preferences diverge. Suppose at t_1, A lays out in an advance directive in which situations he does (respectively, does not) want life-sustaining treatment. Tragically, such a situation comes about at t_2. Now he renounces his former decision. Suppose further that A acquired all his new and relevant preferences through control-bypassing routes (medication, traumatic near-death experiences, and, if you like, add a manipulative inheritor). Should the directive of t_1 be binding? Certainly not.

This illustrates a deeper point: we cannot let the past prevail over the present. Legal competence and corresponding responsibility for actions in this sense are not "essentially historical notions." They are rather *ahistorical;* what matters is the present preference structure, irrespective of its genesis. If there is any time-related dimension, competence is *future oriented.* The only arguable reason for an interference with a decision at t_2 is not that the person talked differently at t_1 but that, foreseeably, it might be too deeply regretted at t_3.

We take these arguments to be decisive against historical responsibility accounts; and presumably, historicists and internalists do not endorse these ramifications of their position either.[24] In response, they could concede that "not his own will" is merely a figurative way of speaking and does not apply to entirely self-regarding affairs, only to other kinds of responsibility. Fair enough. Then let us take a look at forms of responsibility, starting with contractual liability.

RESPONSIBILITY IN PRIVATE LAW

Contractual Liability

Contractual responsibility is instructive because it illustrates the widely underappreciated interdependence of freedom and responsibility. Do manipulated agents have enough "will of their own" to enter into contracts? As civil law metaphorically says, contract formation requires meeting of minds. Is this possible with a history of manipulation? Suppose brainwashed transhumanist S takes a huge credit loan to finance cryonic conservation of his brain, to be washed and frozen after his death. Is the contract valid?

History-sensitive minds might be inclined to deny its validity because it imposes obligations on S he would not have consented to but for the manipulation. Then, however, he would be deprived of his legal freedom to enter into contracts and therewith of all factual possibilities to participate in social life. Without competence to enter contracts, he cannot even go out and have lunch. In order to *enable* incompetent persons to participate, the law needs to assign legal guardians who make decisions on their behalf. This is surely anything but a prosperous perspective for manipulated persons. And some legal reality: Having the legal power to form contracts is a fundamental human right. Affected persons often fight desperate legal struggles to have their legal competence restored.

Presumably, historicists do not want to deprive S of this freedom. Then they might argue that manipulated agents do have the power to form binding contracts but are not liable for consequences, that is, the other party cannot enforce contractual obligations against them. This, however, violates general principles of contract law, the synallagmatic relation between parties, *quid pro quo, do ut des,* or *tit for tat*: a party has to fulfill its obligation only if the other does so, too. One-sided obligations are, by any standard, unfair. So if there is a lunch for manipulated persons, it cannot be free. We suggest that contractual obligations reveal a principle that is quite fundamental: *freedoms correlate with responsibilities.* One cannot have the fruits of freedom without carrying its burdens.

In light of this, historicists might be inclined to affirm manipulated agents' powers to enter into contracts and restrict their argument to other forms of responsibility. However, this is a major concession at the heart of their argument. Generally, the convincing part of their reasoning also applies to contract law. What may very well turn out to become the most worrisome field of application for neuroscientific knowledge and tools may not be remote-controlled criminals and Manchurian candidates but instead the mundane exchange of money and goods in stock- and supermarkets (Levy 2007). "Oxytocined" gamblers highlight this problem. Although their evaluation of chances and their emotional attitudes toward risk taking have been manipulated, they satisfy all conditions for legal competence. Nonetheless, their spending more money than they would have without manipulation makes one wonder whether they freely entered into contracts and have to fulfill corresponding obligations (payment) vis-à-vis the casino. Provided the influence is above a threshold of intensity, the answer must be negative. And this is precisely the intuition historicists' arguments draw upon.

Only, the argument is better formulated differently. Instead of locating the source of concern in "inauthentic" wills or "un-owned" springs of action, it is to be found in the relation between casino and gambler. More concretely: the casino has illegitimately interfered with the gambler's right of mental self-determination, and this unlawful act results in financial expenditures. Compensation for the harm inflicted on the gambler includes compensation for the additional money he spent. This becomes evident if the case is slightly modified to a three–party case. Suppose it was not the casino but his wife who secretly administered oxytocin to him in order to make him a more bearable husband. With the gambler's mind in exactly the same internal state, he surely has to pay the casino (and may then sue his wife for compensation). Any internalist strategy must resign here. It is, again, not a property of brain or psyche but considerations of the duties persons owe to each other and compensation for their violations that provide the solution.

CRIMINAL RESPONSIBILITY

Finally, let us turn to criminal responsibility. Is the absence of a brainwashing defense for manipulated agents a fallacy, a gap in legal doctrine in need of reform, or are there good reasons to hold such agents responsible? What we have said so far is more or less a logical deduction that we take to be largely beyond dispute. The following, however, is arguable, because no accepted theory of punishment and excuse exists that provides consensual answers to questions about the responsibility of manipulated persons. Still, there are better and worse arguments.

As a starting point, the fact that interventions have *caused* the agents' preferences is not an excuse. It has been convincingly argued that causation per se cannot be an excuse (Morse 2006). Furthermore, the law would be well-advised to not resort to such abysmal notions as free will in the sense of "could have done otherwise" (PAP) and to avoid talking of people having "chosen evil over rightfulness," or similar vague but suggestive phrases still so abound in legal reality. Any honest answer to unresolved metaphysical matters of that kind must be agnostic. But by its very nature, criminal law needs a conceptually (and metaphysically) stable basis, which cannot be found in issues eternally *in dubio*. Avoiding PAP is, of course, just a start with the profound problems still ahead. Worries over fairness and legitimacy of punishment are anything but resolved (Greene and Cohen 2004). And the still fashionable strategy among many legal scholars to declare criminal law (largely) immune to findings of cognitive and behavioral sciences because free will or its equivalents are no conditions of legal responsibility simply evades the problem. Such a purely positivist stance can only apply to decision within the current legal framework but poorly fails to address external critiques.

For criminal law, the fascinating aspect of manipulation cases is that they present a prima facie compelling case for an excuse. Thereby, they transform the external critique of punishment into an internal problem. They urge the law to separate exculpatory causal routes from others while observing the internal constraints of systematic coherence. If not all causal factors are normatively equal, where to draw the lines? A compatibilist, determinism-friendly theory of excuses explaining (and justifying) why we punish some but not others is needed.

Character-Based Excuses

What seems to partially account for the intuition that manipulated persons should be excused is their character or personality. Manipulated agents are, at heart, not the bad guys their deeds portray. In a sense, this is plausible. However, whether character is a reason for legal excuse is a controversial matter. In criminal systems based on actors and their attitudes, broadly in line with a Humean view of responsibility, persons are only responsible for deeds "in", not for ones "out" of character (Fletcher 2007:35). For act-based systems, by contrast, character evaluations are confined to the sentencing stage.

Quite generally, we are skeptical about ideas of "true" or "deep selves" and related concepts, which are necessary to define what is in and what is out of character. Whatever they exactly are, "selves" are not fixed but constantly redeveloping entities; there is no lasting quintessence of being oneself (Metzinger 2011). Therefore, we might cherish ideas such as authenticity in our personal self-conceptions, but they should not be taken too seriously in determining the adequate legal assessment of persons and their actions (Bublitz and Merkel 2009). Even if such notions could be rendered more plausible, punishing (and excusing) for character blurs the line between law and morality. What constitutes a "good" or "bad" character is, in the last instance, a moral matter. We also wish to remind that even a thoroughly antisocial personality enjoys legal protection[25] and cannot as such warrant distinctions between excusable and culpable behavior. Although character evaluations certainly do implicitly influence legal and moral reasoning (Lacey 2011), they are not helpful in formulating theories of excuse.

Punishment as Stabilization of Normative Expectations

Any general theory of excuses has, at some point, to confront the aims of punishment. What the proper and justifiable aims are, is, of course, highly controversial. Strict retributivist views may come to very different conclusions about punishing or excusing manipulated agents than rehabilitative or deterrence-oriented theories. We have reservations with all of these positions, at least in their mutually exclusive versions. Instead, we want to sketch what we take to be the core of a theory of punishment not undermined by neuroscientific findings. It plausibly explains many currently accepted excuses and provides good arguments against their extension to cases of manipulation.

Functions of Criminal Law

Abstractly, criminal law is a means of conflict solution and prevention. Breaches of norms cause two types of conflict: between offenders and victims and between offenders and society at large, or more precisely, the normative order regulating interpersonal relations. Negative sanctions are means to appease and perhaps solve both these conflicts. They cannot restore the factual status quo ante, but they may compensate former and prevent future harms. Vis-à-vis victims, offenders owe financial compensation for harms inflicted, mostly achieved through tort law. But at least in regard to grave cases, many would argue that financial compensation alone does not sufficiently redress the injustice and that some kind punishment has to be inflicted. There is arguably some truth to this, but nothing in our argument hinges on it. In the remainder, we want to focus on the other conflict, the one between offender and the normative order. For this, we like to draw attention to an aspect of punishment providing an instructive perspective on excuses: norm stabilization.

NORM STABILIZATION

Lawful conduct, excuses, and punishment are related to expectations. At times, the law cannot reasonably expect compliance and hence exculpates wrongdoers. To John Gardner, the gist of excuses is "that the person with the excuse lived up to our expectations" (2007:121). These expectations are basically normative, that is, their content, *what* is expected, is determined neither through empirical observations of what people in fact expect from each other nor by actual capacities of persons, but rather by judgments about their roles in social life. Gardner's observation is certainly correct. But it does not explain *why* expectations should play such a crucial role in criminal law. And, as far as we can see, none of the traditional theories of punishment does so either. Thus, what we need is a theoretical foundation for the legal importance of expectations.

Drawing on Niklas Luhmann's (2004:150) observations about law as a social system, the prominent German legal scholar Günther Jakobs (2004; 2008) has devised a theory of punishment centered on expectations. Broadly, he suggests that normative expectations structure society. Here is one of his examples: Walking alone in a park at night, knowing that others are not *allowed* to harm you is not sufficiently comforting. What is necessary for you to make use of your legally guaranteed freedom to walk in the park is the reliance that most likely others will not *do* so, which is the expectation that the norm to not harm you is accepted and obeyed by everyone else.

Norms correlate positively with expectations about other people's conduct over time. These expectations are destabilized by breaches of these respective norms. The primary function of punishment, then, is to counterfactually (re-) stabilize these expectations. Violations of a norm do not only harm the party protected by it but additionally convey a crisis of trust in the validity of the broken norms. In the reading of most modern theories of law, a legal norm is valid if and only if it has been enacted by authorized legislators, has not been formally repealed, and exhibits at least a certain minimum of factual efficacy, which is demonstrated either by (sufficient) obedience of those subordinated under the norm or by formal sanctions in cases of its breach.[26] Hence, legal norms are or become invalid when their specific function to guide human social behavior falls below a minimum threshold of factual efficacy. Because norms are only concepts, thoughts, and symbols, they do not exist in any physical or material sense but only in the specific mode of possessing validity. A substantial dwindling of their factual efficacy in guiding and constraining the factual behavior of their addressees would induce a process of gradual fading of their force and hence of their sheer existence as norms altogether. This in turn would be bound to lead to a massive loss of trust in the reliability of mutual expectations of norm compliance among citizens—and hence to the gradual vanishing of one of the most important premises of a well-ordered and peaceful society: the confidence that one is sheltered by *valid* norms in whatever one (lawfully) does, and that one's expectations in the (by and large) lawful behavior of others are substantiated and reliable.

Against the backdrop of these interrelations, the basic function of punishment in criminal law becomes clear. Breaches of norms of the criminal law are, as it were, the first step of an erosive process with regard to their validity. Would the state as the guarantor of that validity ignore or knowingly tolerate this first step, it would perform the second: symbolically demonstrate its abandonment of the broken norm. That is exactly what penal sanctions are supposed to counter. They symbolically invalidate the violator's disobedience against the norm and thus "repair" its damaged validity. To put it briefly, penal sanctions cannot abolish the past and undo the breach of a norm. But they can defend the status quo of the normative order itself and thus secure its future existence. They restore the validity of broken norms by verifying the normative consequences of such breaches in imposing sanctions. And this, to all our knowledge, appears to be the only way of ensuring mutual trust (reliable mutual expectations) in basic norm compliance among citizens and thus the precondition of a peaceful society. This also lays the groundwork for a compatibilist theory of excuse. When norm-violating behavior does not undermine expectations and, hence, its factual validity, the law may not need to counteract the contravention through punishment. This is the reason that situations in which norm compliance is an "unreasonable expectation," an idea found throughout the law, may have exculpating effects.

The idea of punishment as norm stabilization is perhaps just a "modern label for the Hegelian idea of validating the legal norm in the face of crimes" (Fletcher 2007:57) or a variation of what has become known as expressive or communicative theories of punishment. As everything in this field, it is highly controversial. Much more needs to be said, especially whether it, by itself, sufficiently justifies punishment (Gardner 2007:201), its empirical assumptions, or the "just" amount of punishment it implies (Sanchez 2007).[27] In more familiar terms, norm stabilization contains an element of deterrence and an element of symbolic norm restoration. Both these functions may be flagged with the label "*positive* general prevention" (Gardner 2007). Importantly, punishment is directly linked to the harm caused to the normative order. Hence, offenders are held liable for compensation—are made to pay for what they did, not merely treated instrumentally as a means to serve society's ends by deterring others.

Neurointerventions and Excuses

If this functional theory provides part of the rationale for punishment, how do manipulated agents fare in its light? A more fine-grained look at *what* is being manipulated is in order. Of course, mind interventions can bring about mental effects in the addressee that rule out his responsibility for the usual reasons: no voluntary act; lack of knowledge; impaired cognitive or control capacities. Irrespective of their origins, such deficits exculpate agents. The question is whether the realm of excuse should, as Delgado suggests, be expanded to agents who satisfy all conditions of responsibility, but whose preferences for action have been manipulatively implanted.

Norm Responsiveness and Momentary Disorientation in Values

A first category that should be interpreted slightly wider than it currently is concerns reason responsiveness, requiring that agents are, at the moment of action, responsive to those normative reasons that apply to the respective situation. When interventions induce volitions (as in case 3) that directly usher into actions, one may argue about reason responsiveness. Even if the volition is resistible but has been brought about by stimuli bypassing all faculties of practical reasoning, one may deny reason responsiveness. These are, however, very exceptional cases.

The more interesting type of cases the law has yet to come to terms with concerns rapid changes in preference structures as in case 5. Is P reason responsive just after the DBS is turned on? Hard to say. His preferences change suddenly and extensively. One may be tempted to consider such Jekyll/Hyde scenarios as changes of personal identity, but we are reluctant. As the law in many respects depends on the continuity of persons and because it does not have a doctrine of diachronic identity, it should resort to such notions only in extremely rare and clear-cut cases of fairly complete subjective fissions between a present and a former state of mind of one and the same biological human being (Merkel 1999). Preferrably, rapid preference changes should be regarded as instances of deficient reason responsiveness. To be responsive to reasons requires some kind of stable (not fixed) preference structure in light of which varying situations are evaluated. It is a truism that decision making needs time, and all the more time if it involves complex factors. Reason responsiveness is a capacity not just of passively absorbing and understanding information but also of actively processing and weighing arguments. Therefore, reason responsiveness may be disputed if agents have only a very limited time span to make decisions. The same should, we suggest, apply if the complexity is due to rapid inner upheavals of preferences. Agents then may need some time for inner exploration and adjustment. And for such situations, we propose to introduce a novel excusing condition: *profound momentary disorientation in values*.

These two minor suggestions are more or less concretizations of accepted excuses. Further down the line of deficiencies in mental control, we find manipulated agents. They are norm responsive and act according to their preferences, but their formation was somehow interfered with. Such deficits are not incapacities in the immediate action sequence but prior impairments; they are, as it were, historical phenomena. This characterizes the typical philosophical manipulation cases. Should we not, in the name of fairness, expand the realm of excuses just a bit to include those agents? Can we really sustain a principled line between the approved excuses and the kinds of defects just sketched? And why should it run exactly here?

Manipulation Defense and Rationale of Excuses

At first glance, expanding excuses may indeed appear as a small step, yet normatively it would be a huge leap. Manipulated agents are not incapacitated but rather, given their (implanted) preference structure, law-abiding behavior is simply no genuine option for them. To appreciate why they cannot be excused, let us recall the aims of punishment.

Common to all aforementioned excuses is that agents do not undermine the validity of the norm albeit their actions contravene them. They do not cast doubt on the norm as a binding constraint of behavior, nor do their actions express a deliberate denial of obedience to it. Instead, persons insensitive to normative reasoning, incomprehensive of the scope of a norm, or incapable of turning better judgments into action can, by and large, still be portrayed as agents abiding by the law. Their actions do not seriously unsettle the confidence of others that the norm is in effect.[28] Legal sanctions may aim to "repair" (treat) the agent by conferring sufficient cognitive or normative competence, but no need to repair the validity of the norm arises. Therefore, the law can grant excuses without harming the normative order.

By contrast, manipulated agents are aware of the norms commands, can feed them into their reasoning processes, and are not generally unreactive to their orders—they would abide by them when the breach of the norm would come at a much higher price. However, all things of the *actual* context considered, other preferences prevail. Bluntly, their excuse would amount to nothing but a remorseful, "Sorry, I did not *want* to comply—given my preference structure, norm compliance was just not a viable option for me." If there is one excuse a legal system cannot accept, it is precisely this. It would dramatically undermine the validity of the norm in question, turning its claim to universal compliance into something akin to a recommendation at the individual's (and their preference structure's) disposal.

In more familiar categories, consider the preventive effects of excuses based on unwillingness. Why should P contemplating to harm V not be deterred? Just as he is about to form the intention to harm V, the "motivating" force of the norm should, ideally, intervene—and that is achievable only by connecting deeds with negative consequences. If agents' histories rendered them unsuitable targets for sanctions, criminal law would abandon one of its most salient aims—crime prevention through the sheer existence of sanction-threatening norms—at the very moment for which they are designed. Again, there is a future-directed aspect to the praxis of responsibility that historical accounts are bound to overlook.

Reformulate as Incapacity?

In the final step, let us tie this together by considering the following objection. The argument for the nonresponsibility of manipulated agents might be reformulated in terms of incapacities. Because their preferences have been implanted (and not "freely" chosen), one might say that they lack "preference formation control." Yet, this cannot constitute a condition of excuse because it contradicts an indispensable premise of liberal states.

Liberal Premise: Sufficient Capacities for Norm Compliance

Liberal states, primarily based on fundamental negative rights against the state, by default guarantee individuals a substantive realm of personal freedom. Inherent in the idea of such systems is the assumption that behavior of citizens can and should be regulated through norms. Instead of factual police force, norms are prime regulators of social interaction, presuming the expectation that social behavior be guided in line with their stipulations. Such systems do and must presuppose a general ability of compliance, that is, the "regular" person's capacity to sufficiently motivate herself to abide by those norms.

If this premise were abandoned—a theoretically arguable option—society would need to devise other instruments to prevent conflicts and secure its order, particularly the rights of citizens not powerful enough to protect themselves. How to sustain a well-ordered and peaceful community in the absence of the idea that social behavior be guided through norms? A host of protective measures of state surveillance and police strategies would have to be introduced, attacking the liberal presumption of freedom at its core.

Here, a facet of responsibility we previously hinted at resurfaces: *liberty implies accountability.* As long as individuals' self-governance over their affairs is institutionally guaranteed, they must be expected to refrain from norm-violating behavior. Self-governance presupposes abilities to norm-complying behavior and, likewise, responsibility for failures.

The tension between freedom and security is vividly demonstrated by the problems many liberal states currently face in engaging with purportedly dangerous persons (habitual offenders).[29] Beyond any doubt, unless they put others at risk, they are entitled to as much freedom as everyone else. Nonetheless, it is equally true that they cannot simply be treated as if they were not dangerous. A minor step is their exclusion from sensitive social activities (e.g., pedophiles from kindergartens), more infringing are measures like supervision or, ultimately, and with all due concern for their rights, preventive detention. These are unfortunate, but sometimes unavoidable, restrictions of freedom precisely because society can demonstrably not rely on these people living up to the expectation of abiding by the law.[30]

If a defendant complains about the unfairness in holding him responsible because he was truly unable to align his preferences with the requirements of the law, he declares he cannot be expected to act in accordance with the liberal premise. But then, on the reverse, he cannot be treated in accordance with it, either. Thus, there is a fiction in the law's view of persons, and this is one of the founding fictions of liberal states: citizens are deemed to have sufficient internal resources to guide their behavior according to norms. Admittedly, this has a Kantian spin: you ought, and therefore you can.[31] And, as Schopenhauer already objected, we cannot want to want what we want. Hence, by treating each other as if we could, we treat each other on the basis of a fiction.

Instead of concealing this foundational premise (behind talk about "will" and metaphysical powers of its freedom), the law should openly declare and defend it. It is, in the end, a matter for political decision whether we want to live by it or pay the price of abandoning it. Other forms of social control and security are conceivable, but undesirable. Thus, the fiction does not derive—as some philosophers argue—from an "illusion", the laymen's belief in free will, which we should uphold as it is helpful to maintain order (Smilansky 2003). Instead, we treat each other on the basis of a legal-political arrangement. It assigns burdens, not personally chosen by anyone, as the reverse side of liberty.

This interdependence has implications for currently discussed (future) interventions into minds of criminals for rehabilitative purposes and the public good. Neuroscience may enable interventions to alter behavior, thought patterns, and emotional dispositions that are sources of delinquent tendencies (Greely 2008; Shaw 2012). We strongly oppose the use of such measures without informed consent by the persons concerned. In principle, the mental sphere of the individual has to enjoy strong legal protection. Unfortunately, legal systems have not yet attempted to formulate concepts such as freedom of thought or mental self-determination and define its contours more precisely. But it seems a common tacit premise that the inward workings of one's mind, thoughts, and preferences are beyond the purview of legitimate state action (Bublitz and Merkel 2012). The corollary of the intangibility of the inner sphere is that persons have to be expected to arrange their mental life in a way that respects the social order. Again, freedom and responsibility are inextricably intertwined.

Gardner calls it a "condition of legitimacy of the liberal state" to not bypass rationality (2007:209). This easily connects to the no-bypass criterion developed by Mele; but the normative consequences are different: control-bypassing interventions should be illegitimate, even constitute a criminal offense, but they do not rule out the responsibility of the victim for harms inflicted on third parties. Although P has been harmed by M, he remains obligated to not harm V.

For skeptics, some courtroom reality: more often than not, defendants desperately attempt to avoid being treated as incapacitated, although they do suffer from excusing deficits. Many affected persons perceive the security regime of preventive detention—such as therapy or supervision—as decidedly more constraining and prefer being normal prisoners because then they retain the full status of a legal person embedded in mutual normative expectations as sketched previously. Imagine the law were to introduce a two-track system giving everyone a choice of either being treated—and punished—on the liberal premise, or being considered devoid of the required capacity, with all preventive consequences. We suspect that not even the most determined determinist would wish to enroll.

Exceptional Excuse?

Nonetheless, fairness problems emerge. Isn't it unfair to punish manipulated persons only because of the fictitious assumption that they could have formed their preferences differently? Is not a special excuse limited to manipulation cases as proposed by Delgado warranted?

The notion of fairness, of course, needs finer contours. Its main legal formal element is the right to equal treatment, prescribing to treat like cases alike and prohibiting to treat unlike cases alike. To invoke this right, manipulated agents need to show that punishing and subjecting them to the liberal premise—the attribution of the ability to align preferences in harmony with the law—is unjust because their factual inability qualitatively differs from those of other persons.

Here, the philosophical debate has a lot to contribute. It's time to resort to the above premise (inc 2): there is no normatively relevant difference between manipulation and all the other innumerous factors shaping thoughts, emotions, and preferences. To a large extent, we agree. In respect to our preferences and personality, we are all shaped by forces ultimately beyond our control.

The law must recognize that no one is self-created; the very idea seems incoherent as every preference formation would require an anteceding or higher order preference, regressing ad infinitum. Looking at the factual development of humans, the springs of anyone's preferences lie in genetic, socioeconomic, and cultural conditions. We all begin as entirely dependent beings and are conditioned, indoctrinated, and shaped by a plethora of control-bypassing influences. In this sense, we are simply made what we are, "inevitably fashioned and sustained, after all, by circumstances over which we have no control" (Frankfurt 2002:28).

There are two lines of objections against the no-difference argument. The first distinguishes between manipulations of existing, full-grown preference structures and those that take place early in life (Fischer 2006). However, we do not see a normative difference. Take cases 2, 3, and 6. Why should persons being mistreated from early on be more responsible than, say, Patricia Hearst, having been indoctrinated for the first time at the age of 20? Somewhere behind such arguments, clandestine images of real, authentic evil people, in contrast to unlucky victims of circumstances, seem to be in effect. We think this is a nonstarter.

Others point to the fact that one can make gradual differences between manipulation and other influences. Delgado (1979) defends his point by emphasizing the exceptional character of coercive persuasion. Nonetheless, a lack-of-preference-formation defense would have to equally apply to many other persons who populate courtrooms. Consider terrorists growing up in refugee camps and indoctrinated by religious fanatics; violent offenders with their own history of childhood abuse or emotional mistreatment; kids from rotten social backgrounds, and without real opportunities, rioting warehouses; persons suffering from debilitating diseases; survivors of traumatic events; war veterans; or, with respect to neuroscience, people with brain tumors, unfortunate genetic dispositions (MAOA), reduced prefrontal cortex activity, and so forth.

We cannot find categorical differences.[32] Rather, we conceive of *manipulation cases as time-condensed, fast-forwarded stories and mishaps of ordinary human life*. What happens there in one treacherous act is what otherwise may happen in an extended biographical course of human development. Because there is no such thing as an ultimate self-creation, granting some persons excuses because of their history is not a postulate, but more a contravention of the principle of equal treatment.

Fairness

Manipulation cases, however, seem to be a particularly vivid instance of a more general problem of fairness in the practice of punishment. Commanding everyone to defend the normative order and submitting everyone to the same universal expectation despite the overtly manifest inequalities of *factual* opportunities of their fulfillment is, at least, worrisome. Some persons have a good chance of living a lawful life without having to sacrifice much, whereas others are prone to conflict with the legal order. None of these dispositions stems from anything ultimately controlled or deserved by the individual. The law has to treat persons alike, but we are not alike.

Equal treatment is an emancipatory achievement. It entitles and empowers people, enables social participation, and eliminates some factual injustices. When it comes to obligations, however, its reverse side is that everyone is measured by the same standards irrespective and in spite of factual differences. Obliging some to carry heavier burdens in stabilizing the legal order appears at least prima facie unfair. This aggravates due to the empirical fact that those carrying heavier burdens are typically not the ones who benefit most from a legal order primarily protecting the status quo. It is the gradual differences in factual freedoms from which the apparent unfairness in manipulation cases derives, and in regard to ordinary situations, the very same unfairness often strikes observers from outside the legal profession, and at times seems to be forgotten by insiders.

These worries cannot be brushed aside easily, for instance, with reference to social contract theories or utilitarian we-are-all-better-off-with-punishment arguments. And they won't go away through granting exceptional excuses. They are systematic in nature. Surely, this issue needs to be addressed in a wider framework of justice and moral luck. But it overshadows the legitimacy of all harms that criminal justice systems inflict on persons, none of whom can be said to bear ultimate responsibility. A society preserving its order by formally equal, yet potentially unfair, measures seems at least to be obliged to counteract unequal distribution of opportunities. Moreover, convicted persons should be relieved of some of their burdens by more meaningful and constructive forms of punishment than the ones currently employed and essentially rooted in practices of medieval times. This is not a luxury, suspiciously eyed by the public as it may be, but a demand of justice. After all, "the idea of distinguishing the truly, deeply guilty from those who are merely victims of circumstances" is not only pointless (Greene and Cohen 2004:1781)—it is unfair itself.

CONCLUSION

If interventions impair control over bodily movements or reasoning capacities, agents are not responsible. This is not, however, because such deficits are caused by manipulations, but rather because they are impairments of a sufficiently grave kind. Criminal responsibility is, essentially, a personal matter, and manipulation does not in principle differ from other forms of the genesis of preferences. However, as we have indicated, a new criminal offence against the mind threatening to penalize severe manipulations should be introduced. Then, manipulators would have to bear the foreseeable consequences of their illegitimate deeds, which include harms inflicted on third parties by the manipulated. Similarly to civil law, costs can be shifted to a considerable extent onto offenders. Nonetheless, fully excusing manipulated agents only because of their mental history would run counter to established principles and well-founded doctrine. It would also be (largely) incoherent with the legal treatment of coerced persons who commit crimes (and are held responsible) and most importantly, would alleviate the manipulated from their prime obligation: to mobilize all possible inner means to avoid violations of others and of norms. On occasion, of course, it may suffice to subject manipulated agents to a symbolic condemnation only.

(Partial) transference of blame is the only arguable way to exempt manipulated agents. Whether and to what extent responsibility can be transferred is an issue entirely independent of metaphysical or ontological truths. Therefore, *if* manipulated persons are not responsible, it is neither because of personal identity, authenticity, or ownership considerations, nor because of deterministic events in their brains, but because of the fact of their manipulation. What does this mean for the philosophical quarrel? Well, we think it undermines the relevance of the manipulation case debate. No matter how crafty the incompatibilists, they cannot present their story without manipulation (Mele 2008:272). And if that is doing all the work, not much follows for ordinary life in a deterministic universe.[33]

Moreover, some philosophers seem to infer from the arguable result that manipulated agents are not responsible that something must be wrong in their minds or brains. This bears some striking resemblances to the *Zeitgeist*, the contemporary tendency to locate the source of mental illness, life problems, and even the current financial crises in individual deficits and psychological or neuronal processes. But with this perspective, we run a risk for losing sight of social and structural levels.

Looking into the brain, after all, may often be looking in the wrong direction. The literature is full of attempts to search for responsibility conferring causal neuronal routes. This not only sometimes leads to wild speculations over mechanisms and their individuation,[34] but also is taken up by neuroscientists trying to figure out which parts of the brain confer responsibility and where neuronal correlates of autonomy are located (Felsen and Rainer 2011). You don't have to be clairvoyant to foresee that once "identified", lawyers will deny responsibility of clients because they have some significant neuronal peculiarity around those same spots. To be sure, this is fascinating research. But it is important to appreciate the nature of one's explanandum, to understand what one is looking for and what remains unobservable. Even if we knew all mental states of a person and all neuronal correlates, "mechanisms", and physiological processes, along with their complete history, we would not be able to assess responsibility simply from that. As argued elsewhere, just as knowledge may not be in the brain, responsibility isn't there, either (Bublitz and Merkel 2009).

Responsibility is a judgment about a person's conduct in respect to his freedoms and the duties he owes to others. To partake in the social structure as free, equal, and responsible agents, persons must have some basic mental capacities. Neuroscience may help determine whether agents possess them. When they pass this entrance bar, they are considered free and likewise accountable. If they fail, they may be relieved from blame but also lose, unfortunately but unavoidably, some legal powers and factual freedoms. Where these bars are set depends less on neurological than on social-political considerations.

At least, it seems that conditions of excuses in normative systems can, to a large extend, only be inferred from the system-level perspective, not from the perspective of the individual to whom they are applied. From the latter view, not much more can be gained than a moral intuition, a felt unfairness that may serve as a departure point for more general principles of a fair distribution of freedoms and obligations. Even in nonlegal theories of responsibility, we think, the view of victims and society has to come in somewhere. Responsibility is always a functional concept. Changing its conditions directly affects others and, as freedoms are interrelated to responsibilities, the structure of reciprocal freedoms on a larger scale.

Perhaps philosophers may, after all, not be much interested in legal responsibility, its functional relations to freedom, and the architecture of the normative order. They may dismiss the foregoing arguments as affirmative apologies trading individual justice against demands of a social system. Even Fletcher suggests that functional theories "lose the capacity to engage in serious moral criticism of the system" (2007:331). We think this is misguided. On the contrary, by reconstructing criminal law and its aims in light of objections from empirical and philosophical sides and by discarding all the straw-men theories of punishment so often needlessly fought about, one paves the way for a more profound critique by exposing sources of real concern.

Surprisingly, on a deeper level, manipulation cases allude to the social dimension of punishment and the entanglement of retributive with distributive justice. Although we can—and possibly ought to—continue to expect each other to abide by the law and treat each other accordingly, we should be aware of the fiction and the residue of unfairness to individuals it implies.

Even with more meaningful sanctions, such a residue in punishment ineluctably remains. Without ultimate desert, criminal law cannot evade the hunch of what the famous German legal philosopher Gustav Radbruch (1929) reminded lawyers of, long before the age of neuroscience: in criminal law, a good jurist can only be one with a bad conscience.

NOTES

1. Oxytocin seems to influence choice and risk-taking behavior (Kosfeld et al. 2005).
2. DBS has not been used for such purposes, but the nucleus accumbens was targeted by DBS for major depression with automatic stimulators over which patients do not have control (Schlaepfer et al. 2007). Whoever controls the device has access to one of the central areas mediating pleasure and motivation. In the 1970s, bizarre conditioning experiments were conducted with electric stimulation of homosexual men in order to initiate heterosexual behavior (Moan and Heath 1972).
3. That violent dispositions, previously alien to the person concerned, may emerge as disturbing side-effects of DBS has been documented (Appleby et al. 2007).
4. Scientist warned about security risks of DBS (Kohno et al. 2009).
5. In an Italian murder case, sentencing was reduced because of defendant's genetic dispositions toward violent behavior (MAOA) (Baum 2011).
6. For considerations where indeterminism is placed best, see Clarke (2003); from a critical perspective, Walde (2006).
7. For example, Haji (2008); Fischer (2006); Berofsky (2003).
8. Even extrapolating shared legal principles in comparative (criminal) law has only just begun (Fletcher 2007, is a good introduction), primarily as a result of new international criminal law institutions such as the ICC.
9. Testimonies and court decisions in the Hearst case are documented in Anspacher (1976).
10. Delgado stipulates the following conditions: (1) that coercive persuasion has occurred, (2) that the defendant's unlawful action was the proximate result of that coercive persuasion; and (3) that exculpation for the act committed is morally justified (1979).
11. The concepts of excuse, exculpation, exemption from blame, impunity, factors negating responsibility, and similar terms do not have a universally shared meaning—we use them synonymously.
12. For example, § 2.01 Model Penal Code.
13. Art. 9 European Convention on Human Rights (ECHR); Art. 10 European Charter of Fundamental Rights (ECFR), Art. 18 Universal Declaration of Human Rights (UDHR).
14. Decisions of the German Federal Court of Justice (BGHSt), Vol. 39, p. 1.
15. We wish to address an objection Shaw (2012) has leveled against our responsibility-shift argument (2009). To her, the manipulator would then function as a "blame magnet" leading to the "paradoxical conclusion" that the responsibility

of one person depends on the responsibility of another. Rather, both should be "fully accountable". Her argument surely merits a thorough response; here we can only affirm that we hold it to be a facet of both moral and legal reasoning that the relation between offenders affects assessment of autonomy, blame, culpability, and sentencing.

16. This is, however, methodologically dubious when philosophers test their theories with fictional cases but on ordinary moral intuitions. What responsibility might they allude to if not the kind we usually apply in real-life to real persons with the usual real negative consequences? Despite all differences between morality and law, there seems to be a common core at least partially tracked by the kinds of legal responsibility.
17. Also, it is useful to indicate some common ground: everyone agrees that the issue at stake is not personal identity in a strong sense (agents remain "themselves"); moreover, no one thinks that actions of manipulated agents are justified (i.e., they remain wrongful).
18. Interestingly, many philosophers explicitly deny that they are internalists because they take into account the history of preferences. But this does not make them externalists in our sense. As long as their argument is confined to the history of internal states, they may be *internal historicists,* but not externalists interested in agents' social situations.
19. For clarification of Frankfurt's view and an unsuccessful attempt to rescue purely internal hierarchical accounts, see Taylor (2003).
20. We borrow this example from Anderson (2011).
21. Cf. § 3.02 Model Penal Code.
22. Let us add that, from a legal perspective, harms to life or sufficiently grave harms to physical integrity can *never* be outweighed, no matter how grave the threat of the harms averted may have been. That follows from the basic principle underlying such justifications: that of (legally compelled) solidarity (cf. our remarks above). In liberal constitutional orders, there can be no such thing as a legal duty of solidarity to have one's life or significant features of one's body or health sacrificed for the sake of others that one does not threat with harm. As legal philosopher Ronald Dworkin once remarked in a different context: "His life is the only one he has." To have it sacrificed for reasons of solidarity with others cannot be a legitimate legal duty.
23. EU: Riera Blume v. Spain (ECHR App. 37680/97). US: Peterson v. Sorlien 299 N.W.2d 123 (Minn. 1980).
24. Fischer and Ravizza propose that under certain conditions, agents can "take ownership" of implanted mechanisms (1998:232). But instead of solving the problem, it puts it back a step. In legal terms: How can incompetent agents make a legally binding decision to take ownership? This presupposes what it seeks to establish: their competence. For arguments in our vein, see Judisch (2005), Zimmerman (2003), and Shaw (2012).
25. Personality rights are to be found, e.g., in Art. 22 UDHR, Art. 8 ECHR, and Art. 2 I German Basic Law. For the scope of personality protection under the ECHR, see Marshall's (2009) comprehensive analysis.

26. This is the classic positivistic stance as seminally formulated by Kelsen (1960), approved by most other modern theories of law as well.
27. We wish to note that norm stabilization does not necessarily lead to what is called "enemy law", in which persons of whom we cannot expect norm compliance (such as terrorists) are stripped of basic rights.
28. At this point, we have to reemphasize that these expectations are not simply social-psychological phenomena but, again, are normatively constituted. The law only refers to those general expectations that are in line with legal principles. For instance, violence against certain minority groups may—unfortunately—enjoy wide popular support. Even if it were proved empirically that these acts do not destabilize expectations about conduct in other situations, the law should neither refrain from punishing nor sentence lightly. Moral psychology shows that moral intuitions are prone to bias. Vox populi and legal judgment often diverge, especially in matters of punishment and sentencing. Even though the law protects expectations, it cannot yield to the ever-changing and highly suggestible public opinion. Instead of tracking "shared intuitions of justice of communities" (Robinson 2011:10), it has to influence social dynamics based on adequate public reasons.
29. See the judgments of the ECHR striking down German security prevention laws (M v. Germany, App. 19359/04 and Haidn v. Germany, App. 6587/04).
30. This is not to say that these persons should be imprisoned. Every other, softer police measure must be taken before that, even if it is much more costly for society and cannot avert all risks.
31. Cf. Kant's famous remark (1908:4).
32. Robinson seems to suggest a different treatment in view of diverging public moral intuitions about personal histories (2011:19). Yet, as said, criminal law is not an exercise in tracking and enforcing moral intuitions. Excuses cannot be grounded in public opinion about blameworthy personal developments.
33. And because of this, manipulation cases are a problem for both compatibilists and libertarians (Haji and Cuypers 2001; Mele 2006).
34. Broad or narrow—what if replaced by functionally equivalent implanted mechanisms? See Fischer (2006) and Judisch (2005).

REFERENCES

Anderson, S. (2011). Coercion. In: *Stanford Encyclopedia of Philosophy.* Retrieved January 20, 2012, from plato.stanford.edu.

Anspacher, C. (1976). *The Trial of Patricia Hearst.* San Francisco, Great Fidelity Press.

Appleby, B., P. Duggan, A. Regenberg, and P. Rabins (2007). "Psychiatric and neuropsychiatric adverse events associated with deep brain stimulation: A meta-analysis of ten years' experience." *Movement Disorders* 12: 1722–1728.

Baum, M. (2011). "The monoamine oxidase A (MAOA) genetic predisposition to impulsive violence: Is it relevant to criminal trials?" *Neuroethics* DOI 10.1007/s12152-011-9108-6.

Berofsky, B. (2003). "Identification, the self and autonomy." *Social Philosophy & Policy* 20(2): 199–220.

Bublitz, C., and R. Merkel (2009). "Authenticity and autonomy of enhanced personality traits." *Bioethics* 7: 360–375.

Bublitz, C., and R. Merkel (2012). "Crimes against minds." (forthcoming in *Criminal Law and Philosophy*). DOI 10.1007/s11572-012-9172-y.

Clarke, R. (2003). *Libertarian Accounts of Free Will*. Oxford, UK, Oxford University Press.

Clausen, J. (2009). "Man, machine and in between." *Nature* 457: 1080–1081.

Degaldo, R. (1978). "Ascription of criminal states of mind: Toward a defense theory of the coercively persuaded ("brainwashed") defendant." *Minnesota Law Review* 63: 1–33.

Delgado, R. (1979). "A response to Professor Dressler." *Minnesota Law Rev* 63: 361–365.

Dressler, J. (1979). "Professor Delgado's brainwashing defense: Courting a determinist legal system." *Minnesota Law Review* 63: 335–360.

Felsen, G., and P. Rainer (2011). "How neuroscience informs our conceptions of autonomy." *American Journal of Bioethics—Neuroscience* 2: 3–14.

Fischer, J. M. (2000). "Responsibility, history and manipulation." *Journal of Ethics* 4: 385–391.

Fischer, J. M. (2006). *My Way: Essays on Moral Responsibility*. New York, Oxford University Press.

Fischer, J. M. and M. Ravizza (1998). *Responsibility and Control*. Cambridge, UK, Cambridge University Press.

Fletcher, G. (2007). *The Grammar of Criminal Law*, Vol. 1. Oxford, UK, Oxford University Press.

Frankfurt, H. (1969). "Alternate possibilities and moral responsibility." *Journal of Philosophy* 66(3): 829–839.

Frankfurt, H. (1988). Coercion and moral responsibility. In: *The Importance of What We Care About*. Cambridge, UK, Cambridge University Press.

Frankfurt, H. (2002). Reply to Fischer. In: *Contours of Agency*. S. Buss and H. Overton. Cambridge, MA, MIT Press.

Gardner, J. (2007). *Offences and Defences*. Oxford, UK, Oxford University Press.

Greely, H. (2008). "Neuroscience and criminal justice: Not responsibility but treatment." *University of Kansas Law Rev* 56: 1103–1138.

Greene, J., and J. Cohen (2004). "For the law, neuroscience changes nothing and everything." *Philosophical Transactions of the Royal Society, B* 359: 1775–1785.

Grey, B. (2011). Neuroscience and emotional harm in tort law: Rethinking the American approach to free-standing emotional distress claims. In: *Law and Neuroscience*. M. Freeman. Oxford, UK, Oxford University Press.

Haji, I. (2008). "Authentic springs of action and obligation." *Journal of Ethics* 12: 239–261.

Haji, I., and S. Cuypers (2001). "Libertarian free will and CNC manipulation." *Dialectica* 3: 221–238.

Husak, D. (2010). *Philosophy of Criminal Law*. Oxford, UK, Oxford University Press.

Jakobs, G. (2004). *Staatliche Strafe: Bedeutung und Zweck*. Vienna, Verlag Ferdinand Schöningh.

Jakobs, G. (2008). *Norm, Person, Gesellschaft*, 3rd ed. Berlin, Duncker & Humblot.

Judisch, N. (2005). "Responsibility, manipulation and ownership." *Philosophical Explorations* 2: 115–130.

Kane, R. (1996). *The Significance of Free Will*. Oxford, UK, Oxford University Press.

Kant, I. (1908). *Kritik der praktischen Vernunft*. Academy Edition Vol. 5. Berlin, Preussische Academie der Wissenschaften.

Kelsen, H. (1960). *Reine Rechtslehre*. Vienna, Deuticke.

Kohno, T., T. Denning, and Y. Matsuoka (2009). "Security and privacy for neural devices." *Journal of Neurosurgical Focus* 27: 1–4.

Kosfeld, M., M. Heinrichs, P. Zak, U. Fischbacher, and E.Fehr (2005). "Oxytocin increases trust in humans." *Nature* 435: 673–676.

Lacey, N. (2011). The resurgence of "character"? Responsibility in the context of criminalisation. In: *Philosophical Foundations of Criminal Law*. A. Duff and S. Green. Oxford, UK: Oxford University Press, pp. 151–179.

Levy, N. (2007). *Neuroethics*. Cambridge, UK, Cambridge University Press.

Levy, N., and T. Bayne (2004). "A will of one's own: Consciousness, control and character." *International Journal of Law and Psychiatry* 27: 459–470.

Libet, B. (2004). *Mind Time: The Temporal Factor in Consciousness*. Cambridge, MA, Harvard University Press.

Marshall, J. (2009). *Personal Freedom through Human Rights law?* Leiden: Martinus Nijhoff Publishers.

McKenna, M. (2008). "A hard-line reply to Pereboom's four-case manipulation argument." *Philosophy and Phenomenological Research* 1: 142–159.

Mele, A. (1995). *Autonomous Agents*. Oxford, UK, Oxford University Press.

Mele, A. (2006). *Free Will and Luck*. Oxford, UK, Oxford University Press.

Mele, A. (2008). "Manipulation, compatibilism, and moral Responsibility." *Journal of Ethics* 12: 263–286.

Merkel, R. (1999). "Personale Identität und die Grenzen strafrechtlicher Zurechnung." *Juristenzeitung* 10: 502–511.

Metzinger, T. (2011). The No-Self Alternative. In: *Oxford Handbook of the Self*. S. Gallagher. Oxford, UK, Oxford University Press, pp. 279–296.

Metzinger, T. (In press). Two principles for robot ethics. In: *Robotik und Gesetzgebung*. E. Hilgendorf and J.-P. Günther. Baden-Baden, Nomos.

Moan, C. and R. Heath (1972). "Septal stimulation for the initiation of heterosexual behavior in a homosexual male." *Journal of Behavior Therapy and Experimental Psychiatry* 3: 23–30.

Morse, S. (2006). "Brain overclaim syndrome and criminal responsibility: A diagnostic note." *Ohio Journal of Criminal Law* 3: 397–412.

Nolan, T. (2004). "The indoctrination defense: From the Korean War to Lee Boyd Malvo." *Virginia J of Social Policy & Law* 3: 435–465.

Pereboom, D. (2003). *Living Without Free Will*. Cambridge, UK, Cambridge University Press.

Radbruch, G. (1929). *Einführung in die Rechtswissenschaft*, 8th Ed. Reprinted in: A. Kaufmann (1987): Radbruch Gesamtausgabe, Vol 1. Heidelberg, C. F. Müller.

Rawls, J. (1971). *A Theory of Justice*. Cambridge, MA, Harvard University Press.

Rehaag, C. (2009). *Prinzipien von Täterschaft und Teilnahme in europäischer Rechtstradition*. Freiburg, Max-Planck Society for International Criminal Law.

Robinson, P. (2011). Are we responsible for who we are? The challenge for criminal Law theory in the defenses of coercive indoctrination and "rotten social background." University of Pennsylvania Law School—Research Paper Series.

Sanchez, B. F. (2007). Positive Generalprävention. Gedanken zur Straftheorie Günther Jakobs'. In: *Festschrift für Günther Jakobs*. M. Pawlik. Köln, Heymanns, pp. 75–95.

Schlaepfer, T., M. Cohen, C. Frick, M. Kosel, D. Brodesser, N. Axmacher, A. Joe, D. Lenartz, and V. Sturm (2008). "Deep brain stimulation to reward circuitry alleviates anhedonia in refractory major depression. *Neuropsychopharmacology* 33: 368–377.

Shaw, E. (2012). "Direct brain interventions and responsibility enhancement." (forthcoming in *Criminal Law and Philosophy*). DOI 10.1007/s11572-012-9152-2

Sie, M., and A. Wouters (2010). "The BCN challenge to compatibilist free will and responsibility." *Neuroethics* 3: 121–133.

Skinner, B. F. (1976). *Walden Two*. New York, Macmillian.

Smilansky, S. (2003). *Free Will and Illusion*. Oxford, UK, Oxford University Press.

Taylor, J. S. (2003). "Autonomy, duress, and coercion." *Social Philosophy and Policy* 2: 127–155.

Thalberg, I. (1989). Hierarchical analyses of unfree action. In: *The Inner Citadel*. J. Christman. Oxford, UK, Oxford University Press.

Vincent, N. (2011). A structured taxonomy of responsibility concepts. In: *Moral Responsibility: Beyond Free Will and Determinism*. N. Vincent, I. van de Poel, and M. J. van den Hoven. Heidelberg, Springer.

Walde, B. (2006). *Willensfreiheit und Hirnforschung: Das Freiheitsmodell des epistemischen Libertarismus*. Paderborn, Mentis.

Wegner. D. (2002). *The Illusion of Conscious Will*. Cambridge, MA, MIT Press.

Wertheimer, A. (1987). *Coercion*. Princeton, NJ, Princeton University Press.

Zimmerman, D. (2003). "That was then, this is now." *Nous* 37: 638–671.

Index

Note: Page numbers followed by "n" refer to notes.